Practice-Based Innovation: Insights, Applications and Policy Implications

Helinä Melkas • Vesa Harmaakorpi
Editors

Practice-Based Innovation: Insights, Applications and Policy Implications

Springer

Editors
Prof. Helinä Melkas
Lappeenranta University of Technology
Lahti School of Innovation
Saimaankatu 11
15140 Lahti
Finland
helina.melkas@lut.fi

Prof. Vesa Harmaakorpi
Lappeenranta University of Technology
Lahti School of Innovation
Saimaankatu 11
15140 Lahti
Finland
vesa.harmaakorpi@lut.fi

ISBN 978-3-642-21722-7 e-ISBN 978-3-642-21723-4
DOI 10.1007/978-3-642-21723-4
Springer Heidelberg Dordrecht London New York

Library of Congress Control Number: 2011938152

Printed on acid-free paper

Springer is part of Springer Science+Business Media (www.springer.com)

Foreword

Today's innovation environment is changing rapidly and new challenges are emerging. The locus of innovation is shifting towards rapidly developing countries and corporate operations are increasingly executed in global value networks. Social structures will change due to urbanisation and ageing of the population in industrial countries. Demand for environmentally sustainable and tailored solutions will grow as ecological issues move to the forefront in most areas of society. Meanwhile technological developments are accelerating, combined with an increasing emphasis on non-technological innovations where usability and user experiences are the main drivers.

To cope with these challenges extensive renewal through a broad-based view on innovation is needed. It will not be sufficient to fund only technological breakthroughs. Service-related, design, business, and social innovations should also be emphasised. Innovations increasingly emerge in practical contexts where different types of knowledge from different disciplines have to be continuously combined. Networking and interaction abilities will be of key importance.

Tekes – the Finnish Funding Agency for Technology and Innovation – is the most important publicly funded expert organisation for financing research, development and innovation in Finland. Tekes aims to boost wide-ranging innovation activities in research communities, industry and service sectors and to work with the top innovative companies and research units to achieve this. Every year Tekes funds some 1,500 business research and development projects, and almost 600 public research projects at universities, research institutes and polytechnics. The activities are targeted to projects that create the greatest socio-economic impacts in the longer-term. Tekes is continuously monitoring developments in its operating environment and supporting innovation policies, among other things, by funding innovation research projects and undertaking foresight schemes.

In 2012, Tekes new strategy will introduce novel operating modes in public research carried out by universities and research institutes. These will help to create new business opportunities and generate areas of expertise in strategic areas which are vital for Finland. The new operating modes aim to create expertise that, on one hand, improves the odds for generating new business activities and companies, and on the other, sets in motion research teams aiming for leading edge expertise that is of key importance for the Finnish business sector. Tekes also strongly supports internationalisation of Finnish business and research.

This book showcases examples of broad-based innovation activities, which is welcome in a situation where the concept of broad-based innovation, and practice-based innovation in particular, still requires improved conceptualisation both in Finland and internationally. The chapters in this book – that focus both on macro-, meso- and micro-level perspectives on innovation through contributions by experts from many different countries – guide us on this journey towards a deeper understanding of broad-based and practice-based innovation. The book has been co-funded through Tekes innovation research activities and I hope that it can provide new insights and viewpoints about innovation and its drivers in an increasingly challenging and complex world.

Dr. Veli-Pekka Saarnivaara
Director General of Tekes

Helsinki, Finland

Contents

Part II Micro-, Meso- and Regional Level Applications

Part III Case Studies and Policy Implications

Contributors

Amara, Nabil Full Professor, Department of Management, Laval University, Québec, Canada

Aula, Pekka Professor, Department of Social Research, Media and Communication Studies, University of Helsinki, Helsinki, Finland

Cooke, Philip Professor, Centre for Advanced Studies, Cardiff University, Cardiff, UK

Cramer, Christina Doctoral Candidate, Brandenburg University of Technology, Cottbus; Druckhaus Cramer, Münster, Germany

Dalziel, Margaret Associate Professor, Telfer School of Management, University of Ottawa, Ottawa, Canada

Darsø, Lotte Associate Professor of Innovation, Department of Learning, Danish School of Education, University of Aarhus/Copenhagen Campus, Copenhagen, Denmark

Ellström, Per-Erik Professor, Helix Vinn Excellence Centre, Department of Behavioural Sciences and Learning, Linköping University, Linköping, Sweden

Hagino, Hiroo Professor, Faculty of Comprehensive Management, Tohoku Fukushi University, Sendai, Japan

Harmaakorpi, Vesa Dean, Faculty of Technology Management; Professor, Lahti School of Innovation, Lappeenranta University of Technology, Lappeenranta/Lahti, Finland

Hennala, Lea Senior Researcher, Lahti School of Innovation, Lappeenranta University of Technology, Lahti, Finland

Herstatt, Cornelius Full Professor and Director, Institute for Technology and Innovation Management, Hamburg University of Technology, Hamburg, Germany

Hipp, Christiane Full Professor and Vice-Dean, Department of Economics and Business Sciences, Brandenburg University of Technology, Cottbus, Germany

Holbrook, J. Adam Associate Director and Adjunct Professor, Centre for Policy Research on Science and Technology, Simon Fraser University, Vancouver, Canada

Høyrup, Steen Associate Professor of Education, Department of Learning, Danish School of Education, University of Aarhus/Copenhagen Campus, Copenhagen, Denmark

Järvensivu, Timo Research Manager, School of Economics, Aalto University, Helsinki, Finland

Konsti-Laakso, Suvi Researcher, Lahti School of Innovation, Lappeenranta University of Technology, Lahti, Finland

Landry, Réjean Full Professor, Department of Management, Laval University, Québec, Canada

Lühring, Norbert Partner, Lischke Consulting GmbH, Hamburg/Munich, Germany

Mahy, Isabelle Professor, Department of Social and Public Communication, Université du Québec à Montréal, Montréal, Canada

Melkas, Helinä Professor, Lahti School of Innovation, Lappeenranta University of Technology, Lahti, Finland

Mäkimattila, Martti Researcher, Lahti School of Innovation, Lappeenranta University of Technology, Lahti, Finland

Nilsen, Per Associate Professor, Department of Medical and Health Sciences, Linköping University, Linköping, Sweden

Nooteboom, Bart Full Professor of Innovation Policy, Tilburg School of Economics and Management, Center for Innovation Research, Tilburg University, Tilburg, The Netherlands

Nykänen, Katri Project Manager, PhD student, School of Economics, Aalto University, Helsinki, Finland

Ogasawara, Koichi Professor, Director of International Affairs, Tohoku Fukushi University, Sendai, Japan

Oikarinen, Tuija Senior Researcher, Lahti School of Innovation, Lappeenranta University of Technology, Lahti, Finland

Parjanen, Satu Researcher, Lahti School of Innovation, Lappeenranta University of Technology, Lahti, Finland

Parviainen, Olli Researcher, Department of Social Research, Media and Communication Studies, University of Helsinki; Entrepreneur, Verkostoanatomia, Helsinki, Finland

Pässilä, Anne Researcher, Lahti School of Innovation, Lappeenranta University of Technology, Lahti, Finland

Rajala, Rika Researcher, School of Economics, Aalto University, Helsinki, Finland

Rantanen, Hannu Vice Rector; Professor, Lahti School of Innovation, Lappeenranta University of Technology, Lappeenranta/Lahti, Finland

Saarnivaara, Veli-Pekka Director General, Tekes – the Finnish Funding Agency for Technology and Innovation, Helsinki, Finland

Salminen, Juho Researcher, Lahti School of Innovation, Lappeenranta University of Technology, Lahti, Finland

Saunila, Minna Researcher, Lahti School of Innovation, Lappeenranta University of Technology, Lahti, Finland

Skiba, Florian Managing partner, ARKTIK GmbH; Institute for Technology and Innovation Management, Hamburg University of Technology, Hamburg, Germany

Ukko, Juhani Senior Researcher, Lahti School of Innovation, Lappeenranta University of Technology, Lahti, Finland

Uotila, Tuomo Professor, Lahti School of Innovation, Lappeenranta University of Technology, Lahti, Finland

Vince, Russ Professor, School of Management, University of Bath, UK

Wixted, Brian Research Fellow, Centre for Policy Research on Science and Technology, Simon Fraser University, Vancouver, Canada

Ogasawara, Koichi Professor, Director of International Affairs, Tohoku Fukushi University, Sendai, Japan

Oikarinen, Tuija Senior Researcher, Lahti School of Innovation, Lappeenranta University of Technology, Lahti, Finland

[illegible]

Chapter 1
Introduction

Helinä Melkas and Vesa Harmaakorpi

1.1 Perspectives into Practice-Based Innovation

The mainstream economic development policy in Europe has until recently relied on a cluster approach and on the power of knowledge and research as the sources of innovation. Innovation policy has been to a great extent equivalent to science and technology policy, and cluster policies have aimed at building competitive advantage with strong regional and national clusters. Recent discussions have, however, emphasised other forms of economic order and origins of innovation. According to some innovation surveys, only a few percent of innovations are based on scientific sources. Cluster policy seems to have its weaknesses, as well. The current science and technology policy is not very effective, partly due to the fact that innovation policy, on the one hand, and science and technology policy, on the other, are not clearly defined but are mixed up in speech. Moreover, the practical context and interaction between the two subsystems of an innovation system (acquisition and assimilation of knowledge; transformation and exploitation of knowledge) seem to offer a lot of unused potential for innovation. This potential remains largely untouched due to lack of policy measures to foster practice-based, networked innovation processes that combine diverse knowledge bases. Advancing open innovation activities also requires new types of policy measures.

It is increasingly considered essential that science and technology-based innovation policy (STI) would be combined with practice-based innovation policy (DUI – doing, using, interacting; cf., e.g., Lundvall 2007). However, the contents and exact meaning of DUI policy are still quite unclear. The present knowledge concerning DUI is sufficient for understanding its importance, but the basis for truly successful policy-making in this area should still be strengthened. There are

---

H. Melkas (✉) • V. Harmaakorpi
Lahti School of Innovation, Lappeenranta University of Technology, Lahti, Finland
e-mail: helina.melkas@lut.fi

H. Melkas and V. Harmaakorpi (eds.),
Practice-Based Innovation: Insights, Applications and Policy Implications,
DOI 10.1007/978-3-642-21723-4_1,

diffusion of innovations. Social networks' role in promoting the diffusion of innovations is widely recognised, but networks are used more as a vague metaphor than an analytic concept. The authors base their findings on a case study of a food industry organisation. They conclude with remarks on how the study of innovation diffusion might benefit from adapting the methods of social network analysis.

In Chapter 5 ("Dilemmas of practice-based innovation policy-making"), *Réjean Landry* and *Nabil Amara* discuss dilemmas that emerge when policy-makers are equally attracted to promoting concepts that are not based on evidence and to developing innovation policies that are evidence-based. Turning policy concepts into evidence-based policy is difficult. The professionals of innovation are very often submitted to tensions where they have to choose between equally attractive but opposite alternative innovation policy measures. The chapter addresses nine dilemmas of innovation policies. It is based on a review of concepts and buzzwords linked to innovation, and the types of data and diagnostics used by policy-makers to develop new national and regional innovation policies. The chapter discusses how policy-makers could resolve the tensions between equally attractive policy alternatives. The results show that policy ideas that appear to provide new policy prescriptions have to be confronted to evidence and diagnostics that take into account the diversity of the situations and needs of firms, industries, and regions.

Cornelius Herstatt and *Norbert Lühring* (Chapter 6, "Coordination in innovation projects") note that many empirically based studies support a positive connection between cross-functional cooperation and successful innovation. The more recent discussion on integrated product development shifts the focus to an overall understanding of the various functional areas, which at times make considerable contributions to the success of an innovation project. There appears to be a need for a higher level of integration and coordination in innovation projects to ensure that the results of individual work packages are aligned to the overall objective of a project and that the implementation of the product development process is effective and efficient. Consequently, we are faced with the question of which mechanisms can be used to best meet coordination needs in innovation projects. Based on an in-depth study of three innovation projects, Chapter 6 attempts to answer this question. The authors first give an overview of diverse aspects of coordination in innovation projects. For this purpose, five dimensions of coordination are derived from organisational theory. A coordination model for innovative projects is then developed to serve as a basis for the empirical study: a comparative case study analysis conducted in the electronics and the automotive supply industries.

Margaret Dalziel and *Satu Parjanen* (Chapter 7, "Measuring the impact of innovation intermediaries: A case study of Tekes") discuss innovation intermediaries that are believed to have a beneficial influence on innovation processes. As there are no universally accepted metrics of intermediary performance, however, it is difficult for these organisations to provide the evidence of their contributions. The authors present a general-purpose methodology for measuring the impact of innovation intermediaries that applies across all types of intermediaries. They then demonstrate the methodology by assessing the impact of the Global Access Program (GAP), which is made available to Finnish firms through

Tekes, an intermediary organisation whose mission is to enhance Finnish industry through technology and innovation. The findings show that the GAP program has had an impact on the performance of participating firms in terms of revenue growth, exports, new international customers, and employment growth. Consistent with their expectation that impacts on firm performance are a consequence of earlier impacts on firm resources and capabilities, the authors find a statistically significant relationship between the immediate impact of strategic information and advice, and information and advice on new markets, and longer term impact on firm performance.

1.3.2 Part II: Micro-, Meso- and Regional Level Applications

Part II focuses on micro-, meso- and regional level applications related to practice-based innovation. The micro-level applications concern day-to-day human interaction. The meso-level is the lesser known of the society groupings; it is the marriage between micro- and macro-level studies, analysing medium-sized groups, such as work communities or municipal citizens. Part II contains eight chapters.

In Chapter 8, *Lotte Darsø* and *Steen Høyrup* ("Developing a framework for innovation and learning in the workplace") develop an analytical framework for conceptualisation and analysis of the interplay between innovation and workplace learning. The analytical differentiation as well as the juxtaposition of preject and project is enabling for comprehending the relationship between innovation and learning. The chapter mainly directs attention to the less developed and discussed concept of preject. Learning theory is examined from the perspective of the individual in his or her social setting, and the concepts of innovative learning, adaptive learning, and reproductive learning are introduced and discussed. By relating these concepts to a problem matrix, the authors are able to, firstly, differentiate between preject, problem solving, and project and, secondly, associate innovative learning, adaptive learning, and reproductive learning with the corresponding arenas. The project mainly affords reproductive learning. Problem solving mainly affords adaptive learning. The preject affords innovative learning as well as adaptive learning.

Per Nilsen and *Per-Erik Ellström* (Chapter 9, "Fostering practice-based innovation through reflection at work") state that practice-based knowledge (experience) is indispensable to obtain expertise, perform many work tasks well and solve different types of problems, but it is not conducive to change or renewal in the workplace. Practice-based knowledge needs to be integrated with research-based knowledge. The development of practice-based innovations can be fostered by creative learning made possible through the deliberate use of research-based knowledge to challenge well-established attitudes, beliefs, and practices. Reflection in the workplace is a mechanism to integrate research-based knowledge with practice-based knowledge to scrutinise and possibly revise prevailing thought and action patterns, thus facilitating creative learning. Reflection is often not allotted formal priority on the management agenda, yet the potential of reflection cannot be fully realised without formalising this as an expected, legitimised activity in the workplace. The authors emphasise that decisions and planned mechanisms, structures, and procedures are thus needed to facilitate and support reflection in the workplace.

In Chapter 10, *Anne Pässilä*, *Tuija Oikarinen* and *Russ Vince* also investigate reflection ("The role of reflection, reflection on roles: Practice-based innovation through theatre-based learning"). According to them, a key issue for practice-based innovation is how organisations can generate innovation in the midst of action. In order to answer this question, they discuss the relationship between learning, reflection, and practice-based innovation. Reflection is seen as a crucial organisational process that can create spaces for innovation. The authors demonstrate how theatre-based learning can offer an effective strategy for the creation of reflective spaces that reveal the dynamics of innovation, both in terms of what promotes and what prevents innovative behaviour and practice. Through research and intervention in three organisations, the authors show that viewing roles and relations 'acted out' in theatre helps to reduce the unconscious acting out of entrenched emotional and political dynamics in practice. The struggle to create innovation in the midst of action can be seen in the reflexive tension between the radical possibility of such interventions and the political purpose they may serve for established power relations. The authors conclude that there will always be a tension in organisations between dynamics that support innovation and dynamics that undermine it.

Isabelle Mahy (Chapter 11, "From the artists to the managers: Responsible collective innovation practices, inspiration flowing through hosting and harvesting profound change") demonstrates in her chapter how artists' creative work can inspire a collective process supporting and leading to practice-based innovation. The underlying hypothesis is that enabling innovation through art is a powerful means to foster practice-based innovation. After having presented innovation as new situated knowledge, relevant at a micro-level work process activity, the concept of *ba* is brought forward in the chapter to describe organisational contexts that invite, sustain, and foster innovation. The art of hosting and harvesting of complex situations are then presented as appropriate and powerful collective facilitation and information gathering processes to nurture innovation. Considering innovation as knowledge emerging from collective intelligence, two cases illustrate how collective intelligence can be nurtured by artistic practices. The results are focused on the specifics of the principles and practices at work, which are creative, artistic, playful, sensible, involving concerns for ethics and aesthetics, and helpful in creating meaningful experiences.

Chapter 12 by *Juho Salminen* and *Vesa Harmaakorpi* also looks into collective intelligence ("Collective intelligence and practice-based innovation: An idea evaluation method based on collective intelligence"). The authors note that users and customers are becoming increasingly important sources of knowledge due to changes in innovation policies and paradigms. Simultaneously innovation is becoming more of a networking activity. New methods are thus needed for processing information and ideas coming from multiple sources more effectively.

For example, the whole personnel of an organisation are seen as a great potential for innovation. The recent development of communication technologies such as the Internet has increased interest towards the multidisciplinary field of collective intelligence. To investigate the possibilities of collective intelligence, the authors used the nest-site selection process of honeybees as a model for an idea evaluation tool, a prototype of which was then tested in a case organisation. The results were promising; the prototype was able to evaluate ideas effectively, and it was highly accepted in the organisation. The authors conclude that collective intelligence can be utilised for some of the tasks at the front end of innovation.

Florian Skiba and *Cornelius Herstatt* study lead users in service innovation (Chapter 13, "Users as sources of radical service innovation: A closer look into opportunities for integrating service lead users in service development"). Although theory and praxis both underline the importance of radical service innovations, until now lead users in the service industry have been excluded from the scope of research. Based on explorative findings on user innovations in services and recent research on service innovations, Chapter 13 investigates whether users can lead trends and independently innovate in the service sector; if so, whether lead users in the service industry show similar characteristics to lead users from other industries, and whether approaches to identification and integration from the consumer and investment goods sectors can be applied in the service sector as well. By illustrating the potential impact of lead users as a source of radical service innovations, the authors hope to raise interest in this powerful concept that is not yet systematically applied in service industry.

Lea Hennala, *Suvi Konsti-Laakso* and *Vesa Harmaakorpi*, the authors of Chapter 14 ("Challenges of bringing citizen knowledge into public sector service innovation") discuss user-driven innovation as an essential part of practice-based innovation. Their qualitative case study aims at finding out what kinds of challenges there are pertaining to a municipal resident-involving service innovation approach in the public sector. This research question is approached from the points of view of both the municipal resident and the public authorities. With the municipal resident point of view the authors investigate how residents as care service users are disposed towards having the possibility to personally participate in the service development activities, and from what kinds of positions the service users produce a voice to support care service development. The public authority point of view examines how the customer-driven service development approach is perceived by developer authorities, and from what kinds of positions they listen to the message compiled from the users' voices. As a result, the authors present six challenges of the service user-involving service innovation approach.

Philip Cooke (Chapter 15, "The increasing use of dramaturgy in regional innovation practice") reports on advances in regional innovation practice. Regional innovation has become a maturing field of economic governance. Regions have become more prominent actors in the innovation field in the past decade. Innovation is widely seen by supranational, national and regional governance bodies and agencies as a mainspring of improved regional economic performance and wellbeing. Leading regional innovation practitioners are increasingly being understood as catalysts of innovation, a development in their earlier role as being supporter or partner in innovation essentially conducted by others. One technique that Cooke's chapter devotes attention to where regional 'orchestration' of innovation occurs is the use of narrative, drama and non-scientific laboratory experimentation to open business and community minds to the constructed regional advantages of innovation. The theoretical context is 'post-cluster' hence platform-minded and using matrix models to induce innovation through stimulating cross-cluster 'transversality'.

1.3.3 Part III: Case Studies and Policy Implications

Part III consists of six chapters. In Chapter 16 ("Service innovation and service design in the German printing industry"), *Christina Cramer* and *Christiane Hipp* note that service innovation has received less attention in science over the last decade as compared to product innovation, although the importance of service innovation is steadily growing. In fact, for the printing industry service innovation and the new approach, service design, have received almost no attention at all. Because of the current crisis in the printing industry that is due to digitalising, there is considerable pressure to find new business opportunities. According to the authors, service innovation and service design can perhaps help to overcome the crisis, so that companies can remain competitive in a dynamic business climate. Therefore, their chapter analyses cross-industry benchmarks in the field of service innovation and service design. The use of the Circle of Service Design is shown in order to highlight possible opportunities for the printing industry to develop better and optimised services in the future.

Brian Wixted and *J. Adam Holbrook* investigate the knowledge system in Vancouver, Canada, and its place on the Pacific Rim (Chapter 17, "Innovation, cities and place: An empirical study of the knowledge system in Vancouver and its place on the Pacific Rim"). The authors argue that human capital intensive clusters emerged surprisingly early in a number of technology fields in Vancouver, but they largely failed to develop beyond entrepreneurial rent seekers. The authors discuss this in the light of Vancouver's international economic spatial position. Vancouver appears to occupy a strategic Canadian innovation pivot point position on the Pacific Rim. As a generalisation, the innovation systems literature has downplayed the overall physical geographic setting of particular places and the connections between cities in national or international urban systems. This chapter examines the innovation history of a few of Vancouver's peculiarly human capital intensive clusters noting how the isolation with no nearby cities and its connectedness (a Pacific gateway point) appear to have shaped its trajectories. The analysis makes sketches of how place and innovation can come together, emerging from a 10-year study of innovation clusters and cities in Canada.

In Chapter 18, *Koichi Ogasawara* discusses user-driven innovation ("User-driven innovation and knowledge integration in elderly care services: A community integration model"). His chapter shows key methodological factors in applying a user-driven co-creation approach to the domain of service innovations in elderly care, based on three field studies and action research conducted in Japan. Through interpretation of the logical structure and service process mechanisms of elderly care, he focuses on the relation between the values of elderly people and the knowledge and skills of experts to fulfil these values. The complexity of dimensions of values and categories of support functions is regulated by a moral game between the service user and experts during the process of service creation. Interpreting the user's status and values during service design by bringing together experts from different disciplines results in services that are easy for users to adopt and facilitates user accommodation of the process of service implementation. This chapter proposes an active cluster net centred on user-driven co-creation as a community integration model that reflects innovative societal structuring in coordinated human service allocation. According to the author, in service innovation, the macro-level care provision system, funding that supports innovation, meso-level clustering of knowledge integration, and the micro-level service interface that takes into account human behavioural factors need to be managed as a whole.

Chapter 19 by *Timo Järvensivu, Katri Nykänen* and *Rika Rajala* ("A holistic model of innovation network management: Action research in elderly health care") introduces a holistic, four stage model for managing innovation networks. The authors argue that the model offers an array of practical advice for managers on how to improve an innovation network's knowledge mobility and network stability through fostering trust and commitment. Network management, in particular innovation network management, is a topic of increasing interest and scope. Research on innovation network management has offered various, but ultimately partial, theoretical and practical contributions. Trust and commitment have been identified as the basic elements of a functioning network, and it is known that there are certain factors that foster or discourage their existence. Networks are different; each network has its particular challenges. Practice-based innovations involve specific challenges for network management. The authors' aim is to look at innovation network management from a holistic perspective, bringing together the relevant but scattered viewpoints and contributions. They use action research to look at what managers can do to manage an innovation network. The resulting holistic model rises from one particular practice-based innovation context—elderly health care in Finland—but the authors argue that it is applicable in other contexts and innovation types as well.

Hiroo Hagino also looks into elderly health care (Chapter 20, "Practice-based innovations at 'Sendan No Oka': Motivation management and empowerment management"). He introduces Sendan No Oka, Tohoku Fukushi University's affiliated nursing home in Sendai, Japan. Sendan No Oka has a strong reputation in Japan for its high quality health care and social services, and its excellent business performance. At its establishment, these were only distant goals. The staff at Sendan No Oka have together tackled many difficult challenges and instituted a wide variety of useful innovations. Management focuses on motivation and empowerment for both clients and staff have been the two pillars of Sendan No Oka's innovation ecosystem. With motivation management and empowerment management based on expertise and seamless professional collaboration, incremental, practice-based innovation has led to the achievement of high-level health care and social services at reasonable cost. Based on these achievements, Sendan No Oka is now stepping on to attempt radical innovation in client-oriented, comprehensive community care.

In Chapter 21, *Minna Saunila, Juhani Ukko* and *Hannu Rantanen* focus on innovation capability ("Innovation capability and its measurement in Finnish SMEs"). The importance of the development of an organisation's innovation capability for its success is highlighted in current literature. According to the authors, this study provides a comprehensive description of the status quo of Finnish SMEs regarding the different dimensions of innovation capability, expertise, work wellbeing, and the measurement of issues related to innovation capability. The study also clarifies whether the results of the studied issues differ depending on the size of the organisation, the industry and the organisational position. The most important insight of the study is that the perceptions of executives and employees differ significantly with regard to all the aspects investigated. The study suggests that both academics and practitioners should focus on the development of new methods and practices for enhancing innovation capability, and especially the measurement of issues related to it.

Finally, in Chapter 22 ("Epilogue: Two modes of practice-based innovation"), we as editors summarise certain aspects concerning practice-based innovation and this book. We introduce our thinking with regard to the meaning and implementation of practice-based innovation in practice.

The book thus informs the reader concerning the essence of practice-based innovation and discusses, for instance, what we should know about it before we can assess its policy implications. The differences between research-based or science-based, on the one hand, and practice-based innovation, on the other, become visible, and prerequisites for sustainable practice-based innovation are investigated. The essence of user-driven innovation is also discussed in several chapters. What is actually new in user-driven innovation; what does it bring with it/demand in innovation activities? What is the new 'place' of the customer; which new methods, competences and attitudes are needed? With what success are they implemented? These issues, for instance, are focused on. The intriguing topic of 'related variety' is studied, and the reader also finds out about how practice-based innovation is communicated, and how innovation systems and communication systems are combined. The relationship between foresight and innovation is investigated.

Creativity in practice-based innovation is given emphasis in several chapters. The reader finds out about how practice-based collective creativity and learning processes are triggered and what kinds of creativity methods are useful in bridging distances in practice-based innovation. After all, the aim is to extend innovativeness to all members of an organisation. New forms of expertise and new roles of actors are needed in practice-based innovation activities; several chapters focus on generation of the necessary expertise and knowledge. Value networks in practice-based innovation and measurement of innovation capability are challenging topics; they are also looked into, for instance, by studying how quality of work life is related to innovation capability.

We extend our sincere thanks to all chapter authors for their important contributions. Without their different fields of expertise and hard work this book could not have been produced. The preparation of this book has been made possible by Tekes, the Finnish Funding Agency for Technology and Innovation, as well as the European Regional Development Fund and the Regional Council of Päijät-Häme. Their financial support is gratefully acknowledged. We thank the publisher, Springer, and all the competent people there, especially Claudia Schmiedeberg, Michael Bursik and Barbara Bethke, who patiently guided us through the book process. We also thank Kristian London for his excellent assistance in polishing the language. The thoughts presented by us in this book have been developed in close collaboration with our great colleagues at Lappeenranta University of Technology, Lahti School of Innovation. Sincere thanks to all of you, too.

Lahti, Finland *Helinä Melkas & Vesa Harmaakorpi*
April 2011

References

Harmaakorpi, V., & Mutanen, A. (2008). Knowledge production in networked practice-based innovation processes – interrogative model as a methodological approach. *Interdisciplinary Journal of Information, Knowledge, and Management, 3*, 87–101.

Harmaakorpi, V., Tura, T., & Melkas, H. (2011). Regional innovation platforms. In P. Cooke, B. Asheim, R. Boschma, R. Martin, D. Schwartz, & F. Tödtling (Eds.), *The handbook of regional innovation & growth*. Cheltenham, UK: Edward Elgar. In press.

Lundvall, B.-Å. (2007). Innovation system research: Where it came from and where it might go. Globelics Working Paper Series No. 2007–01. The Global Network for Economics of Learning, Innovation, and Competence Building System. http://dcsh.xoc.uam.mx/eii/globelicswp/wpg0701.pdf. Accessed 29 December 2010.

Part I
Insights into Practice-Based Innovation and Innovation Strategies

Chapter 2
A Pragmatist Theory of Innovation

Bart Nooteboom

Abstract In this chapter, I develop theoretical foundations for practice-based innovation from the *embodied cognition* school of thought in cognitive science and from the *pragmatist* line of thought in philosophy. From it, I derive the notion of *cognitive distance*. I use the resulting insights to discuss the well-known notions of *absorptive capacity* and *exploitation* and *exploration* from the innovation literature, and the way in which exploitation and exploration are connected. In the analysis I model practice in terms of the notion of a *script*. Cognitive difference ('distance') complicates but also enriches collaboration, and this positive effect is related to the notion of bridging *structural holes*. Among other things, the analysis yields an underpinning of the idea that application is not just a result of research but also a basis for ideas for research, in two-way traffic between theory and practice, and between research and application. The analysis also has implications for innovation policy, in particular for the currently popular principle of *focus and mass*.

2.1 Introduction

In innovation, what is, and what should be, the relation between theory and practice, between science and industry? In the traditional 'linear view' of innovation, science is autonomous and its results are carried into application in business. Since the 1990s, the innovation literature has shifted to a non-linear view, according to which a variety of factors, including not only science and technology but also design, marketing, organisation, training, and entrepreneurship, are parts of a wider innovation system in which feedback also travels from practice to theory, and there

B. Nooteboom (✉)
Tilburg School of Economics and Management, Center for Innovation Research, Tilburg University, Tilburg, The Netherlands
e-mail: b.nooteboom@uvt.nl

H. Melkas and V. Harmaakorpi (eds.),
Practice-Based Innovation: Insights, Applications and Policy Implications,
DOI 10.1007/978-3-642-21723-4_2, © Springer-Verlag Berlin Heidelberg 2012

should be feedback from industry to academia. While this is the 'espoused theory', the actual 'theory in use' often still is one of a linear science and technology push. I propose that one reason for this is that our basic, tacit notions of knowledge favour the linear view, and that for a proper grasp of the non-linear view, we need to reconsider those basic notions.

We talk a lot about the 'knowledge economy', and in innovation, knowledge is central. We should, then, know what we are talking about, and rather than reinventing wheels, let us employ the insights from cognitive science that are available. The problem is that there are different schools of thought in cognition. I opt for the 'embodied cognition' school. I will show that it has deeper roots in philosophy, in the school of American Pragmatism. Hence I call my theory of innovation a pragmatist theory. The crux of it is that action is constitutive of knowledge, that ideas get transformed in their application, which leads to new ideas. The term *application* suggests that what is applied remains what it was, like brushing paint on a surface, while in fact application entails integration with other ideas and transformation of those ideas, depending on the conditions of application. Knowledge and the goals of action shift through action to the extent that one encounters obstacles and novel opportunities. An apt term for such form of action is the anthropologist Levy-Strauss' term *bricolage* (the French term used for 'do it yourself', which also has the connotation of 'muddling through'). This is a key feature of innovative, Schumpeterian entrepreneurship.

After sketching the fundamental ideas, I will use them to discuss and develop a number of concepts that play a role in innovation management and innovation policy.

2.2 Embodied Cognition

The embodied cognition school, in the work of Damasio (1995, 2003) and Lakoff and Johnson (1999), rejects the Cartesian separation of body and mind in the recognition of how cognition is rooted in largely unconscious bodily processes of perception, feelings, and emotions. As a result, cognition is a wide concept, including knowledge as well as feelings and normative evaluations.

The roots of cognition in the body and feelings allow the theory to connect with neural science and with social psychology. In neural science, it connects with the work of Edelman (1987, 1992), who proposed a 'neural Darwinism' where neural structures develop in analogy to evolution, with variety generation via connections between existing structures and selection via reinforcement of structures that generate conduct that is experienced as successful.

In social psychology, embodied cognition connects with the notion of mental framing and with decision heuristics, in which unconscious psychological mechanisms, emotions, and rational evaluation mix (see e.g. Bazerman 1998). Decision heuristics can be interpreted as making sense from a perspective of evolutionary psychology. While they may be substantively irrational by the standards of economic logic, taking into account survival conditions that require rapid perception, interpretation, and response to opportunities and threats, they are adaptive and in that sense procedurally rational.

Cognition and action are situated. People develop repertoires of mental frames, forms of perception, and dispositions to interpretation, judgment, and action, and how selections and combinations from those repertoires are made depends on the situation. From the situation, people adopt cues that trigger mental frames and are assimilated into them to make sense. In this process, people attribute characteristics to situations and people according to mental schemas even when those are not observed. This yields prejudice, but is also functional in fast response and the use of experience.

2.3 Pragmatism

Embodied cognition also connects with the American school of pragmatic philosophy of James, Peirce, Dewey, G.H. Mead and related, more contemporary philosophers, such as Hans Joas (1996). This reference to pragmatism is tricky, since in everyday language pragmatism is misleadingly seen as a shedding of principles and ideals in a muddling through with compromises. Philosophical pragmatism holds that cognition, in a wide sense that includes normative judgments and goals, occurs on the basis of mental dispositions and categories that are developed in interaction with the physical and especially the social environment. Intelligence is internalised practice. Questions of truth soon lead to questions of workability. There is a cycle of interaction between, on the one hand, cognition and language and, on the other, action, in which knowledge and meaning are applied (or better: exercised) in action and there run into limitations and novel challenges that lead to an adaptation or transformation of knowledge and meaning.

For an elaboration of that in an earlier (Nooteboom 2000) and a more recent work (Nooteboom 2009), I have been inspired by the ideas from the developmental psychologist Jean Piaget about how intelligence develops in children. The gist of it is that in the process of assimilating experience into existing cognitive structures (*assimilation*) in different stages, those structures are transformed (*accommodation*). An important element here is that in novel applications (*generalisation*), existing knowledge and competence are subjected to novel challenges and opportunities that yield new insights into the limitations of existing competence and yield inspiration as to how they may be overcome. In the face of failure, one first tries out different options from existing repertoires of competence (*differentiation*), and if that does not suffice, one adopts elements from practices in the novel context that appear to succeed where one's own practice fails (*reciprocation*). This yields hybrid structures of old and new elements that allow one to try out new elements without yet surrendering the basic design or architecture, and to see what elements have promise, to find where in the structure the obstacles lie to the full realisation of their potential, which gives hints as to where to make more radical, architectural change, in *accommodation*. This, I propose, yields a theory of discovery. It connects very well with the pragmatic principle of exploring while engaging in application.

The pragmatic perspective has important implications for notions of rationality, uncertainty, knowledge, learning, innovation, entrepreneurial behaviour, and innovation policy, which I will elaborate on in this chapter. The economic doctrine of rational choice of means to achieve pre-established goals according to given preferences is fundamentally misleading. According to pragmatism, goals, means, and action interact. We do have goals, preferences, and largely subconscious dispositions to action, but they are revised as the result of discovery of means and of results of actions, especially when action encounters problems or new opportunities are found. This happens especially in innovation, but on a lower scale also in ordinary life. The dynamics of cognition and action is not an add-on to statics as the base case; it is the base case. Situations and institutions not only condition goal achievement but also are constitutive of goals.

Entrepreneurs engage in what Levy-Strauss called *bricolage*. They do have some initial view of where they want to go, but on the way they shift targets as they encounter new problems and opportunities. This has important implications for innovation management and innovation policy, as I will discuss.

Since goals, intentions, perceptions, and meanings become determinate and shift in interaction with other people, action and cognition are inherently social, and the methodological individualism of economic theory fails fundamentally.

Embodied cognition and pragmatism explain why collaboration is important for innovation. To elaborate on this, I will next discuss the notions of absorptive capacity and cognitive distance.

2.4 Absorptive and Expressive Capacity

The well-known notion of absorptive capacity requires a conceptual widening to encompass the wider notion of cognition: not only the *competence* side of substantive understanding but also the *governance* side of insight and empathy with respect to styles of thought and action, motives, survival conditions of a partner firm, and moral views and predilections (Nooteboom 2004). On the organisational level, it includes ways of communication and knowledge sharing, organisational memory, and cultural features concerning views and attitudes towards the outside world, in organisational 'cognitive focus' (Nooteboom 2009).

Absorptive capacity refers to the receiver side in communication and needs to be complemented by expressive capacity on the sender side, in the ability to be clear, to explain and to give clever examples and metaphors that trigger understanding. Together, wider absorptive capacity and expressive capacity yield a wider notion of collaborative capacity. In connection with the earlier text, absorption equals *assimilation*.

In terms of the earlier analysis, interaction between people is a source of new knowledge, because the attempt to fit knowledge into the absorptive capacity of the partner may be seen as *generalisation*. Next, misfits, through misunderstanding, yield an incentive to 'put it differently', through *differentiation*, and there are perceived needs and opportunities to try and fit elements of knowledge from the one into the cognitive framework of the other, through *reciprocation*, yielding novel combinations. The process of learning is enhanced because people can help each other to *assimilate, differentiate,* and *reciprocate.*

2.5 Cognitive Distance

Cognitive construction on the basis of interaction with other people entails 'crossing cognitive distance': different people develop different cognitive structures along different life paths in different environments, and to collaborate one must cross that distance. Cognitive distance arises between different people, organisations, and scientific disciplines as well as between theory and practice. Cognitive distance entails difference in cognition in the narrow sense of knowledge but also difference in moral perceptions and views. This combination of the intellectual and the moral has the implication for the study of firms that we should combine perspectives of competence and governance (Nooteboom 2004), while in the literature on firms, they run in largely separate streams, in on the one hand studies of competence, learning, innovation, and the like, and on the other hand studies of collaboration and its governance, as in transaction cost economics.

In economics and society, cognitive distance results in both a problem and an opportunity. The problem is that to the extent that cognitive distance is greater, people understand each other more or less imperfectly, have different normative views and inclinations, and have less empathy, less ability to imagine themselves in the position of the other, which all limit ability to collaborate. The positive side of cognitive distance is that it provides an opportunity for learning and innovation. Hence, cognitive distance can be too small to generate novelty or too large to utilise its opportunities.

This may be modelled as follows. If we model the decline, with cognitive distance, of ability to collaborate as a downward sloping straight line, the increase of novelty potential as an upward sloping straight line, and performance of innovation by interaction as the mathematical product of the two, the result is an inverted U-shaped parabola. This yields the notion of 'optimal cognitive distance'. With this term, I do not wish to suggest that the optimum can be calculated prior to choice of partners, but rather that it is approximated by trial and error.

The optimum depends on how radical the innovation involved in the interaction is. If we adopt the well-known distinction between exploitation, defined as improvements within a basic design, set of principles, or architecture, and exploration as the breaking of such frames, in exploitation the marginal disutility of lack of understanding and agreement (slope of the downward sloping line) is relatively high, and the marginal utility of novelty (slope of the upward sloping line) is lower than in exploration, resulting in a lower optimal distance.

The optimum is not fixed in time. It depends, in particular, on the ability to collaborate at any level of cognitive distance, and this may increase as a function of the accumulation of knowledge and experience in collaboration. That may be modelled as an upward shift of the downward sloping line that represents ability to collaborate as a function of cognitive distance. This causes optimal distance and innovative performance to increase (Nooteboom 2000). This shows how the development of ability to collaborate with people who think differently yields economic advantage. The idea of optimal cognitive distance and its shift on the basis of experience is tested empirically (econometrically) in a study of innovation in inter-firm alliances (Nooteboom et al. 2007).

2.6 Structural Holes

Collaborative relationships can last too long, leading to a reduction of cognitive distance as a result of knowledge sharing, to the point that they no longer have anything new to tell each other if the relationship is exclusive, i.e. excludes collaboration with other parties on the same subject. Thus, there is also an optimal duration for such relationships: long enough to bring cognitive distance down to an optimal level, but not beyond that, to too little distance. However, this principle is broken when the relationship is not exclusive, when both sides of the relationship also engage in other, non-overlapping partnerships, whereby both sides are rejuvenated with new knowledge from mutually unconnected outside partners. This, in fact, reflects the well-known principle of 'bridging structural holes' (Burt 1992). There is innovative potential in building a connection between previously unconnected communities (leaving a 'structural hole' between them). In this case, long-lasting connections (via the bridge) can maintain innovative potential.

2.7 Exploration and Exploitation

A well-known notion in the innovation literature is that of exploration and exploitation (March 1991). Exploitation refers to improvements within basic logics, designs, or architectures, while exploration entails their change. The distinction is related to that of incremental and radical innovation. In terms of the theory of discovery: *generalisation* and *differentiation* are exploitative, while *reciprocation* and *accommodation* are exploratory.

One of the greatest challenges to organisations and management is to somehow combine exploitation and exploration. The first is needed to survive in the short term and the second to survive in the long term, so one should be engaged in both. Exploitation is based on exploration, and exploration must somehow be inspired from the experience of exploitation. But how does one combine exploitation, which presupposes a given basic design logic, given structures of practice, with a given distribution of functions and roles, a shared purpose and clear, established ideas and standards, with exploration, which entails their change, with shifting roles and meanings? This reflects both the need for the link between innovation and practice and the difficulty of it.

The theory of discovery sketched above gives part of the answer. It shows how in exploitation, by moving to a new area of application (*generalisation*), one can keep on exploiting while nevertheless setting out on a path of exploration through the stages of *differentiation, reciprocation,* and *accommodation.*

This sequence does not, however, solve the problem of how to combine exploitation and exploration at the same time and place. One solution to this is the well-known separation of production and R&D. The perennial problem there is that production people blame development people for not being conscious of problems and costs of production, while development people blame production people for being too conservative and missing out on new opportunities in the market or in technology. Exploration and exploitation require different mentalities that can be difficult to hold together in a single 'organisational focus'. A solution may then be to locate the two tasks in different organisations. We find this, for example, in the pharmaceutical industry, with small firms engaging in exploration and moving the outcomes on to exploitation in larger companies that can better deal with clinical testing and large-scale, efficient production and distribution.

In collaboration, optimal cognitive distance is greater for exploration, to generate more radically new 'novel combinations', than it is for exploitation, where the scope of novelty is less and the penalty for misfits and misunderstandings is larger. This was borne out in the empirical study of Nooteboom et al. (2007). There, exploration was defined and measured as innovation outside the technological profile of the firm and exploitation as inside it.

2.8 Scripts

The theory of discovery sketched earlier can be further clarified and made more tangible with the notion of scripts (Nooteboom 2000). A script is a graph of nodes connected by linkages. A script may be abstract, depicting the logical structure of an argument or theory in logical implication or causation. It may also represent a physical process, such as a production process, a service process, or a user process, with the nodes as component activities and the linkages indicating relations of succession in time and transfer of goods or information. These connections may indicate various forms of dependence (Thompson 1967): sequential dependence, where one activity feeds into the next; pooled interdependence, where different nodes derive inputs from a common source or contribute to a common output; or reciprocal interdependence, where they interact. Activities within nodes can themselves be represented as (sub)scripts, and the script as a whole may be inserted in a wider process (superscript). A classic example is a restaurant, with nodes of arrival, seating, ordering, serving, eating, paying, and leaving. The paying node may allow for alternatives of cash, check, debit card, or credit card, and each mode of paying has its own subscript. The superscript is the higher system of supply to the shop, waste disposal, arrival, parking, and departure of customers, etc.

There are connected user and producer scripts. It is a well-known principle in marketing for producers to envisage the user scripts in which their product is to be inserted. Thus, a new device for a car is to be inserted in a driver script, a maintenance script, and a script for the production of the car. Acceptance of the innovation depends on how well it fits into existing user scripts. If the product requires an entirely new user script, its risk of rejection is high. If it does not fit into an efficient production script, then efficiency and cost are at risk.

Now the script concept can be used to clarify different concepts of innovation. Does the innovation entail a change in a single node or in the architecture of nodes? If it is a change of architecture, does it preserve existing nodes? An example is the innovation of a self-service restaurant from a service restaurant. With some internal changes, the nodes of arriving, seating, selecting food, ordering, eating, paying, and leaving are preserved, in the different order of arriving, selecting food, paying, seating, eating, and leaving. If the change concerns a single node, say the node for paying, does it change all alternatives or only one, such as abandoning payment by cheques due to the ubiquity of credit card ownership?

If the innovation changes a node, how wide are the repercussions for other nodes to maintain systemic integrity? Systems with many and densely and tightly connected nodes are more vulnerable to local change than a ‘stand-alone’ system with more-or-less isolated and weakly connected nodes. As a result, in a more loosely coupled, stand-alone system it is easier to combine exploration and exploitation within the same structure than in a densely, strongly coupled system. In the former, there is more room for local variation and experimentation without thereby jeopardising the integrity of the system.

The script notion can be used to clarify the theory of discovery, as follows. *Generalisation* now is insertion of a script into a new superscript. *Differentiation* is the selection of a new form of enacting a node from an existing repertoire (think again of the payment node in the restaurant script), or extension of such a repertoire. *Reciprocation* is adoption of a new node in an existing script. *Accommodation* is a re-arrangement of old and new nodes in a novel script. Exploitation preserves script structure, and exploration breaks it up.

2.9 3rd Spaces

In innovation policy, a hot topic is ‘valorisation’: how to put research to more productive use. Usually, the debate is limited to increasing the application of results from existing research or increasing the share of applied, contract research. The latter should not be taken too far, since fundamental research is of unpredictable value, and focusing only on short-term practical use would kill the goose that lays the golden eggs. However, from the present pragmatic perspective, application (exploitation) should also be a source of inspiration for fundamental research (exploration). Application is just as much a test and source of inspiration as other empirical testing is.

So, in a recent (2008) advisory report to the government, produced by the Dutch Scientific Council for Government Policy (WRR), we proposed the institution of so-called 'third spaces' (see Nooteboom and Stam 2008). These are actual or virtual spaces where applied scholars from university and R&D staff from industry meet to conduct joint projects, such as developing new ideas and developing and testing prototypes. Here, scientists from academia have an opportunity to test their ideas and developers from industry get some time off to reflect on how things may be done differently, away from the short-term pressures in their firms. Two conditions need to be met for this to work. First, this activity must be seen as legitimate and as contributing to an academic career, next to peer-reviewed scientific publications. Conversely, industry should see this as a legitimate activity of staff, which also contributes to their careers in industry, and should be willing to contribute to the funding.

This kind of activity may be seen to resemble what is done at state-supported technological institutes in between academia and industry, but there is one big difference. Third spaces engage in temporary, project-based activities, lasting, say, four months, after which participants disperse again to their home environments. This facilitates variety and turnover of people and ideas, and prevents institutionalisation where technological institutes, after a while, start to compete with both universities and industry and present a barrier between them rather than facilitating interaction between them.

2.10 Focus and Mass?

Makers of innovation policy are tempted by the logic of 'focus and mass'. The argument is as follows. A country cannot excel in every field and must therefore focus on what it is exceptionally good at ('focus'), in designated 'key areas'. Also, research and development in a certain field should be concentrated, to avoid wasteful competition where different researchers at different locations are studying the same field ('mass').

The problem with focus is that current excellence in a field is not necessarily indicative of success in the future. In innovation, future success is inherently unpredictable. If it were predictable, it would no longer be innovation. Also, the most radical innovations tend to be 'competence destroying', in 'creative destruction'. Precisely because one is currently successful, one can get locked into exploitation and fail to get out into exploration, which therefore is typically conducted by outsiders and newcomers. Also, innovation by novel combinations typically occurs across the boundaries of current fields rather than within them. And, finally, innovation occurs everywhere, in every industry, and cannot be seen as locked up in any existing industries. An example is retailing, which has exhibited radical innovation, mostly in the use of bar codes that has revolutionised distribution, marketing, and product development. Yet no one would designate retailing as a 'key sector', because it is not 'sexy'.

The problem with mass is that different, competing approaches to a given field increase variety, and that is good for innovation. This insight is highlighted in an evolutionary approach to innovation, on the basis of the three basic principles of evolution: variety generation, selection, and dissemination. Policy makers tend to adhere to an 'intelligent design' view of innovation policy that, in this area, is as misguided as it is in biology. The favourable view of mass is also based, implicitly, on an argument of economy of scale. Admittedly, some minimum efficient scale of research does apply in some cases. For a laboratory with large and expensive machinery and instruments, with corresponding maintenance, one needs a sufficient number of researchers for efficient utilisation. But in some cases, that can be taken care of by scientists traveling to the facility from abroad, as in the case of observatories on faraway mountains for astronomers, and the huge particle accelerator in Geneva for nuclear physicists. Also, it is not the number of scholars doing certain research at a certain location that counts, but the number of researchers that they interact with, from various locations in the world.

I do admit that there is waste in the R&D system, but it is of a different source. It lies in the fact that people from different disciplines conduct research from different perspectives without taking cognisance of each other's insights. If people knowingly compete, that is good, in my view, but if they unknowingly duplicate, it is not good. Economists talk about trust, for example, while ignoring insights from sociology and social psychology, and talk about 'behavioural economics' while ignoring other behavioural sciences. This has nothing to do with lack of mass, but with myopia.

Such myopia is frustrating for policy, since policy makers cannot afford to focus on one perspective while the phenomena they have to deal with present themselves with all their perspectives. Specialisation is good, but integration of perspectives should also be part of science, and not left as a puzzle for policy makers.

2.11 Conclusion

The arguments for practice-based innovation are not only practical but also philosophical, epistemological. To some, *practice-based innovation* sounds pragmatic in the derogatory sense of ignoring foundations and principles, allowing for ignorance, incoherence, or even inconsistency. However, there is a more respectable notion of pragmatism as a line of thought in philosophy, which recognises that application is part of a learning process where ideas change in their application and yield new ideas, so that application is part of discovery. Not only usefulness but also discovery requires application, to provide the exploitation that inspires exploration, as one moves from one application to another.

References

Bazerman, M. H. (1998). *Judgement in managerial decision making*. New York: Wiley.

Burt, R. (1992). Structural holes: The social structure of competition. In N. Nohria & R. Eccles (Eds.), *Networks and organizations: Structure, form and action* (pp. 57–91). Boston: Harvard Business School Press.

Damasio, A. R. (1995). *Descartes error: Emotion, reason and the human brain*. London: Picador.

Damasio, A. R. (2003). *Looking for Spinoza*. Orlando: Harcourt.

Edelman, G. M. (1987). *Neural Darwinism: The theory of neuronal group selection*. New York: Basic Books.

Edelman, G. M. (1992). *Bright air, brilliant fire: On the matter of mind*. London: Penguin.

Joas, H. (1996). *Die Kreativität des Handelns*. Suhrkamp: Frankfurt a.M.

Lakoff, G., & Johnson, M. (1999). *Philosophy in the flesh*. New York: Basic Books.

March, J. G. (1991). Exploration and exploitation in organizational learning. *Organization Science, 2*(1), 101–123.

Nooteboom, B. (2000). *Learning and innovation in organizations and economies*. Oxford: Oxford University Press.

Nooteboom, B. (2004). Governance and competence, how can they be combined? *Cambridge Journal of Economics, 28*(4), 505–526.

Nooteboom, B. (2009). *A cognitive theory of the firm*. Cheltenham, UK: Edward Elgar.

Nooteboom, B., & Stam, E. (Eds.). (2008). *Micro-foundations for innovation policy*. Amsterdam: Chicago University Press.

Nooteboom, B., Van Haverbeke, W. P. M., Duijsters, G. M., Gilsing, V. A., & V. d. Oord, A. (2007). Optimal cognitive distance and absorptive capacity. *Research Policy, 36*, 1016–1034.

Thompson, J. D. (1967). *Organizations in action*. New York: McGraw-Hill.

References

[illegible] M. B. (1988). [illegible] New York: W[illegible]

Burt, R. (1992). Structural holes: The social structure of competition. In N. Nohria & R. G. Eccles (Eds.), Networks and organizations: Structure, form and action (pp. 57–91). Boston: Harvard Business School Press.

[illegible]

Chapter 3
Combining Foresight and Innovation: Developing a Conceptual Model

Tuomo Uotila, Martti Mäkimattila, Vesa Harmaakorpi, and Helinä Melkas

Abstract Foresight and innovation are activities closely linked with each other, the former providing inputs for the latter. However, there have been few attempts to build conceptual and theoretical bridges between these two activities. In this chapter, we present a conceptual model depicting the connections between foresight and innovation activities and learning. Into this broad model we have combined, in a novel way, much-used and well-known concepts and ideas, such as exploration and exploitation, absorptive capacity, three modes of foresight activities, information quality attributes, and information brokerage.

3.1 Introduction

The success of economic actors is strongly related to their adaptability to the emerging techno-economic environment. Sotarauta and Srinivas (2005: 35–36; 2006) make a distinction between strategic adaptation and pure adaptation; the former refers to an actor's capacity to change its destiny by adapting itself to changes and reshaping its local environment. Pure adaptation is more reactive by nature than this more proactive strategic adaptation, which is also strongly related to an actor's ability to learn.

Strategic adaptation is based on decisions that have to be made in a great uncertainty. This uncertainty can be reduced by creation of future-oriented knowledge. Future-oriented knowledge is often very challenging to use in an actor's renewal process, since (1) the possible futures are hard to outline, (2) future-oriented knowledge is even more abstract than tacit knowledge, and (3) due to its nature, future-oriented knowledge is hard to adopt in an actor's organisational learning processes and strategic routines (Uotila et al. 2006). To make use of

T. Uotila (✉) • M. Mäkimattila • V. Harmaakorpi • H. Melkas
Lahti School of Innovation, Lappeenranta University of Technology, Lahti, Finland
e-mail: tuomo.uotila@lut.fi

H. Melkas and V. Harmaakorpi (eds.),
Practice-Based Innovation: Insights, Applications and Policy Implications,
DOI 10.1007/978-3-642-21723-4_3,

future-oriented knowledge, actors need a special dynamic capability: visionary capability. In this context, visionary capability refers to an actor's ability to outline the potential development directions based on paths already travelled – utilising the opportunities emerging from the changing techno-economic paradigm (Harmaakorpi 2004).

Nowadays, knowledge is generally considered the most important source of competitive advantage, but even the most specialised forms of knowledge are becoming a short-lived resource – due to, for example, the accelerating pace of technological change. This emphasises the need for foresight activities at all levels – national, regional, and organisational – and the integration of foresight, innovation, and learning activities. Eerola and Väyrynen (2002: 15) and Salo (2000: 207) see the need to integrate foresight activities more closely into today's decision-making and (action) planning processes as an important challenge for those carrying out foresight activities. This integration is needed in order to increase the effectiveness of foresight activities.

According to Bootz (2010), an agreement between practitioners and theorists about the strong ties linking strategic foresight and learning processes (especially organisational learning) seems to be emerging, but the true nature of this link still remains vague. In this chapter, we elaborate on a conceptual model depicting the connections between foresight and innovation activities and learning. The proposed model makes use of concepts and ideas presented earlier by March (1991) (exploration and exploitation), Cohen and Levinthal (1990), Todorova and Durisin (2007) (absorptive capacity), Salmenkaita and Salo (2004) (three modes of foresight activities), Melkas (2004) (information quality attributes), and Burt (1997; 2004) (information brokerage). Although the basic elements of the model are already well known and documented, in the proposed model they are combined in a novel and holistic way, making the theoretical contributions presented here valuable. The model emphasises the role of absorptive capacity as an important dynamic capability for an actor's success in carrying out innovation processes and discusses ways of enhancing absorptive capacity in order to better link the results of foresight processes and organisational innovation activities and learning. Absorptive capacity includes features of both visionary and innovation capabilities. In order to use and absorb information and knowledge, an actor needs to be able both to explore the future-oriented knowledge generated through foresight processes and exploit it in its innovation processes.

3.2 Networked Innovation Processes

Adaptability, a crucial factor affecting the economic success of actors, depends on their existing resources and capabilities for renewing these resources in order to keep them competitive (Pihkala et al. 2007). Visionary capability is becoming more and more important in renewing an actor's resource base as well as in network leadership (Harmaakorpi 2004), which again can be defined as action that leads all operations and resources of the network in the desired direction (Pihkala et al. 2007).

This chapter takes a resource-based view as its point of departure. In a shift from the earlier research focusing on internal resources, there is an increasing interest in external resources and capabilities available to the actor through networks (Zaheer and Bell 2005). Accordingly, and in line with the teachings of non-linear, multi-actor innovation processes, economic actors are not seen as isolated islands, but as entities belonging to regional, interregional, and global innovating networks and searching for information and knowledge from these networks. Therefore, competitiveness-securing resource configurations have to be considered at the level of innovation networks – as individual actors are embedded in these networks. The capacity to absorb future-oriented knowledge in a dynamic fashion is seen as a crucial competitiveness factor for individual actors and innovation networks (Uotila et al. 2006).

Social structure, especially in the form of social networks, affects economic outcomes, since these networks affect the flow and quality of information (Granovetter 2005). Granovetter (1973) defines the concepts of strong ties and weak ties in social networks. The strength of a tie is a combination of the amount of time, emotional intensity, intimacy, and reciprocal services that characterise the tie. Strong ties are characterised by common norms and high network density. These strong ties are easier for innovation, since they normally include a relatively high amount of trust, common aims, and the same kind of language for communication. However, weak ties are reported as more fruitful for innovations, because more novel information flows to individuals through weak ties than through strong ties (Granovetter 2005). People in the same strong networks tend to share the same knowledge basis, which prevents the Schumpeterian knowledge-combining innovation processes from emerging. Burt (2004; 2005) has developed the 'strength of weak ties' argument further by arguing that innovations are most likely to be found in structural holes between dense network structures (see also Burt 1992; Walker et al. 1997; Zaheer and Bell 2005). An actor able to span structural holes in a social structure is at a higher 'risk' of having good ideas: new ideas emerge from selection and synthesis across the structural holes between groups (Burt 2004). An innovation system rich in structural holes offers a lot of opportunities for new, networked innovation processes.

The weak links or structural holes enabling the biggest innovation potential are somewhat problematic for innovation processes. In order to be able to utilise the innovation potential in these structural holes, information should often be transferred between partners who have totally different knowledge interests. However, this information transfer is hampered by what is called distances in the innovation network; and in order to promote and facilitate information transfer between partners in the innovation network, it is important to acknowledge these distances and their origins.

3.3 Crossing Distances as an Enabler of Learning and Innovation

Following Burt's structural hole argumentation, several authors – for example Knoben and Oerlemans (2006) and Harmaakorpi et al. (2006) – have analysed distances in innovation networks, drawing on the earlier analysis of Boschma on cognitive distance (2005), but describing the relevant dimensions a bit differently. The concept of cognitive distance (and other distances as well) refers to the fact that people belonging to different communities use different frameworks and language, which makes it difficult, sometimes even impossible, to share knowledge (Cillo 2005). According to Cillo (2005), some researchers have even rejected the idea of knowledge transfer models among people who do not share the same context where the knowledge was created. The distinction between 'knowledge' and 'information', which is brought up later in this chapter, sheds some light on this issue.

Harmaakorpi et al. (2006) have identified six central dimensions of distance.[1] The first dimension is the pure physical closeness of the actors, labelled as geographical distance. Harmaakorpi and his colleagues claim that this dimension should be approached as nothing more than what its name indicates. As argued in many studies concerning proximity and distance, this analytical separation enables us to see the real effect of being geographically close to each other, as opposed to the effects of other forms of distance. Cognitive distance refers to closeness of knowledge bases and forms of expertise of the relevant actors. Cognitive distance contributes to accumulation of knowledge and efficiency of interaction. Cognitively close actors are able to assume a certain common knowledge base that does not have to be defined or negotiated. Organisational distance refers to arrangements to coordinate transactions and enable exchange of information within and between organisations. The importance of organisational distance lies in the need to organise utilisation of knowledge and other resources owned by a variety of actors (Boschma 2005: 64). A close but not identical notion to organisational distance is functional distance. Functionally close actors act in areas of expertise close to each other – for example, in the same industry.

While organisational and functional distances are connected purely to relationships between institutions, social distance is fundamentally about relationships between people. It refers to the intensity of trust-based social relations, such as friendship or kinship. The notion of social proximity comes close to the concept of social capital as defined, for example, by Tura and Harmaakorpi (2005) and Burt (2005). When moving from the micro-level of social capital to the macro-level, Harmaakorpi et al. (2006) also identified an institutional or cultural form of distance that is sometimes seen within the context of social capital. Knoben and Oerlemans

[1] Harmaakorpi et al. (2006) and Knoben and Oerlemans (2006) have used the term 'proximity' instead of 'distance', but here we prefer using the latter term in order to emphasise the challenges of crossing distances.

(2006) have, in their literature review, identified an additional component of proximity and labelled it technological proximity. It is based on shared technological experiences and knowledge bases. Technological proximity between actors facilitates acquisition and development of technological knowledge and technologies.

In addition to these distances presented (see Table 3.1), Parjanen et al. (2011) have proposed yet another variety that they labelled as *temporal distance* – referring to differences in the ability to imagine potential futures and make use of future-oriented information and knowledge generated through, for example, foresight activities. This temporal distance manifests itself in the ways in which actors perceive the future – in a reactive or proactive manner. In terms of the arguments presented in this chapter, it is of special interest (see Table 3.1).

Table 3.1 Types of distance in innovation networks (Harmaakorpi et al. 2006; Parjanen et al. 2011)

Distance type	Source	Innovation potential
1. Geographic	Physical distance between actors	Geographic proximity does not automatically lead to innovations, but it may, for instance, facilitate social proximity.
2. Cognitive	Differences in ways of thinking and knowledge bases	A certain degree of cognitive distance enables creation of new innovations.
3. Communicative	Differences in concepts and professional languages	When making a new idea understandable, concepts from other fields or sciences, for instance, may be utilised.
4. Organisational	Differences in ways of coordinating the knowledge possessed by organisations and individuals	An organisation should have both strong and weak links in its network.
5. Functional	Differences in expertise in different industries or clusters	It is useful to obtain novel information from outside of one's own field of operation. In such cases, the information often needs to be adapted to the field of operation in question.
6. Cultural	Differences in (organisational) cultures, values, etc.	The challenge is to get people working in different organisational cultures to collaborate.
7. Social	Social relationships and the amount of trust included in them	Innovations require interaction among different kinds of actors. Trust helps in creation of radical ideas.
8. Temporal	Differences in ability to imagine possible, potential futures	The challenge is to acquire and assimilate future-oriented knowledge so that it could be exploited in a proactive manner.

3.4 The Interplay between Foresight and Innovation

New, networked leadership tries to promote learning and includes an active interpretation of signals for change (Pihkala et al. 2007). Aside from technological signals, these signals can be political, social, environmental, and so on. In order to receive these signals, central players and decision makers need tools that can provide them with meaningful, future-oriented information and shared visions (or frames of reference) that they can use to anticipate the consequences of their choices and negotiate relevant strategies. Foresight is among the tools considered useful in this respect (Eerola and Jørgensen 2002).

Coates (1985: 30) defines foresight as the overall process of creating an understanding and appreciation of information generated by looking ahead. It goes further than forecasting, including also aspects of networking and preparation of decisions concerning the future. Still, it is not planning, but provides 'information' about the future and is therefore one step in planning and preparation of decisions (Cuhls 2003). It includes qualitative and quantitative means for monitoring clues and indicators of evolving trends and developments, and is best and most useful when directly linked to the analysis of policy implications. Thus, technology (and also other) foresight activities may, at best, have a vital role in this renewal process by 'giving time and direction' to decision-makers and thus supporting, for example, strategy formulation.

Learning is facilitated by questioning of either decision-makers' or more widely all actors' representations in the organisation, and learning phenomenon then may be expressed through collective change, mobilisation, and strategy appropriation (Bootz 2010). Foresight can be used to acquire new knowledge from outside sources or conventional domains of the actor, which further enriches and complements the actor's existing knowledge. In this way, foresight processes facilitate the recombination of explicit knowledge and its absorption and conversion into tacit knowledge (Roveda and Vecchiato 2008). Also Cuhls (2003: 96) sees the communication effect of pre-assessing future options or decisions and mobilising and bringing together the different stakeholders of the innovation system (the 'wiring up' effect) as an important result of foresight processes.

Indeed, many authors have related innovation to the concept of future, and in some cases even consider the use of futures research (in its various forms) in innovation processes crucial for the success of an innovation, or argue that at least it can contribute positively to the quality of an innovation process (Van der Duin 2004; du Preez and Pistorius 1999: 215). But even though this relation is acknowledged and although foresight activities can be regarded as an activity providing inputs to the learning and innovation process, not much literature has been published on how foresight is actually used to facilitate innovation processes and how it relates to organisational renewal processes. According to van der Duin (2004), "futures research seems to have the most impact in the first phases of the innovation process, the fuzzy front-end, where it has the function of inspiring people to think about new innovations, new future developments and to challenge prevailing perceptions".

Although foresight exercises have supplied evidence on claimed benefits – such as the generation of future-oriented knowledge and strengthening of collaborative networks (Salmenkaita and Salo 2004), there are still several challenges related to foresight, especially how to improve the effectiveness of foresight and how to strengthen the dialogue between foresight, planning, and innovation processes (Major and Cordey-Hayes 2000; Salo 2000; Eerola and Jørgensen 2002: 29; Eerola and Väyrynen 2002: 15). One special challenge is related to SMEs, where scarcity of time, money, resources, and personnel cause difficulties in utilising the results of foresight exercises (Major and Cordey-Hayes 2000).

3.5 Elements of the Conceptual Model

3.5.1 Exploration and Exploitation

A central issue in studies of adaptive organisational processes is the relationship between the concepts of exploration and exploitation, the former relating to the search for new possibilities and the latter to old certainties (March 1991). According to March (1991), exploration includes things captured by terms such as search, variation, risk-taking, experimentation, play, flexibility, discovery, and innovation. Exploitation, on the other hand, includes such things as refinement, choice, production, efficiency, selection, implementation, and execution. In his article, March (1991) noted that both exploration and exploitation are essential functions for organisations to prosper in the long run, but problems stem from the fact that they compete for the same scarce organisational resources. Thus, the big challenge at an organisational level is to find a suitable trade-off between exploration and exploitation.

The concepts of exploration and exploitation also include a temporal conflict. According to March, adaptive processes characteristically improve exploitation more rapidly than exploration. However, remaining competitive in the long run depends critically on sustaining a reasonable level of exploration, so these tendencies to increase exploitation and reduce exploration can increase unwanted path-dependant behaviour in organisations – being potentially, as March said, self-destructive. So finding a suitable trade-off between exploration and exploitation also involves conflicts between short-term and long-term gains and between gains to individual knowledge and gains to collective knowledge (March 1991). In the case of practice-based innovation, the overarching theme of this book, when innovation processes occur within a very practical context, it seems reasonable to argue that there is an in-built tendency to emphasise exploitation-type activities at the expense of exploration-type activities.

According to Li et al. (2008), some scholars interpret exploration as distant and exploitation as proximate knowledge searches in an innovation network. They concluded that most studies that employed the idea of local or distant knowledge search interpreted exploitation as activities that search for familiar, mature, current, or proximate knowledge, while exploration was seen to consist of activities that search for unfamiliar, distant, and remote knowledge. This notion relates especially to technological innovation: a local search provides a firm with advantages in making incremental innovations, while a distant search might bring opportunities for a firm to achieve radical innovations.

3.5.2 Absorptive Capacity

Information and knowledge are fuels for innovation, and in sourcing them from networks, organisational absorptive capacity becomes a key issue. Absorptive capacity was originally defined by Cohen and Levinthal (1990) as an organisation's ability to value, assimilate, and apply new external knowledge. Kim (1998) argues that absorptive capacity requires learning capability and develops problem-solving skills; learning capability, again, is the capacity to assimilate knowledge for imitation, while problem-solving skills create new knowledge for innovation. Moreover, Zahra and George (2002) define two different types of absorptive capacity: potential absorptive capacity, which is important in acquiring and assimilating external knowledge, whereas realised absorptive capacity refers to the functions of transformation and exploitation of the knowledge collected. Both are, naturally, important in innovation processes: potential absorptive capacity enables the exploration of knowledge (often) through the weak ties of the innovation network, and realised absorptive capacity secures the exploitation (often) through the strong ties of the networks. Absorptive capacity is crucial when pondering questions about future-oriented knowledge adaptation in innovation networks; higher absorptive capacity enables easier crossing of structural holes in the innovation system.

To better understand the characteristics of absorptive capacity as a dynamic capability, we have to take a closer look at its different parts:

1. Acquisition, referring to an actor's capability to identify and acquire externally generated knowledge that is critical to its operations;
2. Assimilation, referring to the actor's routines and processes that allow it to analyse, process, interpret, and understand the information obtained from external sources;
3. Transformation, denoting an actor's capability to develop and refine the routines that facilitate combining existing knowledge and the newly acquired and assimilated knowledge;
4. Exploitation, which as a capability is based on the routines that allow actors to refine, extend, and leverage existing competencies or to create new ones by incorporating acquired and transformed knowledge into their operations (Zahra and George 2002).

According to these definitions, absorptive capacity is like a funnel, where potential absorptive capacity (visionary capability) secures the newness and diversity of the knowledge needed, whereas realised absorptive capacity (innovative capability) stands for operationalisation of the new knowledge within the existing processes in order to make the actual innovation processes to take place.

Todorova and Durisin (2007) have, in their article, presented an interesting interpretation of the concept of absorptive capacity. They have criticised Zahra and George's (2002) model based on several points, the main being that the phases of absorptive capacity presented by Zahra and George (acquisition, assimilation, transformation, exploitation) are not consecutive but alternative routes (route one being acquisition – assimilation – exploitation, AAE, and route two being acquisition – transformation – exploitation, ATE) in a learning process. Based on findings from cognitive psychology, they propose that when a new idea fits well into existing cognitive schemas, it is only slightly altered to improve the fit and then incorporated into the existing cognitive structures. The existing cognitive structure does not change, and the knowledge is 'assimilated'. On the other hand, a process of transformation takes place when new situations or ideas cannot realistically be altered to fit existing knowledge structures. In the latter case, the cognitive structures of the individuals themselves must be transformed to adapt to an idea or a situation that they cannot assimilate (Todorova and Durisin 2007).

3.5.3 Explicit, Emergent, and Embedded Foresight Modes

A fruitful starting point for constructing a conceptual model linking foresight, organisational renewal, and innovation is offered by Salmenkaita and Salo (2004), who have typified three different kinds of technology foresight exercises – explicit, emergent, and embedded – and explored their interrelationships. Although their analysis is related to the context of shaping and implementing science and technology policies, we think the classification they used is applicable also in a wider organisational context.

The first foresight mode is what Salmenkaita and Salo (2004) call an explicit foresight process, in which the participants in the process represent their own separate interests and areas of expertise. They are also implicitly influenced by their own organisational context. Salmenkaita and Salo see an explicit foresight process as a highly structured approach to the development of shared visions about future priority areas, where systematic methodologies are used in order to integrate expert inputs through voluntary consultations and contributions.

The second foresight mode is what Salmenkaita and Salo call an emergent foresight process. It presumes that the participants have overlapping interests, are already sharing similar views of future opportunities, and are willing to commit resources to 'target activity'. Their perhaps competing and complementary views are, through iterative discussions, synthesised into increasingly comprehensive visions of the future, which may embody competing or even contradictory views, but are nevertheless helpful in identifying avenues for further collaboration. The primary benefit for the participants lies in their enhanced capability to acquire resources for further work and to commit other stakeholders to their future-oriented efforts.

The third foresight mode is called embedded foresight, where participants are already involved in collaborative activities, and the organisational framework for such collaboration is intentionally utilised to facilitate future-oriented discussions beyond the scope of individual projects. This kind of continuous consultation among participants supports the adaptation of activities to perceived opportunities and threats. The characteristics of these different foresight modes are compared in Table 3.2.

Table 3.2 Characteristics and comparison of explicit, emergent, and embedded foresight exercises (Salmenkaita and Salo 2004)

	Explicit	Emergent	Embedded
Interests	Disparate	Overlapping	Within same domain
Approach toward innovation policy	'Policy-makers decide'	'Policy-makers follow'	'Policy-makers adapt'
Mode of consultation	Consultative	Competitive	Continual
Methodology	Systemic	Ad hoc	Structured ad hoc
Organisational influence	Implicit	Both	Explicit
Resource allocation	Batch-mode	Continuous	Staged
Result	Recommendations	Actions	Adaptation
Other observations	'All are equal' (expert contributions)	Less competent viewpoints tend to be discounted ('meritocracy')	Promotes self-guidance
Functions	Establish S&T priorities and support 'five C's' in innovation system	Identify RTD opportunities and seek coalitions for RTD activities	Adaptation and adjustment of RTD activities in RTD programmes
Structures	Project organisation	Adaptive 'ad hoc' networks with some management structure	No separate management structures; embedded in RTD programme
Strengths	Ability to seek nonpartisan view of RTD opportunities across established domains	Close links to RTD efforts and participants' RTD priorities	Tight interaction between policy-making and RTD efforts
Weaknesses	Actionability of recommendations may remain weak	Lead roles required to catalyse networking	Constrained to areas in which RTD activity is already established
Pitfalls	Undue political and other nonmeritocratic pressures affecting priority setting	Corporatism may prevail	'Lock-in' of assumptions and actions

3.5.4 Information Quality and Foresight Activities

Salmenkaita and Salo (2004) also discuss difficulties related to different kinds of foresight processes and challenges in the utilisation of the results of those exercises. According to them, there is "an inherent tension in selecting the appropriate level of detail, or granularity, in analysing key technologies". When defining technologies too narrowly, the list of technologies may become excessively long and hence unmanageable, and during the foresight process one cannot pay much attention to interconnections among technologies. Then again, if technologies are defined very broadly, the results may lack the level of detail that would be needed to derive and implement recommendations in specific settings. There can also be a tension between the 'richness' of informational outputs from a foresight exercise and the usability of such outputs for subsequent action plans. These difficulties are reflected in the outputs of foresight processes – the information produced during the process, which again can be approached from the information quality perspective.

Information quality is not an entirely new concept, but it has gained increasing attention during the last few years (Melkas 2004). Several researchers have defined attributes or dimensions of information quality (for a detailed discussion, see Wang and Strong 1996; Appendix 1). Information quality, as well as data quality, has been studied mainly by researchers interested in computing, management information systems, databases and their management, data security, and data warehouse quality, to mention a few. Researchers have mostly concentrated on company environments and business information (cf., e.g., Wang et al. 1998; Becerra-Fernandez et al. 2004; English 1999; Huang et al. 1999), but for example Uotila and Melkas (2007) have used the information quality approach when analysing the information produced in a regional technology foresight process.

For our purposes, it is useful to make a distinction between data, information, and knowledge, although in many studies these concepts are used interchangeably (Melkas 2004), which creates considerable conceptual unclarity. The situation is further complicated by different types of knowledge, such as explicit, tacit, and self-transcending knowledge (see Nonaka and Takeuchi 1995; Scharmer 2001).

Data – for instance, numbers – are the factual content of information, and only meaningful information can be the basis for purposeful action (Melkas and Harmaakorpi 2008). When data are put into a meaningful context and processed, they become information (Lillrank 1997; 2003). When analysed critically in this context, and when its underlying structure in relation to other pieces of information and conceptions about how the world works is understood, this information is transformed into a component of knowledge (Miller et al. 2001; Melkas and Harmaakorpi 2008). The idea that knowledge is much more than information stems from the view that knowledge consists of an assortment of inputs: information, experiences, beliefs, relationships, and techniques that an individual mentally synthesises to determine what a specific situation means and how to handle it (Becerra-Fernandez et al. 2004).

When taking this discussion back to foresight, for example Horton (1995) discusses the elements that should constitute a successful foresight process. According to her, successful foresight consists of three consecutive phases:

1. Phase one comprises the collection, collation, and summarisation of available information and results in the production of foresight knowledge;
2. Phase two comprises the translation and interpretation of this knowledge to produce an understanding of its implications for the future from the specific point of view of a particular organisation;
3. Phase three comprises the assimilation and evaluation of this understanding to produce a commitment to action in a particular organisation (Horton 1995).

When making a distinction between the concepts of 'information' and 'knowledge', Uotila and Melkas (2007) suggested that phase 1 may not result in foresight knowledge but in foresight information (of a 'lower level'), and the interpretation process in phase 2 is of utmost importance in order for the results to benefit the stakeholders. Assessing the hierarchy and quality of data, information, and knowledge is thus necessary, both in the collection of background materials and in the analysis and reporting of results in order to avoid what Uotila and Melkas call the 'black hole of interpretation and implementation of foresight knowledge' (Uotila and Melkas 2007).

3.5.5 Information Brokering as a Facilitator of Organisational Learning and Innovation

Sorenson et al. (2006: 996) state that even within the supportive infrastructure of an organisation, receiving and building on new knowledge can prove difficult: recipients assimilating new knowledge must actively process it by experimenting with its application in new problem domains and environmental contexts. Hence, the act of receiving and building on knowledge can be regarded rather as the beginning of a trial-and-error process than as the acceptance of a complete, well-packaged gift.

Partners participating in networked innovation processes on different sides of structural holes have different knowledge interests. They also have information of a different quality and generated for their own purposes (Melkas and Harmaakorpi 2008). The difference is often so great that a special interpretation function is needed. Burt (1997) calls this special function 'information brokerage' in a 'structural hole'. A structural hole is an opportunity to broker the flow of information between people and control the form of cooperation that brings together people from opposite sides of the hole. However, brokering means more than just linking together partners involved in an innovation process. It also includes the aspect of transforming the ideas and knowledge being transferred, and – at best – allows the widening of optimal cognitive (as well as social, cultural, temporal, and so on) distance between partners in an innovation process and enhances their absorptive capacity (Howells 2006; see also Nooteboom et al. 2007).

Brown and Duguid (1991) also refer to brokering-type activities but use the term 'organisational translator'. Organisations tend to emphasise routine and 'canonical' practice over improvisation, innovation, and adaptation, although in practice, both are needed. Brown and Duguid further note that allowing practice in organisations to evolve unchecked leads to increasing fragmentation, whereas checking it too firmly leads to rigidity; and that over-formalised relations within a firm can lead to endangering weak or informal ties that build unexpected but productive bridges across organisations. Organisational translators are thus actors who can frame the interests of one community in terms of another's practice-shaped world-view.

3.6 The Conceptual Model Integrating Foresight Activities, Absorptive Capacity, and Innovation

According to Major and Cordey-Hayes (2000), translation and interpretation is the most crucial step in a foresight process, but it is still poorly understood and has only a few theoretical techniques. Our aim is to build a bridge between foresight, information brokering, absorptive capacity, and innovation. The idea behind the proposed model is that technology foresight as such (as well as other kinds of foresight processes) can only produce information about possible future development of certain technologies. When this information is contextualised and embedded back into the environment where it will be utilised, into an organisational context, it will be refined or transformed into future-oriented innovation knowledge to promote innovation processes (Uotila and Ahlqvist 2008). This is done via information brokerage, which requires, among other things, a deep understanding of the users' knowledge interests, prior technological choices, and general knowledge level.

Firstly, the model (see Fig. 3.1) assumes, in line with the reasoning of Salmenkaita and Salo (2004), that different foresight modes – explicit, emergent, and embedded – result, in terms of information quality, in different kinds of information. This variation in the resulting information is reflected in information quality, and this again affects the way the information can be absorbed by organisations and further utilised in learning and innovation processes.

When combining this notion with the new interpretation of the model of absorptive capacity presented by Todorova and Durisin (2007), we propose – as is shown in the model – that information resulting from embedded foresight activities is more easily absorbed by organisations and likely to support AAE-type learning and innovation activities, which can usually be characterised as incremental innovations from the organisational point of view. Especially when talking about practice-based innovation, there can be an inherent tendency to over-emphasise AAE-type learning. In this case, there is also a possible risk of 'path-dependency' and 'mental lock-ins'. Path-dependency, or lock-in effects – whether technological

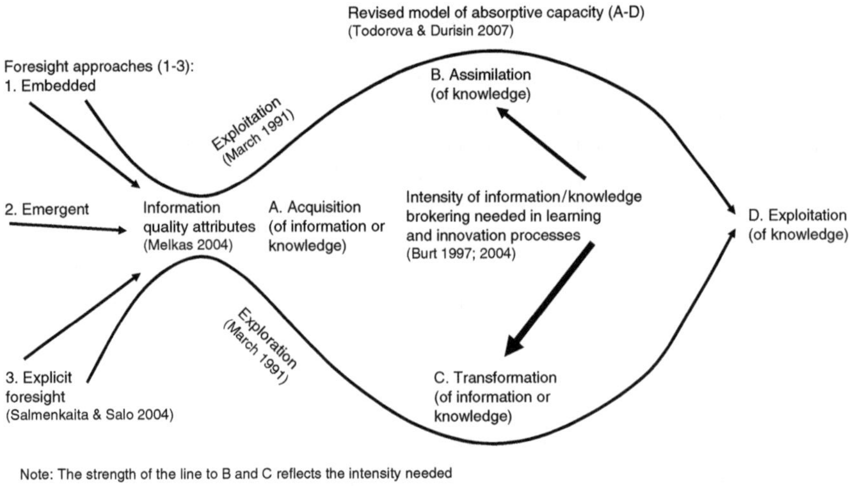

Fig. 3.1 The integrated model of foresight, absorptive capacity, and innovation

or institutional in nature – may support the efficient exploitation of present resources in the short run, but it the longer run they can limit the range of technological options, visions, and value networks, and in so doing reduce innovation capabilities (Könnölä et al. 2007). Using the terminology of March (1991), embedded foresight and AAE-type learning and innovation activities relate to exploitative organisational processes.

On the other hand, explicit foresight exercises result in information that, from the organisational point of view, is more difficult to absorb; but if absorption succeeds, it supports ATE-type learning and more radical innovations. Linking this notion with the terminology used by March (1991), explicit foresight and ATE-type learning relate to explorative organisational processes. When further combining this notion with the concept of information brokering, we propose that the intensity of the necessary brokering is different in different learning or innovation types. This implies that the intensity of brokering needed when an organisation is trying to absorb information resulting from an explicit foresight exercise is greater than that needed in the case of an embedded foresight exercise. When hypothesising further, we suggest that from the point of view of organisational renewal, it is important to find an appropriate balance between, on the one hand, embedded foresight and AAE-type learning and innovations, and, on the other, explicit foresight and ATE-type learning and innovations.

3.7 Summary and Conclusion

Future-oriented knowledge is very tightly centred around the 'knowing self' (see Howells 2002) and is often difficult to articulate and communicate, as well as very sticky and resistant to knowledge-transfer efforts. In his study on the use of future-oriented knowledge in regional innovation processes, Uotila (2008) concluded that the more a foresight process relies on the use of outside expertise (referring here to explicit foresight), the more emphasis also has to be placed on knowledge brokering.

In this process of transforming 'foresight information' into 'innovation knowledge', knowledge brokers play an important role as facilitators of information transfer and new knowledge creation. Parjanen et al. (2011) state that when actors in networks communicate along strong ties and across short distances, what is communicated is more 'knowledge-like, ready-to-use inputs' for learning and innovation processes. On the other hand, when communication takes place across greater distances and along weak links, what is communicated is more 'information-like' inputs, and much greater effort and resources are needed in the interpretation process in new contexts before these inputs can provide support for learning, new knowledge generation, and innovation. People who act as brokers can provide the necessary extra resources, and thus they may help also the innovating actors of a network to cross greater distances.

In the proposed conceptual model, knowledge brokering is seen to play a crucial role when the results of foresight – future-oriented information and knowledge – are used to facilitate innovation processes in organisations. Brokering is an activity promoting information and knowledge flow in innovation networks by helping to decrease different kinds of distances, including temporal distance, between actors in networks. Thus, knowledge brokering and knowledge brokers help also to increase the absorptive capacity of different actors in innovation networks. How they actually do it is still very much uncharted territory. According to Reichert (2006), a new knowledge broker profile is very similar to the "old eighteenth century host or hostess of a salon: smart intellectuals who love to discover nearly as much as they love the sharing of discovery, who not only have the talent for both, but also the communicative disposition and generosity to develop this combination into a human art form, a celebration of shared knowledge development". But were these hosts or hostesses born with their talents and competences, or can they be developed? If one ends up with the latter alternative, then further research is needed concerning the search mechanisms of knowledge, personal incentives and motivating factors of knowledge brokers, personal competencies, and so on, in order to fully understand knowledge brokering and its critical role in facilitating knowledge transfer, learning, and innovation processes.

In this chapter, our aim was to build a holistic model integrating different kinds of foresight approaches, and organisational innovation and learning modes. This was done by combining established and tested theoretical concepts in a novel way. Still, one has to bear in mind that – however conceptually appealing the model might seem – further research is needed to empirically test and validate it.

Appendix 1: Definitions of Information Quality Dimensions (Wang and Strong 1996)

Believability: The extent to which information is accepted or regarded as true, real and credible.

Value-added: The extent to which information is beneficial and provides advantages through its use.

Relevancy: The extent to which information is applicable and helpful for the task at hand.

Accuracy: The extent to which information is correct, reliable, and certified as being free of error.

Interpretability: The extent to which information is in an appropriate language and units and the information definitions are clear.

Ease of understanding: The extent to which information is clear, unambiguous, and easily comprehended.

Accessibility: The extent to which information is available or easily and quickly retrievable.

Objectivity: The extent to which information is unbiased (unprejudiced) and impartial.

Timeliness: The extent to which the age of the information is appropriate for the task at hand.

Completeness: The extent to which information is of sufficient breadth, depth, and scope for the task at hand.

Traceability: The extent to which information is well documented, verifiable, and easily attributed to a source.

Reputation: The extent to which information is trusted or highly regarded in terms of its source or content.

Consistent representation: The extent to which information is always presented in the same format and is compatible with previous information.

Cost-effectiveness: The extent to which the cost of collecting appropriate information is reasonable.

Ease of operation: The extent to which information is easily managed and manipulated (i.e., updated, moved, aggregated, reproduced, customised).

Variety of information and information sources: The extent to which information is available from several differing information sources.

Concise representation: The extent to which information is compactly represented without being overwhelming (i.e., brief in presentation, yet complete and to the point).

Access security: The extent to which access to information can be restricted and, hence, kept secure.

Appropriate amount of information: The extent to which the quantity or volume of available information is appropriate.

Flexibility: The extent to which information is expandable, adaptable, and easily applied to other needs.

References

Becerra-Fernandez, I., Gonzalez, A., & Sabherwal, R. (2004). *Knowledge management: Challenges, solutions and technologies*. Upper Saddle River, NJ: Prentice-Hall.

Bootz, J. P. (2010). Strategic foresight and organizational learning: A survey and critical analysis. *Technological Forecasting and Social Change, 77*(9), 1588–1594.

Boschma, R. A. (2005). Proximity and innovation. A critical assessment. *Regional Studies, 39*(1), 61–74.

Brown, J. S., & Duguid, P. (1991). Organizational learning and communities-of-practice: Toward a unified view of working, learning, and innovation. *Organization Science, 2*(1), 40–57.

Burt, R. (1992). *Structural holes: the social structure of competition*. Boston: Harvard University Press.

Burt, R. (1997). The contingent value of social capital. *Administrative Science Quarterly, 42*, 339–365.

Burt, R. (2004). Structural holes and good ideas. *American Journal of Sociology, 110*(2), 349–399.

Burt, R. (2005). *Brokerage and closure. An introduction to social capital*. Oxford: Oxford University Press.

Cillo, P. (2005). Fostering market knowledge use in innovation: The role of internal brokers. *European Management Journal, 23*(4), 404–412.

Coates, J. F. (1985). Foresight in federal government policymaking. *Futures Research Quarterly, 1*, 29–53. Summer.

Cohen, W., & Levinthal, D. (1990). Absorptive capacity: A new perspective on learning and innovation. *Administrative Science Quarterly, 35*, 128–152.

Cuhls, K. (2003). From forecasting to foresight processes—New participative foresight activities in Germany. *Journal of Forecasting, 22*, 93–111.

Du Preez, G. T., & Pistorius, C. W. I. (1999). Technology threat and opportunity assessment. *Technological Forecasting and Social Change, 61*, 215–234.

Duin, van der, P. A. (2004). Linking futures research to innovation processes. A selection of preliminary findings and some possible future directions. Keynote presentation. EU-US Seminar: New Technology Foresight, Forecasting & Assessment Methods, Seville, 13–14 May 2004.

Eerola, A., & Jørgensen, B. H. (2002). *Technology foresight in the Nordic countries*. Roskilde: Risø National Laboratory.

Eerola, A., & Väyrynen, E. (2002). *Teknologian ennakointi- ja arviointikäytäntöjen kehittäminen eurooppalaisen kokemuksen pohjalta* [Development of practices of technology foresight and assessment on the basis of European experience]. Tiedotteita–Research Notes, 2174. Espoo: VTT. (In Finnish.)

English, L. P. (1999). *Improving data warehouse and business information quality*. New York: Wiley.

Granovetter, M. (1973). The strength of weak ties. *American Journal of Sociology, 78*, 1360–1380.

Granovetter, M. (2005). The impact of social structure on economic outcomes. *Journal of Economic Perspectives, 19*(1), 33–50.

Harmaakorpi, V. (2004). *Building a competitive regional innovation environment – The regional development platform method as a tool for regional innovation policy*. Doctoral Dissertation Series 2004/1. Espoo: Helsinki University of Technology Lahti Center.

Harmaakorpi, V., Tura, T., & Artima, E. (2006). *Balancing regional innovation policy between proximity and distance*. Paper presented at the Fifth Proximity Congress, Bordeaux: France, 28–30 June 2006.

Horton, A. M. (1995). A simple guide to successful foresight. *Foresight, 1*(1), 5–9.

Howells, J. (2002). Tacit knowledge, innovation and economic geography. *Urban Studies, 39* (5–6), 871–884.

Howells, J. (2006). Intermediation and the role of intermediaries in innovation. *Research Policy, 35*, 715–728.

Huang, K. T., Lee, Y. W., & Wang, R. Y. (1999). *Quality information and knowledge*. Upper Saddle River, NJ: Prentice-Hall.

Kim, L. (1998). Crisis construction and organizational learning: Capability building in catching-up at Huyndai Motor. *Organization Science, 9*, 506–521.

Knoben, J., & Oerlemans, L. A. G. (2006). Proximity and inter-organizational collaboration: A literature review. *International Journal of Management Reviews, 8*(2), 71–89.

Könnölä, T., Brummer, V., & Salo, A. (2007). Diversity in foresight: Insights from the fostering of innovation ideas. *Technological Forecasting and Social Change, 74*, 608–626.

Li, Y., Vanhaverbeke, W., & Schoenmakers, W. (2008). Exploration and exploitation in innovation: Reframing the interpretation. *Creativity and Innovation Management, 17*(2), 107–126.

Lillrank, P. (1997). *The quality of information in business processes*. Working paper No. 10/1997/ Industrial Management, 2nd revised edition. Espoo: Helsinki University of Technology.

Lillrank, P. (2003). The quality of information. *International Journal of Quality & Reliability Management, 20*(6), 691–703.

Major, E. J., & Cordey-Hayes, M. (2000). Engaging the business support network to give SMEs the benefit of foresight. *Technovation, 20*, 589–602.

March, J. G. (1991). Exploration and exploitation in organizational learning. *Organization Science, 2*(1), 71–87.

Melkas, H. (2004). *Towards holistic management of information within service networks: Safety telephone services for ageing people*. Doctoral dissertation, Espoo: Helsinki University of Technology, Department of Industrial Engineering and Management. Also available at http://lib.hut.fi/Diss.

Melkas, H., & Harmaakorpi, V. (2008). Data, information and knowledge in regional innovation networks: Quality considerations and brokerage functions. *European Journal of Innovation Management, 11*(1), 103–124.

Miller, B., Malloy, M. A., Masek, E., & Wild, C. (2001). Towards a framework for managing the information environment. *Information and Knowledge Systems Management, 2*, 359–384.

Nonaka, I., & Takeuchi, H. (1995). *The knowledge creating company: How Japanese companies create the dynamics of innovation*. New York: Oxford University Press.

Nooteboom, B., Van Haverbeke, W., Duysters, G., Gilsing, V., & Van Den Oord, A. (2007). Optimal cognitive distance and absorptive capacity. *Research Policy, 36*(7), 1016–1034.

Parjanen, S., Melkas, H., & Uotila, T. (2011). Distances, knowledge brokerage and absorptive capacity in enhancing regional innovativeness: A qualitative case study of Lahti region, Finland. *European Planning Studies, 19*(6), 921–948.

Pihkala, T., Harmaakorpi, V., & Pekkarinen, S. (2007). The role of dynamic capabilities in breaking socio-institutional inertia in regional development. *International Journal of Urban and Regional Research, 31*(4), 836–852.

Reichert, S. (2006). *The rise of knowledge regions: Emerging opportunities and challenges for universities*. EUA Publications 2006. Brussels: European University Association.

Roveda, C., & Vecchiato, R. (2008). Foresight and innovation in the context of industrial clusters: The case of some Italian districts. *Technological Forecasting & Social Change, 75*, 817–833.

Salmenkaita, J.-P., & Salo, A. (2004). Emergent foresight processes: Industrial activities in wireless communications. *Technological Forecasting & Social Change, 71*, 897–912.

Salo, A. (2000). Teknologian arviointi ja ennakointi [Technology assessment and foresight]. In T. Lemola (Ed.), *Näkökulmia teknologiaan* (pp. 193–212). Helsinki: Gaudeamus. (In Finnish.)

Scharmer, C. O. (2001). Self-transcending knowledge: Organizing around emerging realities. In I. Nonaka & D. Teece (Eds.), *Managing industrial knowledge: Creation, transfer and utilization* (pp. 68–90). London: Sage Publications.

Sorenson, O., Rivkin, J. W., & Fleming, L. (2006). Complexity, networks and knowledge flow. *Research Policy, 35*, 994–1017.

Sotarauta, M., & Srinivas, S. (2005). *Co-evolution of policy and economic development: A discussion on innovative regions*. Special Working Paper Series on Local Innovation

Systems, MIT-IPC-LIS-05-001. Cambridge, MA: Massachusetts Institute of Technology, Industrial Performance Center.
Sotarauta, M., & Srinivas, S. (2006). Co-evolutionary policy processes: Understanding innovative economies and future resilience. *Futures, 38*, 312–336.
Todorova, G., & Durisin, B. (2007). Absorptive capacity: Valuing a reconceptualization. *Academy of Management Review, 32*(3), 774–786.
Tura, T., & Harmaakorpi, V. (2005). Social capital in building regional innovative capability. *Regional Studies, 39*(8), 1111–1125.
Uotila, T. (2008). *The use of future-oriented knowledge in regional innovation processes: Research on knowledge generation, transfer and conversion*. Dissertation. Acta Universitatis Lappeenrantaensis 318. Lappeenranta: Lappeenranta University of Technology.
Uotila, T., & Ahlqvist, T. (2008). Linking technology foresight and regional innovation activities: Network facilitating innovation policy in Lahti region, Finland. *European Planning Studies, 16*, 1423–1443. November.
Uotila, T., Harmaakorpi, V., & Melkas, H. (2006). A method for assessing absorptive capacity of a regional innovation system. *Fennia, 184*(1), 49–58.
Uotila, T., & Melkas, H. (2007). Quality of data, information and knowledge in regional foresight processes. *Futures, 39*, 1117–1130.
Walker, G., Kogut, B., & Shan, W. (1997). Social capital, structural holes and the formation of an industry network. *Organization Science, 8*(2), 109–125. March-April.
Wang, R. Y., Lee, Y., Pipino, L., & Strong, D. (1998). Manage your information as a product. *Sloan Management Review, 39*(4), 95–105.
Wang, R. Y., & Strong, D. M. (1996). Beyond accuracy: What data quality means to data consumers. *Journal of Management Information Systems, 12*(4), 5–34. Spring.
Zaheer, A., & Bell, G. G. (2005). Benefiting from network position: Firm capabilities, structural holes, and performance. *Strategic Management Journal, 26*, 809–825.
Zahra, S. A., & George, G. (2002). Absorptive capacity: A review, reconceptualization and extension. *Academy of Management Review, 27*(2), 185–203.

Systems. (No. IPC-LIS-03-001). Cambridge, MA: Massachusetts Institute of Technology, Industrial Performance Center.
[illegible], M., & [illegible], S. (2006). Co-evolutionary policy process: Understanding innovative economies and future resilience. *Futures, 38*, [illegible].
Todorova, G., & Durisin, B. (2007). Absorptive capacity: Valuing a reconceptualization. *Academy of Management Review, 32*, 774–786.
Tura, T., & Harmaakorpi, V. (2005). Social capital in building regional innovative capability. *Regional Studies*, [illegible].
[illegible]

Chapter 4
Communicating Connections: Social Networks and Innovation Diffusion

Pekka Aula and Olli Parviainen

Abstract The role of social networks in promoting the diffusion of innovations is widely recognised, but networks are used more as a vague metaphor than an analytic concept. In this chapter, we study the possibilities that social network analysis (SNA) offers to promote the diffusion of innovations. In addition, we investigate the roles of opinion leaders and opinion brokers in the networks of innovation diffusion. We base our findings on a case study of a food industry organisation. We conclude with some remarks on how the study of innovation diffusion might benefit from adapting the methods of social network analysis.

4.1 Introduction

According to Harmaakorpi (2006), a valuable source of innovations are factors like the ability to interact, learn collectively, and build relationships of trust between the innovating partners. Consequently, an organisation's innovation system ought to be understood as a social infrastructure that consists of various networks with different kinds of social relationships and social ties (Granovetter 2005; Burt 2004; Harmaakorpi and Melkas 2005) as well as technological, information, and business models (Carlsson et al. 2002; Gelsing 2010). The emergence and diffusion of innovations, by definition, stem from situations where individuals or organisations engage in communication processes (Rogers 2003). The communication system

P. Aula (✉)
Department of Social Research, Media and Communication Studies, University of Helsinki, Helsinki, Finland
e-mail: pekka.aula@helsinki.fi

O. Parviainen (✉)
Department of Social Research, Media and Communication Studies, University of Helsinki; Entrepreneur, Verkostoanatomia, Helsinki, Finland
e-mail: olli.parviainen@helsinki.fi

H. Melkas and V. Harmaakorpi (eds.),
Practice-Based Innovation: Insights, Applications and Policy Implications,
DOI 10.1007/978-3-642-21723-4_4,

needed to enable this interaction has been traditionally understood as the physical system composed of technical devices and communicative linkages that form the means of distributing messages and the corresponding organisational actions. The emphasis has been on the quantity – not the quality – of the channels used.

We argue that the communication perspective is essential to understanding the complexity and multidimensionality of the processes involved in diffusion of innovations. We will support this argument with evidence from a case study in which we identified central actors and defined channels of communication and innovation diffusion in a company's social network. We are interested in the possibilities *social network analysis provides at the inter- and intra-organisational levels in promoting the diffusion of innovations.* Instead of focusing on institutionalised communications channels in an organisation's innovation system, we propose that a more holistic view of the communication system involved in innovation diffusion is needed. Placing emphasis on social relations might yield a more diverse and complete picture of the communication system that enables diffusion of innovations within an innovation system. Research on opinion leaders, for example, has revealed that the biggest influence on individuals' decision to adopt an innovation is the opinion of a trusted few (Valente and Davis 1999; Rogers 2003). This power to influence others, opinion leadership, is based on the degree of expertise and the opinion leader's position within the particular social network. We suggest that this influence over others means that content of communication and position within the communication network matter as much as (or even more than) channel of communication. Social network analysis as a method has provided a workable instrument for studying these channels and key position holders, and it also provides an instrument for studying the channels with specific content. More importantly, social network analysis enables the detection of *innovation potential* by detecting so-called structural holes (Burt 1992).

Social network analysis has been used to facilitate the adoption of innovations in a variety of settings. However, it has primarily been used to detect individuals in central network positions who are able to efficiently disseminate existing innovations (Valente 1995; Valente and Davis 1999; Rogers 2003). Though the term 'network' is widely used in innovation research (Parjanen et al. 2010), the systematic use of social network analysis throughout the innovation diffusion process has been rare at both the micro- and the macro-level. Research combining the inter-organisational macro- and intra-organisational micro-level has been conducted, but the sources used in gathering network information have been archives of institutional relationships, such as patent applications (for example Ahuja 2000; Cantner and Graf 2006), which has been proved problematic (Carlsson et al. 2002).

4.2 Diffusion of Innovations in Social Networks

According to Rogers (2003), an *innovation* is an idea, a practice or an object that the unit of interest experiences for the first time. Innovations are conceived when people exchange information on different needs and solutions to them. The diversity of skills, talents, knowledge, and connections inside and outside of

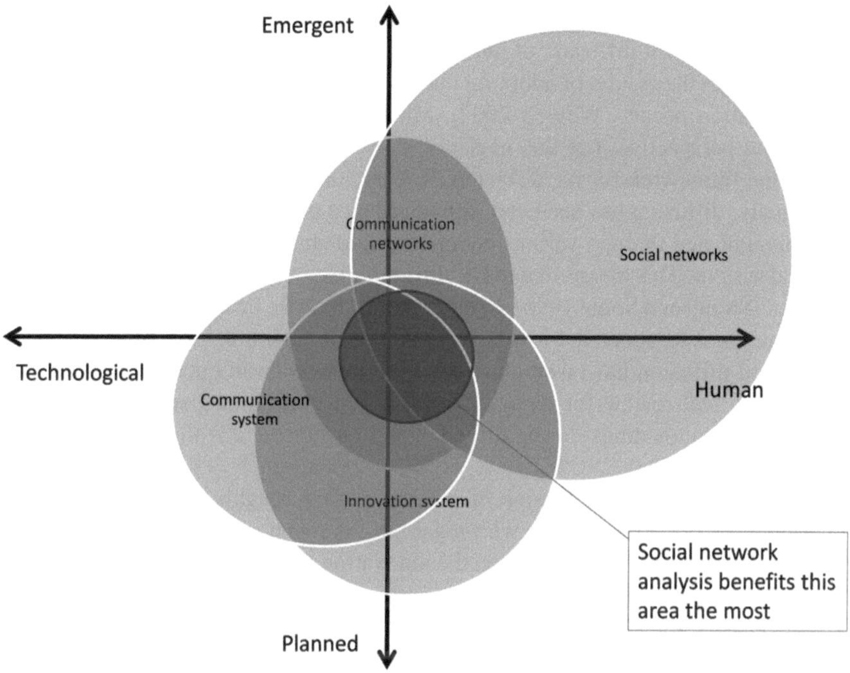

Fig. 4.1 The relationships between the central concepts

the organisation increases the innovativeness of the organisation. Thus, the instrumental element in innovation is connections to other people and organisational entities. In the following sections, we approach our research question from three overlapping perspectives: *diffusion of innovations* happens through *social networks* that are based on *communicative interactions*. Figure 4.1 explains the relationships between these central concepts, most of which are overlapping but not synonymous. We positioned these concepts in two dimensions: the technological-human dimension and the planned-emergent dimension. Social networks are more human than technological and emerge from interaction between people. Communication networks overlap with social networks but also possess a technological aspect. For example, the internet-based social networking service Facebook has a social and a technological aspect. A communication system is a planned technological structure, and an innovation system is a planned structure that usually consists of both technological and human elements. In this study, we use social network analysis to study the area of intersection formed by the central concepts.

4.2.1 Explaining the Diffusion of Innovations

The study of diffusion of innovations has its origins in the agricultural research of the 1940s and 1950s. Since then, the theory has been used in sociology, communications, marketing, management, and health care (Wejnert 2002; Rogers 2003). The study of diffusion of innovations can range from the characteristics of innovations to the results of adopting innovations and to the environment in which the innovation occurs (Wejnert 2002). Much of the research has focused on the individual perspective, but this neglects, somewhat ironically, the social context that is the framework for the diffusion (Rogers 2003).

Usually, diffusion has been conceptualised as a process in which an innovation is communicated through various communication channels between individuals in a social system. This means that to be 'diffused', the innovation does not have to be adopted (Strang and Soule 1998). In other words, one can be aware of an innovation but decide not to adopt it for use. As a matter of fact, much of the research on innovation diffusion has taken place after the innovation in question has diffused (for a historical review for research diffusion on innovations, see Rogers 2003). Research on innovations during diffusion and on innovations that did *not* diffuse is rare (Rogers 2003).

Explanations for the diffusion of innovations are roughly derived from either internal or external standpoints with regard to the social system (Strang and Soule 1998). The external view focuses on the innovation itself: the relative advantage of the innovation, its complexity, trialability, observability, and compatibility (Rogers 2003). That is, the characteristics of the innovation are the critical factor as it is adopted in a system of rational individuals. From an internal perspective, on the other hand, diffusion of an innovation can stem from social pressure, learning, or contagion (Young 2009), from structural equivalency (Burt 1999), or from prestige, spatial proximity, or culture (Strang and Soule 1998).

A social system can be described as any stabilising framework in which communication and human behaviour occur (Parsons 1991). Traditionally, the social system within an organisation has been divided in to two parts or subsystems: the official structure, consisting of hierarchical structures and established communications channels, and the unofficial structure, consisting of social relations between the members of the social system. It is important to keep in mind that social systems are not the same as social networks and vice versa. Social systems are a combination of the official structures mentioned above and informal social networks. Though this divide is ubiquitous, the two structures overlap and are very difficult to distinguish from each other (Flap et al. 1998). Official structures are usually visible and well documented, making them the obvious subject of research. Unofficial social structures are more dynamic, and thus laborious to capture without the proper tools.

4.2.2 Social Network View on Diffusion of Innovations

The term *network* is used in various ways as a metaphor for decentralised, complex and multiple-actor systems. 'Network' can be used as a part of a theory (e.g. Castells 1996), when it is used to model actors' relationships to each other.

'Network' can also be used as an analytic concept that is formed from methods of gathering, handling, and modelling action (Johanson et al. 1995). In addition, 'network' is used in identity construction and cognitive conceptualisation of organisations (Ibarra et al. 2005).

Network theories do not form a cohesive and uniform set. In contemporary network studies, there are two different but converging traditions: the first tradition, the social sciences approach, applies network theories with an emphasis on quantitative natural sciences, and second tradition, the social anthropology approach, uses it more as a metaphor (Knox et al. 2006). In addition, in the studies the unit of interest varies greatly; for example, from cognitive semantic networks to trade networks between nations. This same diversity can also be found in communication network theories. Monge and Contractor (2003) offer a multi-theoretical, multilevel model (MTML) that tries to bring together various theories and interest levels. They argue that, instead of one level of interest, one should take in to account multiple levels and theories at the same time. The theory of diffusion of innovations provides one such MTML approach. Networks form conduits for attitudes and behaviours to spread among the network members. This contagion can occur through being in a similar position within a network or by direct contact (Burt 1999). In this study, we will use the MTML as a guideline for involving networks at the organisational, group, and individual level.

Social network analysis (SNA) is a set of various methods for describing and studying human interaction, and it is used to study repeating patterns in connections linking social actors (Wasserman and Faust 1994). This means that the unit of analysis is not the individual actor, but the entity consisting of these actors and the connections connecting them (Wasserman and Faust 1994). Usually the aim of social network studies is to locate a set of actors forming a distinct subgroup and actors filling a specific role or position (Freeman 2005).

SNA carries a few central concepts. Subjects interacting within a network are referred to with various terms (nodes, points, actors), as are the relationships that connect them (links, ties, arcs, connections). In this study, we use the terms *actor* and *connection* for clarity's sake. Actors in networks can be people, groups of people, or larger organisations. Connections between actors can describe, for example, family ties, the flow of support or resources, emotional valuing, interactive behaviour or membership in a group (Wasserman and Faust 1994). Depending on the relationship, connections can be directed or undirected. A connection describing e.g. 'discusses with' is undirectional, whereas in 'asks advice from someone', the direction of the connection is important. In a directed connection, the sender and the receiver can be distinguished. The strength of the connection can vary as well. It can be either dichotomous (connection is present or absent) or valued (Wasserman and Faust 1994). The value of the connection represents its strength, intensity, or frequency (Monge and Contractor 2003).

SNA has been used, in form or another, as a research method in the social and behavioural sciences since the 1930s. The first famous study is from 1932, when Jacob Moreno stated that personal relations more than personal characteristics were responsible for some of people's actions (Moreno 1934). The popularity of network analysis grew slowly until the 1990s, when the number of studies using SNA started to grow rapidly. The reasons for this growing popularity lie in three phenomena: the increased use of the network metaphor (e.g., Castells 1996), the rise of social networking services (such as Facebook, Twitter, etc.), and increased computing power and more developed computer software (Carrington et al. 2005).

Usually, quantitative social science relies on statistical tools designed to study the background information (age, gender, education, etc.) – or attributes – of subjects. These statistical methods assume that all attributes are independent from the attributes of other subjects. In social network analysis, the individual study subjects are considered interdependent. Specialised social network analysis tools have been developed by combining social sciences, empirical studies, mathematics, and traditional statistical science (Wasserman and Faust 1994).

Though the attributes of actors are also used in various statistical tools in social network analysis, they are usually connected to specialised network statistics. Network statistics describe either the actors' network properties or the properties of the whole network. Individual actors' network statistics describe one's position within the network in relation to others. The most important of these measures include *degree*, *betweenness*, and *closeness*[1] (Everett and Borgatti 2005). Statistics describing the whole network can describe values such as density, reciprocity, or centralisation.[2] A central node within a network is an actor whose position allows her to access, spread, or control information, attitudes, and communication. While a central position benefits the actor (see, for example, Burt 1992), it also embeds the actor more deeply in the network (Wasserman and Faust 1994). This allows the actor to monitor and control the network, while at the same time limiting her options if she wishes to remain in the central position (Valente and Pumuang 2007).

4.2.3 Communication as Social Networks

Monge and Contractor (2003) have performed a comprehensive review of network theories in communication. Most network theories have a *structuralistic* approach. According to the perspective of constructive structuralism, people's values, beliefs, and attitudes are the result of their position within the organisation (Monge and Contractor 2003). Burt's (1992) theories of structural equivalence are a great example of constructive structuralism in action (see also Knoke 1994).

[1] Degree centrality measures the number of connections the actor has. Betweenness measures the number of shortest paths (lines of connection) that go through the actor. Closeness measures the actor's distance to other members of the network.

[2] Density is figure between 0 and 1, with 0 meaning none of the possible connections are present, and 1 meaning every possible connection is present. Reciprocity indicates what percentage of the directed relationships in a network is mutual. Centralisation indicates how much structural power within the network is centralised in a single actor.

The relativist structuralist approach argues that communication structures are formed dynamically through action and therefore connect people regardless their official roles and positions (see, for example, Rogers's (2003) extensive research on diffusion of innovations in networks). However, many network theories do not fit clearly in either the constructive or relativist structuralistic approach. Burt's theory (1992) of structural holes and social capital is a good example of this. According to Johansson (2000), a person's fitting into a role is the result of his position within official networks and direct unofficial communications.

Human communication by definition requires more than one participant. When the number of participants exceeds two, the communication event can be approached as a network. According to Monge and Contractor (2003), communication networks are recurring patterns that are formed by the flow of messages in time and place, where a message is something that can move between the actors in a network or be created by them. These messages form a relationship between the communicating actors. Relationships can be described as information exchange, knowledge of something, or communicating with someone. Through these relationships, people are embedded in several different but often overlapping networks.

However, in a social network bound together by communication, the number of connections between the actors is not evenly distributed (Barabasi 2003); that is, some actors have more and more diverse connections than others. According to the structuralistic approach, this inequality of connections means that two groups of actors have more influence in social networks than others: opinion leaders and opinion brokers (Burt 1999; Valente and Davis 1999; Rogers 2003).

Opinion leaders are people who affect other peoples' opinions, attitudes, beliefs and behaviour (Rogers 2003; Valente and Pumuang 2007). Opinion leaders are central figures in social networks and usually adopt innovations early (Rogers 2003; Valente and Pumuang 2007). The term *opinion leadership* originates from the two-step flow of communication (Lazarsfeld et al. 1948), which states that mass media do not influence people directly but through certain key people. Other people ask these opinion leaders for advice, and their opinion is well regarded. Opinion leadership is more often directed at some specific area of interest instead of taking the form of all-encompassing leadership in all walks of life (Merton 1957; Goldsmith and Hofacker 1991; Rogers 2003; Weimann et al. 2007). Studies (e.g., Childers 1986; Rogers 2003) have revealed that opinion leadership is not a strict division into followers and leaders, but more of a continuum of a person's ability to influence others that is part of everyday interaction (Katz and Lazarsfeld 1955). With this distinction, it is possible to use opinion leadership as metaphor for influence in a network with specific content.

Opinion brokers bring new information to groups and act as go-betweens between groups of similar composition (Burt 1999). A broker can influence or regulate the flow of information within a network (Burt 1999; Monge and Contractor 2003). A broker fills a structural hole in a network. *Structural holes* refer to an absence of strong ties within a network. Structural holes are not always a negative phenomenon, as they allow more efficient functioning of the group, insulating it from continuous and time-consuming communication with every other member of the network (Burt 1999). Burt's (1992) theory of structural holes describes how an actor connecting separate groups gains a competitive advantage by making the groups dependent on the brokers' connections. The theory of structural holes is related to Granovetter's (1973) famous theorem on the 'strength of weak ties'. Granovetter divides people's connections into strong ties and weak ties. Strong ties are the connections that define us: family, friendship and spousal relationships, whereas weak ties are the less-emotional ties: people that one knows but has not invested lot of emotions in (acquaintances, etc.). According to Granovetter, weak ties determine a person's success, not strong ones.

Opinion leaders and opinion brokers are an important and highly influential part of a network's properties (Barabasi 2003). Opinion leaders and opinion brokers share regulation of communication within the social network. Opinion brokers bring new information and innovation to the network, but opinion leaders determine if the innovation is adopted by and diffused within the social network (Valente and Pumuang 2007). Using opinion leaders to diffuse new products and innovations has been proved effective (Rogers 2003; Chan and Shekhar 1990), as people tend seek advice from friends and other influential people in uncertain situations (Black 1982).

Opinion leadership and brokership are ways of describing influence within a network, while mapping the structural holes makes it possible to detect untapped innovation potential within it. Human interaction is not easily simulated, so to best evaluate the possibilities of SNA in diffusion of innovations research, some practical measures are needed.

4.3 Case Study

In order to reflect our ideas on innovation diffusion and to analyse the roles of the central actors in social networks, we conducted a case study of a company. The practical aim of the study was to find the central actors and to define the channels of communication and innovation diffusion within the company's social network. We followed the theoretical framework outlined above, where the central actors within a social network are categorised into opinion leaders and opinion brokers.

Analysing communication patterns requires determining of the level and unit of analysis, defining the set of actors, and choosing the right metrics to describe the networks' and actors' properties. The level of analysis can be the organisational level, but at the same time the unit of interest can be an individual. In this study, the level of analysis varies, but the unit of interest is always an individual. Predefining the actors in the network facilitated conducting of the analysis. We focused on the staff and management of the case company. Selection of network metrics depends on the level of analysis: analysing the position of individuals within the network requires a different set of metrics than analysing the entire network.

In our study, we approached the diffusion of innovations and social and communication networks from the three different perspectives or levels of analysis suggested by the MTML (Monge and Contractor 2003): organisational, group, and individual. The first of these, the organisational level, focuses on the general properties of the network. The second, the group level, concentrates on the kinds of groups that are formed within a network and how they communicate with one another. The third level, the individual, focuses on the network's central actors. We incorporated these three levels of analysis into the insights described earlier in light of the theories of innovation diffusion, social networks, and communication networks in order to investigate the possibilities of social network analysis in diffusion of innovations.

4.3.1 Research Design

The study case is a food-industry plant and research unit (henceforth 'the Company') located in Lahti, Finland. It has 43 employees (including temporary workers and a few researchers from the Company's parent company). Data gathering was conducted through a survey questionnaire administered at the Company's facilities; a researcher was onsite to assist in this process. The purpose of the study was explained in a pre-survey letter circulated throughout the Company two weeks prior to the survey. The letter proved to be very successful, as almost all respondents were anticipating the study and little explanation was required during the study. The response rate was 82%. The data were analysed by using the network analysis program Ucinet v.6 and visualised using Netdraw.

Three types of networks were of interest in the study. The first network type was formed on the basis of the listing of organisational connections outside the Company. The Company is a part of an industry cluster situated in the Lahti region, and in the questionnaire the names of members of the industry cluster were given as options for outside connections; a free text field was also offered for respondents to list other connections. In addition, the respondents were asked to report the importance of the connection. Most of the connections reported were from the pre-defined list of names from the cluster. The second network was formed from the answers to the question, "From whom you have received information about new methods or techniques that have been useful in your work?" In other words, the answers provided the basis for the Company's innovation network. The third network was formed based on the answers to the question, "How often do you contact the following people in work-related matters?" This network forms the setting for the two previous questions, as it describes the daily and weekly interactions happening in the Company.

The Company's three social networks are for the most part structured around the organisational chart. Most of the central actors in the networks are also central in the organisational hierarchy. In order to get a better sense of the subgroups within the Company, the actors were divided roughly into three groups according to communication pattern and position. The first group, 'management and experts' (Group I), communicated mainly outwards, while the second and third groups, 'technical support and experts' (Group II) and 'production' (Group III), were oriented inward. All of the groups communicate with each other. The tendency to communicate repeatedly with the same actors was pronounced. The Company is a production plant, and the task-based organisation of work positions actors in fixed relationships to each other (see, for example, Burns and Stalker 1961).

4.3.2 *Three Levels of Investigation*

Organisational perspective. In this study, no concrete innovations were the subject of innovation diffusion. Instead, we focused on the possible connections for innovations to enter the Company and diffuse within it. The Company members' connections to other organisations are numerous. Almost all of the actors whose connections form the only link from the Company to another organisation are also actors central to the inner social network. These actors fill structural holes that bring new innovations into the network. There are seven actors that fill 17 structural holes (Fig. 4.2). This may cause challenges, as an average of more than two structural holes are recreated when one of those actors is not present. All of the actors that fill the structural holes belong to Group I.

Our research focused primarily on the internal view of innovation diffusion, as the research was conducted solely within one organisation. Furthermore, the content of the channels was broadly defined as work-related matters. More could have been accomplished by asking a more specific network question and including more than one organisation in the analysis.

Group perspective. All of the networks analysed were connected, i.e. every actor could reach every other actor within the network. The social networks within the

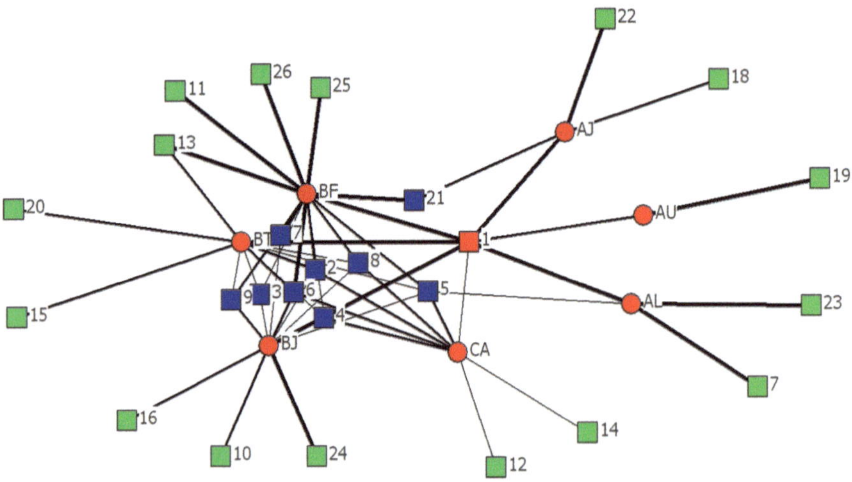

Fig. 4.2 The company's external connections (*Notes*: Round nodes are actors and square nodes are organisations. The green organisations are connections that only one actor has.)

Company form a connected structure that makes the diffusion of innovations possible. The amount of interaction is about one third of the network formed of daily work-related interactions. In the innovation diffusion network, one tenth of the relationships were reported as mutual, suggesting a one-way mode of communication was more prevalent. The average distance for an innovation to spread through the network is three steps, and the maximum distance nine steps. This means that for an innovation to spread to every corner of the social system, it takes an average of three communicative actions, and a maximum of nine. In this kind of elongated network structure, communication from one side of the organisation to the other takes time. The clustering coefficient[3] of the innovation network is rather high. The combination of low density and high clustering indicates that information on innovations spreads within small, tight-knit groups. According to structural equivalence (Burt 1999), people in a tight cluster are most likely to have similar information and require opinion leaders and opinion brokers to spread information between them.

The degree centrality in the innovation network is about the same as in the daily interaction network, but the betweenness centrality is more than 50% higher in the innovation network than in the daily interaction network. This means that the innovation network relies more on brokers than the daily interaction network.

Groups I and II receive information from outside of the group, while members of Group III mainly share information on new innovations within the group. Group I also is well connected inside the group, and its members have the most connections outside the Company.

Individual perspective. There were clear central actors in the Company's social networks. These actors have better possibilities to influence the opinions of other actors, the flow of communication, and the diffusion of innovations. The same actors occupied similar network positions in all of the networks. These key actors were almost all part of the Company's upper hierarchy or experts and could be deduced from the organisational chart.

During the research, it became clear that a strict division into opinion leaders and opinion brokers is not as feasible and clear-cut might be assumed. Degree centrality and betweenness centrality can be used to identify key actors (Fig. 4.3), but no single metric can be used to identify opinion leaders or opinion brokers for innovation diffusion purposes. A combination of metrics offers a much more detailed picture. For example, a high degree combined with low clustering coefficient indicates that the actor has many alters[4] that are not connected to each others' alters (Valente 1995). This indicates brokership status in a social network. If degree

[3] The clustering coefficient is the network's general tendency to form triangles: for example, if A is connected to B, who in turn is connected to C. The clustering coefficient in the example situation is the overall situation ranging from 0 to 1 in networks where A is also connected to C.

[4] In social network analysis, the term *ego* is used to describe the actor whose connections are under scrutiny. The ego has connections to *alters*, other actors that may have connections with each other.

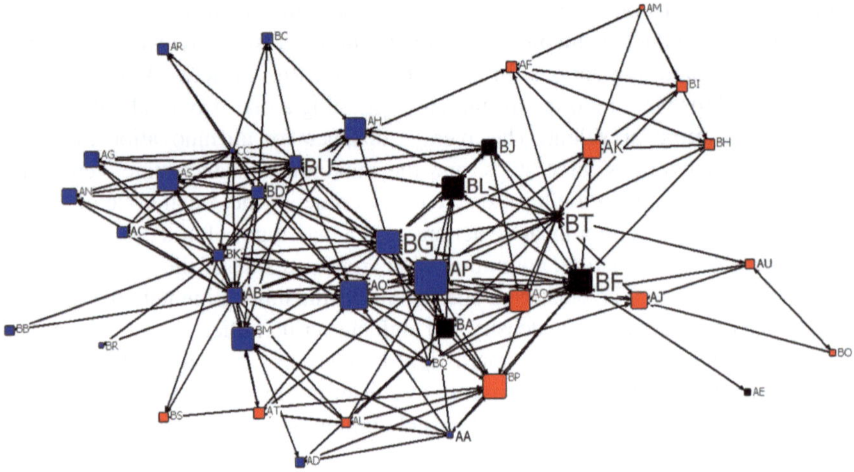

Fig. 4.3 Innovation network within the company (*Notes*: The size of the node represents the degree centrality, and size of the label indicates betweenness centrality. Black nodes belong to Group I, red to Group II and blue to Group III. The actors' names are coded.)

centrality and the clustering coefficient are both high, the actor has the potential for opinion leadership. In the Company, the central actors were found in all of the groups, though Groups I and III had more central actors than Group II.

All of the Company's networks tend to follow Burns and Stalker's (1961) notion that the form follows the function: when working procedures and processes are strictly defined, there is little room for improvisation and authority figures are more central. Also, as Barabasi (2003), among others, has noted, connections among actors are not distributed randomly. Several actors were highly central with regard to the network, but their centrality type (opinion leadership/brokership) varied. Networks are dynamic, and their composition changes according to the environment (Monge and Contractor 2003). Comparison to other networks is, therefore, difficult, as the study subject changes constantly.

4.4 Summary and Conclusion

In this chapter, we have been interested in the possibilities social network analysis can offer to the study of innovation diffusion. In addition, and more precisely, we wanted to ponder the roles of opinion leaders and opinion brokers in innovation diffusion networks. We also argued that the communication perspective is essential to understanding the complexity and multidimensionality of the processes involving diffusion of innovations.

Based on the theory and a case study, we conclude that social network analysis can be utilised in active diffusion of innovations in at least three ways. Firstly, it can

Table 4.1 Utilising social network analysis to promote diffusion of innovations

	Diffusion of innovations	Social networks	Communication
Organisational level	Number of formal ties between organisations	Amount of informal connections between organisations	Processes supporting communication across organisational boundaries
Group level	Structural holes between groups	Physical distance between groups (units, functions, etc.) Detecting communities	Bridging gaps between groups
Individual level	Inter-organisational actor interface plurality	The role of key individuals (opinion leaders and opinion brokers)	Enabling individual network positions to diffuse innovations

be used to monitor to current inner social network structure. Measuring the width, density, centrality, and reciprocity of the network reveals issues that hinder or promote diffusion of innovations. Innovations in, for instance, knowledge-based organisations should have alternative routes (Tsai 2001). Secondly, at the group level, measuring communication densities between groups can show potential places for innovation within the organisation. Connecting two units or groups that were previously less connected provides new interactions and possibilities for innovation. Thirdly, social network analysis can be utilised to identify people central to innovation diffusion. These opinion leaders and brokers can be identified if the innovation type is specified. For example, using a health innovations network to diffuse IT-related innovations may not yield good results. There is no clear division into opinion leaders, opinion brokers, and followers. Actors have different opportunities for influencing the network. The methods of social network analysis provide ways of guaranteeing a thorough spread if the central actors approve the innovation. Table 4.1 is used to summarise the benefits of utilising social network analysis in diffusion of innovations.

The results of this study are limited, as the study did not focus on any particular innovation and its diffusion process. Furthermore, we included only one organisation, which limited inter-organisational results. However, expanding the scope of the study to include an entire innovation and communication system and systemically analyse the social interactions enabling them could provide some practical results for managing diffusion of innovations.

References

Ahuja, G. (2000). Collaboration networks, structural holes, and innovation: A longitudinal study. *Administrative Science Quarterly, 45*(3), 425–455.

Barabasi, A. L. (2003). *Linked.* New York: Plume.

Black, J. (1982). Opinion leaders, is anyone following? *Public Opinion Quarterly, 46*, 169–176.

Burns, T., & Stalker, G. (1961). *The Management of Innovation.* London: Tavistock.

Burt, R. (1992). *Structural Holes*. Cambridge: Harvard University Press.
Burt, R. (1999). The social capital of opinion leaders. *The Annals of the American Academy of Political and Social Science, 566*, 37–54.
Burt, R. (2004). Structural holes and good ideas. *The American Journal of Sociology, 110*(2), 349–399.
Cantner, U., & Graf, H. (2006). The network of innovators in Jena: An application of social network analysis. *Research Policy, 35*(4), 463–480.
Carlsson, B., Jacobsson, S., Holmén, M., & Rickne, A. (2002). Innovation systems: Analytical and methodological issues. *Research Policy, 31*(2), 233–245.
Carrington, P. J., Scott, J., & Wasserman, S. (2005). *Models and methods in social network analysis*. New York: Cambridge University Press.
Castells, M. (1996). *The rise of the network society*. Cambridge: Blackwell.
Chan, K., & Shekhar, M. (1990). Characteristics of the opinion leader: A new dimension. *Journal of Advertising, 19*(3), 53–60.
Childers, T. (1986). Assessment of the psychometric properties of an opinion leadership scale. *Journal of Marketing Research, 23*, 184–188.
Everett, M., & Borgatti, S. (2005). Extending centrality. In P. J. Carrington, J. Scott, & S. Wasserman (Eds.), *Models and methods in social network analysis* (pp. 57–76). New York: Cambridge University Press.
Flap, H., Bulder, B., & Völker, B. (1998). Intra-organizational networks and performance: A review. *Computational and Mathematical Organization Theory, 4*(2), 109–147.
Freeman, L. (2005). Graphic techniques for exploring social network data. In P. J. Carrington, J. Scott, & S. Wasserman (Eds.), *Models and methods in social network analysis* (pp. 248–269). New York: Cambridge University Press.
Gelsing, L. (2010). Innovation and the development of industrial networks. In B. Å. Lundvall (Ed.), *National systems of innovation: Toward a theory of innovation and interactive learning* (pp. 119–132). London: Anthem Press.
Goldsmith, R., & Hofacker, C. (1991). Measuring consumer innovativeness. *Journal of the Academy of Marketing Science, 19*(3), 209–221.
Granovetter, M. (1973). The strength of weak ties. *The American Journal of Sociology, 78*, 1360–1380.
Granovetter, M. (2005). The impact of social structure on economic outcomes. *Journal of Economic Perspectives, 19*(1), 33–50.
Harmaakorpi, V. (2006). The regional development platform method as a tool for regional innovation policy. *European Planning Studies, 14*(8), 1085–1104.
Harmaakorpi, V., & Melkas, H. (2005). Knowledge management in regional innovation networks: The case of Lahti, Finland. *European Planning Studies, 13*(5), 641–659.
Ibarra, H., Kliduff, M., & Tsai, W. (2005). Zooming in and out: Connecting individuals and collectivities at the frontiers of organizational research. *Organization Science, 16*(4), 359–371.
Johanson, J. E. (2000). Intraorganizational influence: Theoretical clarification and empirical assessment of intraorganizational social influence. *Management Communication Quarterly, 13*(3), 393–425.
Johanson, J. E., Mattila, M., & Uusikylä, P. (1995). *Johdatus verkostoanalyysiin [An Introduction into Network Analysis]*. Helsinki: Kuluttajatutkimus. (In Finnish.)
Katz, E., & Lazarsfeld, P. F. (1955). *Personal influence*. Glencoe, Illinois: Free Press.
Knoke, D. (1994). *Political Networks: The Structural Perspective*. New York: Cambridge University Press.
Knox, H., Savage, M., & Harveyn, P. (2006). Social networks and the study of relations: Networks as method, metaphor and form. *Economy and Society, 35*(1), 113–140.
Lazarsfeld, P., Berelson, B., & Gaudet, H. (1948). *People's choice* (2nd ed. [1st ed. 1944]). New York: Columbia University Press.
Merton, R. (1957). *Social theory and social structure*. Glencoe, Illinois: Free Press.
Monge, P., & Contractor, N. (2003). *Theories of communication networks*. New York: Oxford University Press.

Moreno, J. (1934). *Who shall survive?* Washington, DC: Nervous and Mental Disease Publishing Company.

Parjanen, S., Harmaakorpi, V., & Frantsi, T. (2010). Collective creativity and brokerage functions in heavily cross-disciplined innovation processes. *Interdisciplinary Journal of Information, Knowledge, and Management, 5*, 2–21.

Parsons, T. (1991). *The social system*. London: Routledge.

Rogers, E. (2003). *Diffusion of innovations* (5th ed. [1st ed. 1962]). New York: Free Press.

Strang, D., & Soule, S. (1998). Diffusion in organizations and social movements: From hybrid corn to poison pills. *Annual Review of Sociology, 24*, 265–290.

Tsai, W. (2001). Knowledge Transfer in Intraorganizational Networks: Effects of Network Position and Absorptive Capacity on Business Unit Innovation and Performance. *Academy of Management Journal, 44*(5), 996–1004.

Valente, T. (1995). *Network Models of the Diffusion of Innovations*. Cresskill: Hampton Press.

Valente, T., & Davis, R. (1999). Accelerating diffusion of innovations using opinion leaders. *The Annals of the American Academy AAPSS, 566*, 55–67.

Valente, T., & Pumuang, P. (2007). Identifying opinion leaders to promote behavior change. *Health Education & Behavior, 34*(6), 881–896.

Wasserman, S., & Faust, K. (1994). *Social network analysis: Methods and applications*. New York: Cambridge University Press.

Weimann, G., Deon, H., van Vuuren, D., & Joubert, J. P. R. (2007). Looking for opinion leaders: Traditional vs. modern measures in traditional societies. *International Journal of Public Opinion Research, 19*(2), 173–190.

Wejnert, B. (2002). Integrating models of diffusion of innovations: A conceptual framework. *Annual Review of Sociology, 28*, 297–326.

Young, P. (2009). Innovation diffusion in heterogeneous populations: Contagion, social influence, and social learning. *The American Economic Review, 99*(5), 1899–1924.

Chapter 5
Dilemmas of Practice-Based Innovation Policy-Making

Réjean Landry and Nabil Amara

Abstract This chapter discusses dilemmas that emerge when policy-makers are equally attracted to promoting concepts that are not based on evidence and to developing innovation policies that are evidence-based. It addresses nine dilemmas of innovation policies by discussing the following questions: What is the capacity of theories and new policy concepts linked to innovation to provide useful policy prescriptions to foster innovation? To what extent are these theories and new policy concepts supported by solid evidence? The chapter is based on a review of concepts and buzzwords linked to innovation, and the types of data and diagnostics used by policy-makers to develop new national and regional innovation policies. The chapter is also based on the consulting experience of the authors regarding the production of evidence-based diagnostics and derivation of policy measures at both the national and regional levels. The chapter defines policy issues in terms of dilemmas, it provides concrete examples, and it discusses how policy-makers could resolve the tensions between equally attractive policy alternatives.

5.1 Introduction

Turning concepts into innovation policies raises many dilemmas. Hence, ideas of policy concepts emerge from multiple sources such as market failure theories, economic development theories, and cluster theories to name a few. Furthermore, ideas of policy concepts also emerge from various types of diagnostics as to how firms perform with respect to innovation. This chapter discusses dilemmas that emerge when policy-makers are equally attracted to promoting concepts that are not based on evidence and to developing innovation policies that are evidence-based.

R. Landry (✉) • N. Amara
Department of Management, Laval University, Québec, Canada
e-mail: Rejean.landry@fsa.ulaval.ca

H. Melkas and V. Harmaakorpi (eds.),
Practice-Based Innovation: Insights, Applications and Policy Implications,
DOI 10.1007/978-3-642-21723-4_5,

There are two perspectives on evidence-based policymaking. There is a narrow perspective that promotes the use of systematic reviews and meta-analyses to generate evidence on the effectiveness of policy interventions. In this narrow perspective, the evidence is translated into evidence-based policy interventions that should be imposed to policy-makers and practitioners. In this chapter, we endorse a broader perspective to consider evidence-based policy-making as an approach that aims to help policy-makers and practitioners base the development of their policy interventions on the best available evidence (Nutley et al. 2009; Bogenschneider and Corbett 2010). Without evidence, policy-makers and practitioners must rely on intuition, ideology, conventional wisdom, policy buzzwords, and untested theories. We claim that many innovation policies are made this way because the intensity of competition has induced policy-makers at both national and regional levels to continually adapt their innovation policies in order to take into account new policy concepts and diagnostics that pinpoint deficiencies and failures in existing policies. The review of the innovation policy literature suggests major difficulties regarding the capacity of theories to provide useful policy prescriptions. Furthermore, the existing policy diagnostics tend to provide data at levels of aggregation that do not take into account the diversity of the situations and needs of firms, industries, and regions. In this chapter, policy diagnostics have to do with ascertaining the nature of unsatisfactory innovation performances by looking at symptoms and signs of shortcomings and failures in innovation policies. As for evidence-based policy diagnostics, they refer to the type of evidence used to prepare diagnostics. In such a context, practice-based innovation policy-making refers to the idea that the making of policies is more or less based either on the best available evidence concerning the innovation performances of companies or on untested intuitions, concepts, and buzzwords related to innovation performances. Hence, dilemmas of practice-based innovation policy-making arise because policy-makers are equally attracted to the reliance on the best available evidence and on untested concepts, intuitions, and buzzwords.

This chapter addresses dilemmas of innovation policies by discussing the following questions: (1) What is the capacity of theories and new policy concepts linked to innovation to provide useful policy prescriptions to foster innovation? (2) To what extent are these theories and new policy concepts supported by solid evidence?

The chapter is based on a review of concepts and buzzwords linked to innovation, and the types of data and diagnostics used by policy-makers to develop new national and regional innovation policies. The chapter is also based on the consulting experience of the authors regarding the production of evidence-based diagnostics and derivation of policy measures at both the national and regional levels. The chapter defines policy issues in terms of dilemmas, it provides concrete examples, and it discusses how policy-makers could resolve the tensions between equally attractive policy alternatives.

The discussion of practice-based innovation policy-making is organised around nine policy dilemmas:

1. Theory vs. practice-based innovation
2. The magic ingredient concept vs. the multiple ingredients practice
3. Global innovative firms vs. local innovative firms
4. Open vs. closed innovation
5. Inward vs. outward view of innovation
6. Innovations that are first on the world market vs. incremental innovations
7. Policy imitation vs. customised solutions
8. Innovations in knowledge-based firms vs. innovations in traditional firms
9. Producing national innovation diagnostics vs. producing innovation policy diagnostics

5.2 Dilemma 1: Theory Versus Practice-Based Innovation

The dilemma: Policy-makers are equally attracted by the idea of implementing innovation policies based on concepts derived from policy theories and theories of innovation than by its opposite alternative which is to implement innovation policies based on the diversity of situations and needs of firms, industries, and regions.

5.2.1 The Theoretical Argument in Four Parts

How are innovation policies justified by the market theory? In theory, there are market deficiencies when private firms invest less in research and development (R&D) and innovation than is socially beneficial (Lerner 2009). Experts classify the economic justifications following four series of factors. First, market deficiencies pertaining to R&D are related to three factors (Lerner 2009; Reback 2009; Caracostas 2007; Usher 1992): the first factor, which is related to the concept of indivisibility of the innovation process, supports that the interdependence of the different phases of the innovation process entails an indivisibility of fixed costs to the different phases of the innovation process that makes the efficient exploitation of scale costs difficult. The second factor is related to the concept of imperfect appropriability of the innovation results. It claims that firms cannot appropriate the total profits of their investments in R&D and innovation activities, because these investments generate intangible and immaterial public benefits that cannot be commercialised. Furthermore, it is argued that in some cases, the public benefits can be more important than the private benefits. The third factor is related to the concept of uncertainty of innovation. In theory, R&D and innovation processes involve scientific, technical, and commercial uncertainties that render difficult the forecast of results and the calculation of profits of private investments. The intangible character of R&D and innovation investments complicates the life of suppliers of capital by rending the R&D and innovation investments irrecoverable.

Second, markets may also fail due to deficiencies pertaining to diffusion (MacVaugh and Schiavone 2010). First, it is argued that the indivisibility of the diffusion process entails the same type of indivisibility as the R&D process. Second, the diffusion of new technologies and innovation can be impeded because of three deficiencies of the absorption capacity of firms: deficiencies in the mastering of techniques required to adopt new technologies; organisational deficiencies of the firm that prevent the best possible exploitation of new technologies; deficiencies in the abilities of the infrastructure that the firm depends on (commercialisation structures, banks, local institutions, etc.) which impede the best possible exploitation of new technologies.

Third, the diffusion of new technologies and innovation can be impeded because of two deficiencies of information: inadequate availability of information concerning opportunities for entrepreneurs to be familiar with new technologies and innovation, and presentation of the required information in a way that is useless and non-credible for the potential user firm.

Fourth, the innovation process can also be impeded by deficiencies pertaining to the infrastructure that the firm depends on. In this matter, experts tend very often to highlight the following shortages (Peters 2006; Schmoch et al. 2006; Lundvall 1992): lack and inadequacy of technical and professional training; lack of collaboration between public research institutions and firms; lack of institutions of economic intermediation involved in innovation promotion; lack of intermediation institutions dedicated to the promotion of creation and consolidation of networks, partnerships, and horizontal collaborations between actors; lack of social values like trust and reciprocity that raises the costs of knowledge exchange.

5.2.2 Implications for Practice-Based Innovation Policy-Making

Four implications derive from the above theoretical arguments (Lerner 2009; Reback 2009; Caracostas 2007; MacVaugh and Schiavone 2010): first, state intervention is justified by the idea of compensation of the difference between private benefit and social benefit. Second, certain R&D and innovation projects imply fixed costs and investment levels that are so important that they cannot be financed by private firms without governmental grants or fiscal incentives. Third, the imperfect appropriability of innovation results renders the intervention of the state necessary to compensate the gap between the benefits for the community as a whole and those pertaining exclusively to firms. Fourth, the importance of these uncertainties pushes firms to reduce their risks. This propensity justifies state intervention to share the risk with firms in different ways, notably by creating or facilitating the creation of new financing mechanisms, particularly in the field of venture capital.

At first sight, these theoretical arguments are attractive, but, upon further examination, they neither provide specific theoretical diagnostics, nor clarify policy implications regarding innovation. The market failure arguments are formulated at a very high level of aggregation that prevents policy-makers from deriving policy prescriptions that would take into account the diversity of situations and needs of firms, industries, regions, and countries in the matter of innovation. Furthermore, the market failure arguments do not diagnose the innovation process in a way that is disaggregated enough to be able to produce specific policy implications that would contribute to foster innovation in firms.

As a consequence, in its current formulation, the theory of market failures does not help significantly to turn conceptual diagnostics into innovation policies based on the diversity of situations and needs of firms, industries, regions, and countries.

5.3 Dilemma 2: The Magic Ingredient Causal Concept Versus the Multiple Ingredients Practice

The dilemma: Policy-makers are equally attracted by concepts that suggest the implementation of innovation policies based primarily on one or two ingredients – such as R&D and patents – than by their opposite alternative that is to implement innovation policies based on evidence showing that, in practice, innovation depends on multiple factors (ingredients) used in a complementary manner by firms.

5.3.1 The Theoretical Argument

In spite of all the criticisms addressed to the linear science push theory of innovation, policy-makers of many countries, including Canada and many of its provincial governments, tend to actually assume that the innovative performance of firms might be improved by developing innovation policies that rely primarily on increasing R&D and patenting activities. The rationale for such an argument is that increases in R&D activities will increase the development of product and process innovations and, in turn, increasingly relying on patents will increase protection against imitation by competitors of the newly developed innovations. A variation of this rationale argues that increases in industry-university research and increases in the commercialisation of the intellectual property rights derived from the discoveries and inventions of university research will foster the transfer of university inventions and discoveries to industry and, in turn, foster innovation in industry. Such a rationale induces policy-makers to rely on one or two factors of innovation theories upon which to base their policies.

5.3.2 What Is the Evidence on the Determinants of Innovation?

The strong lesson learned from the empirical literature on innovation tells us that multiple ingredients or determinants are required to ensure the development and commercialisation of technological innovations. Hence, Becheikh et al. (2006) have documented in a detailed manner the determinants (explanatory factors) of innovation in a systematic review of the empirical literature on innovation. Many systematic literature reviews now exist on innovation at the national level (Becheikh et al. 2006). The development of product and process innovations is explained by the use of firms' internal and external resources. No firm, even the biggest, has at its disposal all the necessary internal resources to develop or improve, on its own, its products, processes and practices. The empirical literature on innovation shows that it is possible to increase the innovation performance of firms by improving the external factors and resources that contribute to increase innovation. Innovation depends in fact on internal and external determinants or resources, as well as hybrid determinants. The internal determinants refer to the internal capacities of the firms to create and use the necessary knowledge to develop and improve products and processes. The external determinants refer, for their part, to the capacities of the firms to identify, obtain and use the knowledge coming from external sources, and to the capacities of the firms to adapt themselves to the competitive environment. The development of product and process innovations requires the combination of a large number of ingredients as illustrated in the following summary list of determinants that explain innovation in a statistically significant manner.

The determinants linked to the internal resources of firms relate to (Becheikh et al. 2006):

- Size of the firm;
- Strategies (including commercialisation strategies);
- Investments in strategic assets, that is:
  - Knowledge creation: R-D;
  - The use of knowledge embodied in advanced technologies;
  - The use of knowledge embodied in manufacturing practices linked to added value production;
  - The use of knowledge embodied in manpower.

The determinants linked to the external resources of firms relate to:

The use of external sources of knowledge (ideas and information, technical support) to complete and improve internal resources, that is:

- Market sources: clients, suppliers, competitors, consultants;
- Research sources: universities, public centres of research, centres of technology transfer, patent documentation;
- Formal and informal collaborations with other business and research organisations;

- Services of economic development agencies and intermediary organisations helping firms to innovate;
- Variables linked to the firm's competitive environment.

The hybrid determinants of innovation that are partly linked to the firm's internal factors and partly linked to external factors mainly refer to the obstacles to innovation such as lack of information on technology relevant to the development or introduction of new or significantly improved products or processes, lack of financing for the development of development or introduction of new or significantly improved products or processes, lack of skilled personnel to develop or introduce new or significantly improved products or processes, inability to qualify for government assistance programs or R&D tax credits (Amara and Landry 2005; Amara et al. 2008).

The evidence shows that around 15% of small and medium-sized enterprises (SMEs) rely on the use of patents to protect their innovations against imitation from rival firms. By comparison, 40% of SMEs rely on secrecy to protect their innovations from imitation, and other informal strategic mechanisms such as lead time over competitors (51% of cases) and complexity of production processes (25% of cases) (Landry and Amara 2004). Furthermore, recent data on the commercialisation of inventions and discoveries developed in Canadian universities show that investments of 10.9 billion dollars in university research (Statistics Canada 2010a) generated 53 million dollars of economic returns in 2008 (Statistics Canada 2010b), thus confirming in a complementary manner that the patenting and commercialisation of university intellectual property rights is not really a winning strategy when used in isolation from other explanatory factors of innovation. The available evidence shows similar trends in European countries.

5.3.3 Implications for Practice-Based Innovation Policy-Making

Hence, the quantitative evidence on innovation also shows that patenting is used by a small percentage of firms as a tool to protect technological innovations, and that the linkage between university research and economic return is much more indirect than assumed by many policy-makers. The empirical literature on innovation shows that many ingredients are required and combined in a complementary manner in order to ensure the successful commercialisation of technological innovations. Following the concept of evidence-based policy-making, we argue that practice-based innovation policy-making should rely only on variables that influence innovation in a statistically significant way. Variables that do not explain innovation in a statistically significant manner have no impact on innovation and, therefore, should not be considered as innovation policy levers. In practice-based innovation policy-making, a statistical analysis thus allows to identify strengths and weaknesses of firms, and to generate ideas to improve innovation policies.

However, the evidence provided in a statistical analysis entails three main limits that prevent their efficient use in practice-based innovation policy-making: first, available evidence usually does not link the determinants of innovation to specific policy instruments supporting innovation; then, when they tackle the question of policy instruments of innovation support, these studies usually do not tackle systematically the transferability conditions of these instruments; finally, another important limit of empirical studies on the determinants of innovation is to implicitly postulate that countries and regions are homogeneous. Practice-based innovation policy-making should assume, following Asheim et al. (2003), that in order to achieve their goal, innovation policies must take into account the variety of regional contexts, the diversity of firms' capabilities and attitudes, and the diversity of factors facilitating or impeding innovation. Nauwelaers and Wintjes' (2003, in Asheim et al. 2003) thesis asserts that these elements of diversity render impossible the establishment of policies that could be valid and efficient for each of the encountered regional situations. They however add that this does not mean that nothing can be done to develop general policies which could contribute to improve the efficiency and effectiveness of the policy instruments used to support innovation in firms, particularly in SMEs. Hence, based on their empirical studies, they suggest that general policies should embody three characteristics: first, matching the context and companies' needs with the policy instruments in each region; second, confronting the policy tools with the lessons of theory; third, comparing results achieved with a range of policy instruments in different environments (Nauwelaers and Wintjes 2003: 219).

5.4 Dilemma 3: Global Innovative Firms Versus Local Innovative Firms

The dilemma: Policy-makers are equally attracted by the idea of implementing innovation policies based on the concept of globalisation of innovative firms than by its opposite alternative which is to implement innovation policies based on the evidence that shows that about half of the firms buy and sell locally.

5.4.1 The Argument Supporting the Concept of Globalisation

Globalisation describes a process by which firms, regional and national economies have become integrated through trade, capital flows, migration, and the spread of technology. Globalisation is an external force driving change in the business environment. It is argued that no amount of opposition will make globalisation go away and in a rapidly changing global economy, innovation has to be part of company business models and public innovation policies. As a consequence, business leaders and policy-makers argue that all firms, including SMEs, should aim to become global, and start by increasing their exports as a tool for fostering growth and innovation.

5.4.2 What Is the Evidence on Globalisation and Innovation?

On the one hand, policy-makers and business leaders often single out three major components of innovation globalisation: (1) the international exploitation of nationally generated innovations; (2) the global generation of innovations by SMEs; and (3) global techno-scientific collaborations. On the other hand, in practice, the data show that in countries like Canada, with a few exceptions, only large resource firms and large science-based firms are global. Actually, nearly half of the SMEs operating in the manufacturing industry buy and sell locally. In the service industry, again with a few exceptions, only the largest engineering and ICT consulting firms are global. The rest of the firms operating in the service industry also buy and sell locally and at best, provincially or nationally.

Hence, in a typical industrial region of Canada like the Chaudière-Appalaches region in Québec (Landry and Amara 2004):

- 46.5% of sales are generated from clients located within 100 km
- 31.5% from elsewhere in Québec
- 9.2% from elsewhere in Canada
- 11.6% from the USA
- 1.0% from elsewhere in the world

However, the evidence shows that the propensity of firms to sell outside their region and to export increases with their propensity to innovate and with the degree of novelty of their innovations.

Furthermore, the evidence on the acquisition of input factors shows, for the same firms, that:

- 42.8% of purchases are generated from suppliers located within 100 km
- 37.3% from elsewhere in Québec
- 9.2% from elsewhere in Canada
- 8.0% from the USA
- 2.5% from elsewhere in the world

However, the propensity to acquire input factors from outside the region increases with the propensity of firms to innovate and with the degree of novelty of their innovations.

5.4.3 Implications for Practice-Based Innovation Policy-Making

Such evidence carries two policy implications. One might either focus its attention on designing policies that would help firms to increase their exports, or help firms to become more innovative. Both types of policies are appropriate. However, the capabilities of firms to export are not primarily linked to the existence of export support programs, but to the capabilities of firms in the matter of innovation. Hence, the implications of the above evidence for the implementation of practice-based innovation policy-making are to help firms to graduate progressively from non-innovative to innovative firms, and from developing product innovations that carry low degrees of novelty to developing product innovations that embody higher degrees of novelty. Such a practice-based approach would be based on the fact that about 50% of firms are engaged in product development, while the development of product innovations is completed in less than 3 months for about one third of the firms (Landry and Amara 2004). Helping firms to graduate from non-innovative to innovative, and from low to higher degrees of innovativeness, might not be very appealing for policy-makers, but it would be better aligned with the innovative capabilities of firms. Furthermore, such a practice-based perspective would take into account the fact that the globalisation forces are more and more challenged by opposite forces promoting local production and local consumption as a tool to promote environmental protection.

5.5 Dilemma 4: Open Versus Closed Innovation

The dilemma: Policy-makers are equally attracted by the idea of implementing innovation policies based on the concept of open innovation than by its opposite alternative which is to implement innovation policies based on the protection of innovation (closed innovation).

5.5.1 The Argument Supporting the Concept of Open Innovation

The concept of open innovation has caught the attention of policy-makers and business leaders who have transformed this buzzword into a new policy paradigm for the management of innovation. The starting point for the argument supporting this concept is that no firm, even the largest ones, can develop product innovations in isolation by relying only on its internal resources. Therefore, firms have to forge relations with external partners to acquire external ideas, information, knowledge and resources in order to improve their innovation capabilities and performances. The literature on open innovation stresses the benefits of openness in innovation, leaving aside the examination of the costs and disadvantages of openness (Dahlander and Gann 2010). In some policy circles, open innovation is understood to be extremely open, and treats openness as a dichotomous variable: completely open versus completely closed (Balka et al. 2010).

From the standpoint of firms, the innovation process has always been an open process. The degree of openness of the innovation process might have increased due to the rising specialisation of R&D, advanced technologies, and patents whose exploitation requires access to codified knowledge in combination with tacit knowledge. In such a context, open innovation refers to various types of inflows and outflows of knowledge. According to Chesbrough et al. (2006: 1), open innovation is "the use of purposive inflows and outflows of knowledge to accelerate internal innovation, and to expand the markets for external use of innovation, respectively". In their systematic review of literature on open innovation, Dalhander and Gann (2010) have further divided the literature on inbound and outbound flows into interactions that are pecuniary from those that are non-pecuniary, and proposed four emblematic forms of openness:

Sourcing: non-pecuniary inbound flows of resources refer to the exploration and use of external sources of ideas, information, knowledge, R&D and technologies to enhance internal innovation capabilities.

Acquiring: pecuniary inbound flows of resources refer to the acquisition of expertise and technologies (license-in of patents, acquisition of advanced technologies) on the market in order to enhance internal innovation capabilities.

Revealing: non-pecuniary outbound flows of resources refer to how internal ideas, information, knowledge and technologies are revealed to external organisations without any short term financial rewards, thus leveraging the innovative capabilities of external organisations outside the boundaries of the firm.

Selling: pecuniary outbound flows of resources refer to how firms sell or license out their inventions and technological innovations, thus contributing to improve the internal innovation capabilities of other organisations.

The argument for open innovation supports that the internal innovation capabilities of firms are maximised by increasing, in combination, inflows and outflows of resources.

5.5.2 What Is the Evidence on Openness of Innovation?

Despite the attention paid to open innovation, empirical studies on its different forms are still scanty. Hence, most empirical studies have examined non-pecuniary (sourcing) and pecuniary (acquiring) inbound flows of resources.

There is an abundant quantitative literature on the external sources of ideas and information used by firms in the development of product and process innovations (non-pecuniary inbound flows of knowledge) (Amara and Landry 2005; Laursen and Salter 2006). Hence, a study of Canadian SMEs about the frequency of use of external sources of ideas and information used for the development of product and process innovations indicates that traditional market sources of ideas and information dominate in the development of technological innovations (Landry and Amara 2004):

1. Clients (used by 80.3% of SMEs);
2. Exchange of ideas and information with other business leaders (64.3%);
3. Suppliers (62.3%);

4. Industrial and trade fairs (61.7%);
5. Internet (60.0%);
6. Local economic development agencies (34.0%);
7. Patent documentation (21.0%);
8. National Research Council (NRC/IRAP) (19.9%);
9. Community colleges (18.6%);
10. Community college tech transfer offices (18.2%);
11. Universities (18.2%).

Other studies arrive at similar levels of use of non-pecuniary inbound flows of ideas and information in the development of product and process innovations (Amara and Landry 2005; Vega-Jurado et al. 2009). Such data suggest that ideas, information and knowledge are voluntarily exchanged without any pecuniary immediate rewards in order to foster the development of product and process innovations. To the extent of our knowledge, the pattern of use of external sources of ideas and information in the development of innovations has not changed drastically over the last decades, with the exception of the increasing use of the Internet. Landry et al. (2002) showed that firms developing innovations that carry higher degrees of innovativeness are more likely to use research sources of ideas and information, whereas the use of market sources such as clients and suppliers plays a more important role in the development of incremental innovations.

There is also a quantitative literature on the acquisition of advanced technologies that firms use for the development of process innovations (pecuniary inbound flows of technologies) (Roulet et al. 2010).

The evidence on pecuniary outbound flows of resources regarding how 615 innovative SMEs sell or license out their inventions and technological innovations to other firms shows that (Landry and Amara 2004):

- 2.2% of SMEs have licensed out;
- 16.7% have conducted R&D projects for other firms;
- 26.2% have sold equipment to other firms;
- 16.7% have had technicians or engineers lent to other firms.

There are few studies on open sources (Henkel 2006), and case studies on firms like Apple, IBM and Sun (West 2003) on inbound flows, revealing ideas, information, knowledge and technologies to other organisations. However, data on how internal ideas, information, knowledge and technologies are revealed by SMEs to external organisations without any short term financial rewards are almost nonexistent.

The available data on SMEs suggest that they devise open strategies of innovation in a complementary manner with closed strategies to protect their inventions and innovations from imitation by rival firms. Hence, the survey of 615 Canadian SMEs, conducted by Landry and Amara (2004), shows the extent of use of these different mechanisms:

- Patents: 14.5%
- Design registration: 8.9%

- Trademark: 20.3%
- Secrecy: 39.2%
- Complexity of production processes: 24.5%
- Lead-time advantages over competitors: 51.3%

The same study also provides evidence of use of combinations of non-pecuniary inbound and outbound flows of knowledge in the development of product and process innovations (Landry and Amara 2004; Landry et al. 2009). Hence,

- 21.0% of innovative SMEs collaborated with other firms in projects regarding product and process innovations;
- 8.7% of SMEs collaborated with universities in projects regarding product and process innovations;
- 15.8% of SMEs collaborated with public research organisations in projects regarding product and process innovations.

5.5.3 Implications for Practice-Based Innovation Policy-Making

The existing evidence convincingly shows that firms which increase sourcing (non-pecuniary inbound flows of knowledge) and acquiring (pecuniary inbound of technologies and expertise) contribute to improve their innovation performances. However, there is as yet no convincing quantitative evidence regarding the causal linkage between innovation and revealing (non-pecuniary outbound flows of knowledge), and between innovation and selling (pecuniary outbound flows of technologies and expertise). Furthermore, we suggest that in practice, it is very difficult to figure out how firms can avoid revealing ideas and information while sourcing (non-pecuniary outbound flows of knowledge), and vice-versa. Overall, the practical implementation of open innovation strategies is costly and depends in large part on absorptive capacities linked to the current use of advanced technologies, and the current level of skills and expertise of employees.

The available evidence suggests that, in practice, firms devise complex strategies combining elements of openness and closeness. In practice, complete openness is a strategy that would make value capture and exploitation of inventions by innovators difficult, especially when ideas and information are revealed to rival firms which can rely on better complementary assets and production facilities to make use of the revealed information in the development of product or process innovations.

Hence, the implications of the above evidence for the implementation of practice-based innovation policy-making are that policy-makers should develop measures helping firms to progressively increase their degree of openness, in particular sourcing and acquiring, because there is solid evidence of the positive influence of such forms of openness on innovation. However, the lack of convincing evidence regarding the correlation between revealing and innovation calls for policy caution. We have learned in our consulting mandates that SMEs selectively reveal, to other parties, information regarding their expertise, inventions and innovations. Practice-based innovation policy-making should be based on the practice of firms and should aim to help them gradually evolve with respect to sourcing and acquiring rather than transforming a buzzword into policies.

5.6 Dilemma 5: Inward Versus Outward View of Innovation

The dilemma: Policy-makers are equally attracted by the idea of implementing outward-looking approaches to innovation policies fostering niches of excellence than by its opposite alternative which is to implement inward-looking innovation policies that primarily focus on internal factors of regions while neglecting the forging of linkages with the rest of the world.

5.6.1 The Argument Supporting the Outward Concept of Innovation

Policy-makers and business leaders aim to foster innovation and competitiveness by developing niches of excellence and clusters competing in global markets. Hence, innovation policies based on cluster approaches emphasise the impact of the following external factors of firms to explain their success in innovation (Cooke et al. 2000; Morgan and Nauwelaers 2003; Asheim et al. 2003):

- Presence of sophisticated local clients;
- Presence of local clients who anticipate needs that will emerge elsewhere;
- Presence of sophisticated local suppliers;
- Presence of sophisticated local knowledge-intensive business service firms;
- High quality local research infrastructure;
- Strong density of highly qualified human resources in the region;
- Abundant supply of venture capital in the region;
- A local context that encourages investment in innovation;
- A tradition of cooperation between actors;
- The existence of intermediary organisations that support the interaction between firms, and between firms and other actors;
- Close relationships between clients and suppliers that favour trust;
- A regional culture of knowledge sharing.

These factors concern the proximity of firms to their partners, and experts such as Weterings and Boschma (2009), and Boschma (2005) tackle them more and more often by distinguishing between three types of proximity that could contribute to increase the propensity to innovate at the cluster level:

- The cognitive proximity defined by the similarities with which actors perceive, interpret, understand and evaluate their context of action;
- The organisational proximity defined by the extent to which actors share the same relationship space in an organisational system;
- The geographical proximity defined in terms of spatial or physical distance between the economic actors.

5.6.2 Limits of the Cluster Concept

More and more experts (Boschma 2005; Barthelt et al. 2004) are underlying the limits of these diverse forms of proximity by emphasising the necessity for firms of having access to sources of ideas that are different and complementary to those available by cognitive proximity, the necessity for firms of having at their disposal organisational flexibility and training possibilities complementary to those offered by organisational proximity, and finally, the necessity of opening oneself to the outside world and establishing non-local links to avoid a lock-in in geographical local links that prevents innovation.

5.6.3 Implications for Practice-Based Innovation Policy-Making

In practice, innovation policies such as cluster policies/strategies tend to focus their attention on internal factors of countries, industries or regions and to neglect the linkages with the rest of the world. Hence, regional policy-makers are induced to develop their own independent clusters in plastics products, clusters in marine products, etc. by stressing the importance of internal factors at the expense of external sources of knowledge, technologies, and markets.

In such a context, provincial and national policy-makers have the obligation to force regional policy-makers to recognise that the benefits on innovation emerging from the effects of proximity must be counterbalanced by the benefits resulting from the openness to the outside world and the establishment of non-regional links to prevent the effects of lock-in.

5.7 Dilemma 6: Innovations That Are First on the World Market Versus Incremental Innovations

The dilemma: Policy-makers are equally attracted by the idea of implementing innovation policies based on concepts of radical/disruptive innovations than by its opposite alternative which is to implement innovation policies that primarily support incremental innovations in existing firms.

5.7.1 The Argument Supporting the Concept of Innovations That Are First on the World Market

Policy-makers and business leaders are attracted by the concept of radical/disruptive innovations. Radical innovations are innovations that are the first on the world market. As for disruptive innovations, they refer to new products for which there is as yet no market and for which there are no standards. The supporters of these concepts (Christensen et al. 2008) claim that such innovations generate astronomical returns by creating new markets for which there is initially no competition.

5.7.2 What Is the Evidence on Innovations That Are First on the World Market?

The compilation of data of innovation surveys conducted in Canada and Europe reveals that (Becheikh et al. 2006; Amara and Landry 2005):

- Between two-thirds and three-quarters of the manufacturing firms develop product and process innovations.
- Large variations exist from one region to the next and from one country to the next.
- The propensity to develop product innovations varies greatly and in a non-linear manner with the number of employees.
- The development of product innovations also varies greatly depending on the sectors of activity, and also from one region to the next for the same sector of activities.
- The data of the Statistics Canada innovation survey of 2005 show that 10% of the innovative firms introduced, onto the market, at least one product innovation that was a first in the world. However, our own survey data on innovation in manufacturing firms indicate that the propensity to claim the development of innovations that are the first on the world market increases as the number of employees of firms decreases. Such bias is due to the fact that smaller manufacturing firms have less information regarding the state of their competitive environment.

5.7.3 Implications for Practice-Based Innovation Policy-Making

The existing data on innovation in manufacturing industries show that companies usually commercialise incremental innovations. The quantitative evidence on firms that launch innovations that are first in the world, especially on disruptive innovations, is still scanty. The available evidence on spin-offs derived from university research and from venture capital investments suggests that only a small proportion of such firms survive and that a small proportion of this small proportion becomes large firms creating a large number of jobs and large amounts of wealth. In practice, innovations that are the first on the world market make headlines, but not profits.

Rather than focusing all their attention on policy interventions aimed at companies developing innovations that are first on the world market or disruptive innovations, policy-makers should also take into account two more common categories of innovative firms:

1. Firms that do not develop innovations;
2. Firms that develop incremental innovations.

The positioning of policy support measures regarding these three groups of firms represents a good way of improving the efficiency and effectiveness of innovation policy measures that aim to improve the innovative performance of firms. Besides, as we have already mentioned, such a positioning of innovation policies and of their measures would incite to conceive measures around a gradation concept where one would attempt to progressively graduate firms from the non-innovating category towards the incremental innovation category and then, towards the innovation category where firms introduce innovations that are first on the world market.

Hence, we suggest that practice-based innovation policy-making must take into account the degree of novelty that firms bring to their innovations, mostly to product innovations, to have them progressively graduate into innovations that entail higher and higher degrees of novelty in their innovations.

5.8 Dilemma 7: Policy Imitations Versus Customised Solutions

The dilemma: In the absence of good theories and solid evidence, policy-makers and firms are equally attracted by the idea of implementing innovation policies based on the replication of 'models that work' than by its opposite alternative which is to implement innovation policies based on customised solutions that take into account differences in their national contexts (history, institutions, values, R&D intensity, etc.).

5.8.1 The Argument Supporting the Concept

The lack of useful theories incites policy-makers to be inspired by 'models that work', models of successful regions, models of successful firms, models of successful clusters, as sources of orientation for the formulation of practice-based innovation policy support measures. Hence, policy-makers are led to invest resources in attempts to replicate regions like Silicon Valley, the Science Triangle Park, clusters like the Rochester photonic cluster, the Boston biotech cluster, and firms like Nokia and Research in Motion (RIM).

5.8.2 What Is the Evidence on the Concept of Policy Imitation?

The absence of diagnoses based on micro-data that could be appropriate to describe the variety of situations and of firms' needs at the disaggregated level, and the incapacity of theories on innovation to generate precise ideas of innovation support measures at the regional and industry levels, have incited policy-makers to look for ideas of innovation support measures in guidelines and policy papers of exemplary measures of innovation support. A very large number of these guidelines and policy papers now exist. To illustrate this, here are a few examples:

- European Union (2007). European Union Regional Policy. Examples of regional innovation projects. Brussels: European Union.
- US Department of Commerce (2008). Economic Development Administration. A resource guide for technology-based economic development. Washington: US Department of Commerce.
- National Governors Association (2008). Innovation America: A final report. Washington: National Governors Association.
- OECD (2008). Globalization and regional economies: OECD reviews of regional innovation. Paris: OECD.
- Steineke, J. E. and Hedin, S. (2008). Management by objectives and results: Structures and practices in the regional policy field in the Scandinavian countries and Iceland. WP-2008-3. Nordregio. Stockholm: Nordic Centre for Spatial Development.

5.8.3 Implications for Practice-Based Innovation Policy-Making

Such policy guidelines and policy papers of exemplary measures of innovation support in firms are part of the policy-makers' tool kit. They contribute to provide practical ideas on 'how' to foster innovation. However, the major shortcoming of these policy guidelines and policy papers is that they do not provide indications on the contextual factors that ensure the success of the introduction of the listed policy measures. Hence, with the lack of similar conditions on the level of resources (research, human, network capital), of institutions and of history, this tendency to replicate policy measures and approaches of models that work elsewhere incites to copy policy measures that are not adapted to the situations and needs of the imitators.

In practice, however, differences in contexts (history, institutions, values, R&D intensity, etc.) make replication of successful models difficult. Many localised assets are difficult to reproduce in other local contexts. Replication of successful models must be guided by the attributes of the local context of actions and decisions. Practice-based innovation policy-making must be based on history and take into account factors linked to path dependency. Practice-based innovation policy-makers should remember that history is more than the end point of evolution in the matter of innovation performances. Hence, instead of replicating the end point (for instance, supporting the construction of new buildings, new R&D research institutes, etc.), they should pay attention to the take-off process, to the history of the take-off process that was instrumental in generating successful models of innovation. Such a perspective would induce them to develop customised solutions rather than replicating models likely to fail, due to their lack of replicability in different contexts.

5.9 Dilemma 8: Innovations in Knowledge-Based Firms Versus Innovation in Traditional Firms

The dilemma: Policy-makers are equally attracted by the idea of implementing innovation policies based on the concept of knowledge-based firms than by its opposite alternative which is to implement innovation policies based on descriptive reports on innovation which show that so-called traditional manufacturing firms create jobs and wealth, and therefore, that policy-makers are equally incited to help traditional firms become more innovative.

5.9.1 The Argument Supporting the Concept of Knowledge-Based Innovation

Knowledge is considered as the only production factor that creates competitive advantages that are difficult to replicate by rival firms (Zack 1999; Nonaka and Takeuchi 1995; Leonard 1998; Lesser and Prusak 2004). The concept of knowledge-based firms suggests that more innovative firms rely more on knowledge than less innovative ones. The concept of knowledge-based innovation has until now incited policy-makers to emphasise the importance of research based-knowledge as the prime driver of innovation. As a consequence, business leaders and policy-makers argue that all firms, including SMEs, should aim to become research knowledge-based firms by increasing their use of research knowledge such as R&D, patents, and collaborative research projects with external organisations as a tool for fostering growth and innovation. This type of argument incites policy-makers to consider that the only genuine knowledge-based firms are science-based firms operating in industries like, for instance, the photonics industry and the biotech industry.

5.9.2 Why Are the Determinants of Innovations in Traditional Firms also Knowledge-Based Determinants?

As underlined earlier, the development of product and process innovations depends on internal and external determinants or resources, as well as hybrid determinants. The internal determinants refer to the internal capacities of the firms to create and use the necessary knowledge to develop and improve products and processes. The external determinants refer, for their part, to the capacities of the firms to identify, obtain and use the knowledge coming from external sources, and to the capacities of the firms to adapt themselves to the competitive environment. Here is our knowledge-based reinterpretation of the major determinants of innovation in manufacturing firms:

The *determinants* or internal *resources* relate to:

- Knowledge embodied in employees measured by the number of employees according to the level of formal education and technical experience;
- Knowledge about the competitive environment, and strengths and weaknesses of firms embodied in firms' strategies;
- Knowledge creation generated through R-D activities and the number of employees involved in R&D activities;
- The use of knowledge embodied in advanced technologies;
- The use of knowledge embodied in manufacturing practices linked to added value production.

The *determinants* or external *resources* relate to:

The use of external sources of knowledge (ideas and information, technical support) to complete and improve internal resources helping firms to develop product and process innovations:

- Use of market sources of knowledge generated by interactions with clients, suppliers, competitors, consultants;
- Use of research sources of knowledge generated by interactions with universities, public centres of research, centres of technology transfer, patent documentation, technology transfer organisations;
- Use of knowledge generated through formal and informal collaborations with other businesses, and public and private research organisations;
- Use of knowledge generated through interactions with economic development agencies and innovation support organisations.

The *hybrid determinants* that are partly linked to the firm's internal factors and partly linked to external factors mainly refer to the knowledge regarding obstacles to innovation such as the lack of information on technology relevant to the development or introduction of new or significantly improved products or processes or the lack of skilled personnel to develop or introduce new or significantly improved products or processes.

5.9.3 Implications for Practice-Based Innovation Policy-Making

This reinterpretation of the explanatory variables of innovation suggests that, in practice, all the firms and all the sectors, not only those based on science, are in industries and firms which rely on knowledge to innovate. Many firms in so-called traditional manufacturing industries actually use more knowledge that is embodied in their manpower and their advanced technologies than, for instance, firms that make drugs in the pharmaceutical industry or firms that assemble parts in the electronics industry.

Firms' knowledge base varies from one firm to the next, as well as from one industry to the next. Firms' capacity to acquire and use knowledge-based factors in the development of product and process innovations also varies broadly. Knowledge-based inputs must be considered as solutions to solve problems that are specific to the products and production processes of each firm. The goal of innovation policy measures resides in the improvement of the knowledge processing capacities of all firms. In a context of practice-based innovation policy-making, one might assume that this goal can be efficiently achieved only if innovation policies are based on a portfolio of support measures which take into account firms' level: high knowledge processing capabilities, medium capabilities, below-average capabilities and non-existence of knowledge processing capabilities. Hence, the knowledge processing capabilities of firms could be improved gradually with the following types of assistance:

- Coaching, diagnosis and support services;
- Demonstrations of use of advanced technologies;
- Support towards the participation and organisation of equipment trade shows;
- Support towards the creation of technological and knowledge networks;
- Support towards the creation of expertise centres that would provide diagnostics, and support the implementation and optimisation of advanced technologies;
- Support of customised service offering to firms because services offered across the board, irrespective of industries, number of employees, knowledge absorptive capacities, are considered non-pertinent or non-existent by targeted firms.

5.10 Dilemma 9: Producing National Innovation Diagnostics Versus Producing Innovation Policy Diagnostics

The dilemma: National policy-makers are attracted by the idea of producing national innovation diagnostics based on aggregated data, while regional policy-makers are attracted by the alternative concept of producing innovation policy diagnostics based on disaggregated data that take into account the diversity of situations and needs of firms, industries, and regions.

5.10.1 The Argument Supporting the Concept of Knowledge-Based Innovation

In the 1990s, the OECD's Oslo Manual (1992) and its subsequent editions (OECD 1997; 2005) developed a series of methodological guidelines that provide a template used in the various OECD countries to conduct national innovation surveys. The purpose of such guidelines was to generate comparable data on innovation performances at the national level. National innovation diagnostics provide aggregated data useful for benchmarking innovation performances across countries. Hence, they are largely used in OECD reports to compare and assess national innovation performances. Such national diagnostics are also usable to compare the innovation performances of large metropolitan city regions.

5.10.2 The Problem Raised by National Innovation Diagnostics

However, firms of less than 15 employees are excluded from national innovation surveys in countries such as Canada. Hence, Statistics Canada's survey data on innovation include only manufacturing and service firms of more than 15 employees. But most of Canadian regions are made up of more than 90% of firms of less than 15 employees. This limit of available data prevents the production of precise and pertinent diagnostics at the regional level. Furthermore, for legitimate reasons, the data of these surveys cannot be disaggregated at the regional level, owing to the very strict confidentiality rules that Statistics Canada must respect.

The data on innovation produced by national statistical agencies, whether analysed at the industry or firm level, do not allow diagnosing firms' performances by distinguishing innovation performances at the regional level or in terms of industry and firm size. They also do not allow diagnosing firms' strengths and weaknesses at the regional level according to their R&D activities, their capacity to obtain tax credits for R&D activities, the use of advanced technologies, the use of added value production practices, firms' strategies, amongst others, their commercialisation strategies.

5.10.3 Implications for Practice-Based Innovation Policy-Making

National innovation diagnostics are useful for benchmarking innovation performances across countries. However, the implementation of practice-based innovation policies requires disaggregated data and evidence about regions and industries within regions that are not provided in national innovation surveys. Such shortcomings have led policy-makers of many European and Canadian regions to produce regionalised customised innovation diagnostics as a tool to generate evidence-based innovation policies at the regional level or at the industry level.

5.11 Summary and Conclusion

This chapter discusses the dilemmas that emerge when policy-makers are equally attracted to promoting concepts that are not based on evidence and to developing innovation policies that are evidence-based. It addresses nine dilemmas of innovation policies by discussing the following questions: What is the capacity of theories and new policy concepts linked to innovation to provide useful policy prescriptions to foster innovation? To what extent are these theories and new policy concepts supported by solid evidence? The chapter is based on a review of concepts and buzzwords linked to innovation, and the types of data and diagnostics used by policy-makers to develop new national and regional innovation policies. The chapter defines policy issues in terms of dilemmas, it provides concrete examples, and it discusses how policy-makers could resolve the tensions between equally attractive policy alternatives.

The discussion of practice-based innovation policy-making was organised around nine policy dilemmas: (1) Theory vs. practice-based innovation; (2) The magic ingredient concept vs. the multiple ingredients practice; (3) Global innovative firms vs. local innovative firms; (4) Open vs. closed innovation; (5) Inward vs. outward view of innovation; (6) Innovations that are first on the world market vs. incremental innovations; (7) Policy imitation vs. customised solutions; (8) Innovations in knowledge-based firms vs. innovations in traditional firms; and (9) Producing national innovation diagnostics vs. producing innovation policy diagnostics.

The results of this chapter contribute to show that many policy concepts that appear attractive are not supported by robust evidence. Hence, the results of this chapter show that turning policy concepts into evidence-based policy is difficult. The professionals of innovation are very often submitted to tensions where they have to choose between equally attractive but opposite alternative innovation policy measures. The results of this chapter show that policy ideas that appear to provide new policy prescriptions have to be confronted to evidence and diagnostics that take into account the diversity of the situations and needs of firms, industries, and regions.

There are questions about the future of practice-based innovation policy-making. Contrary to what has happened in health policies, drug policies and medical professions, there has not been an institutionalisation of an evidence-based approach in the field of innovation policy-making. As a consequence, there are as yet no organisations which provide leadership and guidance that promote the evidence-based approach in the field of innovation policies. The science and technology councils that exist or existed in many countries were expected to play such a role, but we assume that a lack of appropriate resources prevented them from playing such a role.

References

Amara, N., & Landry, R. (2005). Sources of information as determinants of novelty of innovation in manufacturing firms: evidence from the 1999 statistics Canada innovation survey. *Technovation, 25*, 245–259.
Amara, N., Landry, R., Becheikh, N., & Ouimet, M. (2008). Learning and novelty of innovation in established manufacturing SMEs. *Technovation, 28*, 450–463.
Asheim, B., Iasksen, A., Nauwelaers, C., & Todtling, F. (Eds.). (2003). *Regional innovation policy for small-medium enterprises*. Northampton, MA: Edward Elgar.
Balka, K., Raasch, C., & Herstatt, C. (2010). How open is open source? software and beyond. *Creativity and Innovation Management, 19*(3), 248–256.
Barthelt, H., Malmberg, A., & Maskell, P. (2004). Clusters and knowledge: Local buzz, global pipelines and the process of knowledge creation. *Progress in Human Geography, 28*, 31–56.
Becheikh, N., Landry, R., & Amara, N. (2006). Lessons from innovation empirical studies in the manufacturing sector: A systematic review of the literature from 1993–2003. *Technovation, 26* (5–6), 644–664.
Bogenschneideer, K., & Corbett, T. (2010). *Evidence-based policymaking*. New York: Routledge.
Boschma, R. (2005). Proximity and innovation: A critical assessment. *Regional Studies, 39*, 61–74.
Caracostas, P. (2007). The policy-shaper's anxiety at the innovation kick: How far do innovation theories really help in the world of policy? In F. Malerba & S. Brusoni (Eds.), *Perspectives on innovation* (pp. 464–489). New York: Cambridge University Press.
Chesbrough, H., Vanhaverbeke, W., & West, J. (2006). *Open innovation: Researching a new paradigm*. Oxford: Oxford University Press.
Christensen, C., Horn, M. B., & Johnson, C. W. (2008). *Disrupting class: How disruptive innovation will change the way the world learns*. New York: McGraw Hill.
Cooke, P., Boekholt, P., & Todtling, F. (Eds.). (2000). *The governance of innovation in Europe: Regional perspectives on global competitiveness*. New York: Pinter.
Dahlander, L., & Gann, D. M. (2010). How open is innovation? *Research Policy, 39*, 699–709.
European Union. (2007). *European Union Regional Policy*. Brussels: European Union.
Henkel, J. (2006). Selective revealing in open innovation processes: The case of embedded Linux. *Research Policy, 35*(7), 953–969.
Landry, R., & Amara, N. (2004). *La R&D et les stratégies de protection et de transfert de technologies*. Report prepared for Canada Economic Development and La conférence des élus de Chaudière-Appalaches, Québec, Canada.
Landry, R., Amara, N., & Lamari, M. (2002). Does social capital determine innovation? To what extent? *Technological Forecasting and Social Change, 69*, 681–701.
Landry, R., Amara, N., & Saïhi, M. (2009). Complementarities between strategies to protect inventions and innovations: Evidence from manufacturing SMEs. *International Journal of Intellectual Property Management, 3*(1), 56–78.
Laursen, K., & Salter, A. J. (2006). Open for innovation: The role of openness in explaining innovation performance among UK manufacturing firms. *Strategic Management Journal, 27*, 131–150.
Leonard, D. (1998). *Wellsprings of knowledge: Building and sustaining the sources of innovation*. Boston: Harvard Business School Press.
Lerner, J. (2009). *Boulevard of broken dreams: Why public efforts to boost entrepreneurship and venture capital have failed and what to do about it*. Princeton: Princeton University Press.
Lesser, E., & Prusak, L. (2004). *Creating value with knowledge*. New York: Oxford University Press.
Lundvall, B.-A. (Ed.). (1992). *National systems of innovation*. New York: Pinter.
Macvaugh, J., & Schiavone, F. (2010). Limits to the diffusion of innovation: A literature review and integrative model. *European Journal of Innovation Management, 13*(2), 197–221.

Morgan, K., & Nauwelaers, C. (Eds.). (2003). *Regional innovation strategies: The challenge for less-favoured regions*. New York: Routledge.
National Governors Association. (2008). *Innovation America: A final report*. Washington: National Governors Association.
Nauwelaers, C., & Wintjes, R. (2003). Towards a new paradigm for innovation policy? In B. Asheim, A. Iasksen, C. Nauwelaers, & F. Todtling (Eds.), *Regional innovation policy for small-medium enterprises* (pp. 193–219). Northampton, MA: Edward Elgar.
Nonaka, I., & Takeuchi, H. (1995). *The knowledge-creating company*. New York: Oxford University Press.
Nutley, S. M., Davies, H. T. O., & Walter, I. (2009). Past, present, and possible futures for evidence-based policy. In G. Argyrous (Ed.), *A practical guide: Evidence for policy and decision-making* (pp. 1–27). Sydney: University of New South Wales Press.
OECD. (1992, 1997, 2005). *Proposed guidelines for collecting and interpreting technological innovation data: The Oslo manual*. Paris: OECD.
OECD. (2008). *Globalization and regional economies: OECD reviews of regional innovation*. Paris: OECD.
Peters, S. (2006). *National systems of innovation: Creating high-technology industries*. New York: Palgrave Macmillan.
Reback, G. L. (2009). *Free the market: Why only government can keep the marketplace competitive*. New York: Penguin.
Roulet, N., Dubois, P., Aoussat, A., & Le Coq, M. (2010). The integration of new technologies: The stakes of knowledge. *International Journal of Product Development, 12*(2), 126–140.
Schmoch, U., Rammer, C., & Legler, H. (Eds.). (2006). *National systems of innovation in comparison*. Dordrecht, The Netherlands: Springer.
Statistics Canada. (2010a). *Science statistics: Estimates of research and development expenditures in the higher education sector, 2008–2009*. Ottawa, Canada: Statistics Canada.
Statistics Canada. (2010b). *Survey of intellectual property commercialization in the higher education sector 2008*. Ottawa, Canada: Statistics Canada.
Steineke, J. E., & Hedin, S. (2008). *Management by objectives and results: Structures and practices in the regional policy field in the Scandinavian countries and Iceland*. WP-2008-3 Nordregio. Stockholm: Nordic Centre for Spatial Development.
Usher, D. (1992). *The welfare economics of markets, voting and predation*. Ann Arbor: The University of Michigan Press.
Vega-Jurado, J., Gutierrez-Gracia, A., & Fernando-de-Lucio, I. (2009). *Does external knowledge sourcing matter for innovation? Evidence from the Spanish manufacturing industry*. Valencia: Institute of Innovation and Knowledge Management, Polytechnical University of Valencia.
West, J. (2003). How open is open enough? Melding proprietary and open source platform strategies. *Research Policy, 32*, 1259–1285.
Weterings, A., & Boschma, R. (2009). Does spatial proximity to customers matter for innovative performance? Evidence from the Dutch software sector. *Research Policy, 38*(5), 746–755.
Zack, M. H. (1999). Developing a knowledge strategy. *California Management Review, 41*(3), 125–145.

Chapter 6
Coordination in Innovation Projects

Cornelius Herstatt and Norbert Lühring

Abstract Many empirically based studies support a positive connection between cross-functional cooperation and successful innovation. In this context, the existing literature mainly points to the interplay between the two key functional areas of research and development (R&D) and marketing (Griffin and Hauser 1996; Moenaert et al. 1994). The integration of production (Olson et al. 2001; Kahn 2001; Song et al. 1998) and purchasing (Wijinstra and Stekelenborg 1996) are, however, increasingly mentioned as also being critical for the realisation of innovation projects. The more recent discussion on *integrated product development* (Bullinger et al. 1995) shifts the focus to an overall understanding of the various functional areas, which at times make considerable contributions to the success of an innovation project. From this perspective, there appears to be a need for a higher level of integration and coordination in innovation projects. This is seen as necessary to ensure that the results of individual work packages are aligned to the overall objective of a project and that the implementation of the product development process is effective and efficient (Lawrence and Lorsch 1986; Rueckert and Walker 1987). Consequently, we are faced with the question of which mechanisms can be used to best meet coordination needs in innovation projects. Based on an in-depth study of three innovation projects, our chapter attempts to answer this question. We will first give an overview of diverse aspects of coordination in innovation projects. For this purpose, five dimensions of coordination are derived from organisational theory. Furthermore, a coordination model for innovative projects is developed to

C. Herstatt
Institute for Technology and Innovation Management, Hamburg University of Technology, Hamburg, Germany

N. Lühring (✉)
Lischke Consulting GmbH, Hamburg/Munich, Germany
e-mail: luehring@lischke.com

H. Melkas and V. Harmaakorpi (eds.),
Practice-Based Innovation: Insights, Applications and Policy Implications,
DOI 10.1007/978-3-642-21723-4_6,

serve as a basis for our empirical study: a comparative case study analysis conducted in the electronics and automotive supply industries.

6.1 Introduction

Whenever dependent elements or subsystems are created to achieve a high level of specialisation, the need to align these units through coordination arises in order to minimise dysfunctional behaviour in the units (Frese 1988; Rühli 1992; Staehle 1999; Steven 2001). Coordination complements differentiation; along with work sharing, it is the second basic principle of true organisation (Osterloh and Frost 2000; Rühli 1992; Silber 1986). Generally speaking, the need for coordination increases (Rühli 1992) as:

- The differentiation of the entire system in elements and subsystems caused, for instance, by division of work inside an organisation, advances.
- The interdependence of organisational elements and subsystems grows and the complexity and intensity of relationships develops.
- Great areal, factual, and cultural discrepancies require handling.
- The variety of tasks to be accomplished increases. Unstructured as well as dysfunctional behaviour of organisational elements can negatively influence the success of the entire system.

All these aspects are frequently relevant for innovation projects. The individual relevance of these aspects, however, depends on the specific context of the innovation project.

A cross-organisational approach to innovation projects makes the contributions of diverse departments necessary. The demand for coordination increases due to the fact that departments such as marketing, production and R&D are organised according to the principles of division of work, creating areal, functional and cultural discrepancies between departments. Finally, innovative tasks are characterised by uncertainty and lack of structurability, which increases the need for coordination (Bullinger et al. 1995).

6.2 Dimensions of Coordination and Variables Influencing Coordination of Innovation Projects

The analysis of coordination in innovation projects poses a number of questions. To structure these questions, various approaches from organisational theory have been investigated. Based on these, we have developed a framework that we will use to analyse coordination in innovation projects. This approach is based on five dimensions (see Fig. 6.1).

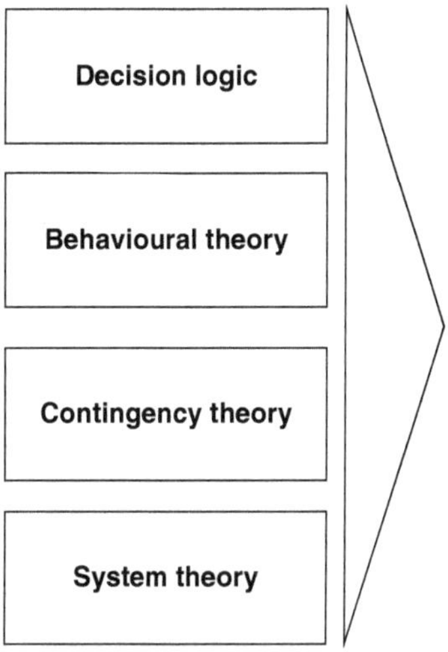

Dimension of coordination	Questions
Content and task	• Who coordinates with whom? • Which parts of a task need to be coordinated in a project?
Reason	• Why are tasks being coordinated? • Which barriers require which coordination mechanisms?
Context	• In which situations is coordination used? • Which factors influence the demand for coordination?
Time	• When is coordination applied? • Which processual aspects and models form the foundation of coordination?
Instrument	• Which tools are used to perform coordination?

Fig. 6.1 Research framework for analysis of different dimensions of coordination in innovation projects

6.2.1 *The Content and Task Dimension: Coordination and Function*

As already noted, integrated product development (Bullinger et al. 1995) is currently seen as a key contributor to the success of any innovation project. An exhaustive overview of empirical findings regarding the dependencies between different functional units like marketing and R&D can be found, for instance, in Griffin and Hauser (1996). From the perspective of interface design, the involvement of production in innovation projects has been further scrutinised by Gerpott (1991), Kahn (1996, 2001), Olsen et al. (2001) and Song et al. (1998).

In addition, the involvement of procurement in the innovation process has been discussed by Wijinstra and van Stekelenborg (1996), Nijssen et al. (2002) as well as Salomo et al. (2003). Furthermore, various authors have analysed the integration of controlling (Stuffer 1994; Ulrich and Eppinger 1995), testing and quality control (Ehrlenspiel 1995), customer service (Stuffer 1994) as well as legal affairs (Ulrich and Eppinger 1995).

Regardless of the requisite functional focus of these works, the literature conveys the impression that the consistent employment of inter-department coordination is to be seen as key factor for success.

The contribution of the different departments results in a high need for coordination of activities and decision-making during innovation projects. This coordination is often hampered by the specific characteristics and differing motivational structures of departments.

6.2.2 The Reason Dimension: Barriers to Coordination

A large number of studies indicate that a focus on aggregation of different departments is not sufficient to guarantee project success (Shaw and Shaw 1998; Song et al. 1997). In particular, personnel-related and structural problems hamper efficient and effective cooperation across interfaces. According to Sommerlatte (1988), about one-third of costs in firms are created by the interface inefficiencies of divided activities. In addition, 70% of the working time per task goes to waiting, transport, or start-up times (Knickel 1997).

Already in 1963, in their work on differentiation in and integration of organisations, Lawrence and Lorsch (1986) identified specific characteristics of highly specialised departments as negatively influencing integration. Souder (1988) indicated in an investigation of 289 development projects that two-thirds of the projects examined showed substantial work climate dissonances, correlating to a total failure of the respective projects.

Various authors (Gupta and Wilemon 1988; Griffin and Hauser 1996; Sherman et al. 2000; Stuffer 1994; Lucas and Bush 1988) indicate that the reason for such dissonances are differences in the individual "thought worlds" of respective departments, which affect efficiency through their high potential for conflicts.

In accordance with function-specific thinking, a function-specific language is often developed through daily work; this language often seems out of reach for other firm members because of its unique methods of wording and interpretation (Seidel 1996; Wiebecke and Tschirky 1987). Collaboration in the case of divided tasks is also important to avoid misunderstandings and frustration. People talk with each other, but often do not fully understand each other (Euringer 1995).

Another integration barrier is apparent in the different time and planning horizons of the departments involved. While marketing departments are usually confronted with fast-changing market situations and customer demands, and are accordingly forced to react quickly, R&D departments often avoid short-term decisions and fast reactions (Euringer 1995). This can possibly cause differences of opinion with regard to the allocation of scarce resources (Staudt et al. 1991).

Potential problems are intensified through inherent conflicts in the goals of different departments, usually dominated by function-related sub-aspects, even if the main goal of the company concerning product development is known and understood by all departments concerned (Griffin and Hauser 1996; Song and Parry 1997; Euringer 1995).

Personnel-related causes for problems at the interfaces between departments often result from the organisational structures involved in project work. By this we are primarily referring to a lack of clearly assigned responsibilities, resulting, for example, from erroneous self-assessments within individual departments and poor organisation. The consequences of this are often unclear internal responsibilities and a hindered (or complicated) perception of responsibility of the department in question (Manns 1992; Schmidt 1996; Specht 2000). The departments involved are often at the same hierarchical level. This may cause difficulties if these departments fail to agree with respect to a certain strategy or decision. Due to the non-existence of a coordinating 'senior' who can bridge the distance between the different departments, there may be no direct way of resolving the problem (Brockhoff and Hauschild 1993).

In addition to group-internal conflicts, team work has the potential to lead to conflicts at the interface between company- and project-level organisations (Stuffer 1994). A lack of coordination related to steering and the planning of necessary capacities and information for the team and the line organisation may put project results at risk (Bullinger et al. 1995; Shanteau and Rohrbaugh 2000).

To sum up, investigating the causes for disturbances in the coordination of innovation projects indicates a great number of entities and barriers that need to be considered in order to ensure project success. A variety of critical aspects must be addressed when considering the choice of certain coordination efforts. As illustrated, the creation of a common understanding regarding concrete tasks within an innovation project, conflict resolution, prioritisation, and exchange of information on the bases for decision-making can all be considered special challenges that create barriers to integration and coordination. When considering the coordination mechanisms relevant to and empirical analysis of an individual project, it is highly important to what extent an individual coordination measure meets these challenges.

6.2.3 The Context Dimension: Moderating Factors in Coordination

In addition to the identification of barriers to coordination, a number of other moderating factors require consideration to determine the need for coordination. In their cause-effect model, Griffin and Hauser (1996) describe the project phase as an important determinant of the need for inter-departmental coordination in an innovation project. Another contextual variable is the specific project management approach applied (Lühring 2003).

6.2.3.1 Impact of Project Phases

Griffin and Hauser support the model developed by Rueckert and Walker (1987) stating that, as marketing frequently lacks the necessary informational, monetary and personnel resources, the resulting dependency has to be seen as a key factor in interaction. Whether such dependency occurs is in itself dependent upon the overlap of jobs of interacting departments and the chosen strategic alignment of the company. The more the functions overlap and the more the strategic alignment of the company demands innovation, the higher the probability of cases in which the use of common resources is necessary. Due to the different types of tasks that need to be completed at different phases of the innovation project, the degree of dependency varies from phase to phase (Griffin and Hauser 1996).

In a study of US-based companies in numerous industries, Song et al. (1998) investigated the relative importance of collaboration among the functional units of marketing, R&D and production across all product development phases. They identified different schemes of collaboration, which were found to be more or less productive, depending on the given project phase.

The remarkable finding from this study was that the collaboration of all three functions in the five phases identified by Song et al. was found to be of either low or even negative productivity. One possible reason could be time intensity. The involvement of different functions increases the number of potential conflicts between departments and individuals (Song et al. 1998; Olsen et al. 2001).

While Olsen et al. (2001), rather than classifying projects into numerous phases, merely categorise these as early- or late-phase projects, they indicate the same demand for integrating marketing and R&D into the early project phases.

The work of Olsen et al. supports the notion of the necessity of integrating R&D and production during the late project phases, to consider the aspects of process capability, quality of production, cost and size of production during new developments. In this, Olsen et al. further show that the degree of integration of the departments during the project life cycle is not constant, but increases with the progress of the project (Olsen et al. 2001; Salomo et al. 2003). The degree of integration increases as soon as the basic technological problems have been solved and routine tasks become more important.

6.2.3.2 Impact of Project Uncertainty

Regarding the external effects of uncertainty, Griffin and Hauser share the interpretation of Rueckert and Walker (1987) as well as that of Gupta, Raj and Wilemon (Gupta et al. 1986), who postulate a causal relationship between complexity, on the one hand, and degree of integration as considered necessary by the organisation, on the other. Depending on the strategic alignment of the company with regard to product development, that is, to what extent the company wants to claim the position of innovator or imitator, the degree of uncertainty and demand for information varies. To decrease uncertainty and the risk of failure, the integration of marketing and R&D is necessary. Gupta and Wilemon assert that the degree of optimum integration decreases from innovator to imitator.

Thom also shares the opinion that, after project phase, degree of newness seems to be the second key factor influencing the demand for integration (Thom 1980). While incremental developments are characterised through a linear development path with low uncertainty and short development time, companies are confronted with an increasing radicality of development when dealing with new technologies, markets and opportunities.

Due to lack of experience, the planning horizon decreases and plan uncertainty increases as the radicality of an innovation projects grows. The existing interfaces become increasingly complex and additional interfaces emerge, which cause an increase in the demand for coordination and integration. Thom states in this context that experiences in this process can only be developed at a process level of development, because of the repetitiveness of the processes. This is not possible on a content level (Thom 1980).

Moenaert and Souder define uncertainty in the context of product development as the difference between the information required to complete a job and the information that already exists within the organisation (Moenaert and Souder 1990). With regard to innovation projects, a division can be made between market- and technology-related sources of uncertainty. Such market-related sources are those involving customers and competitors. Technology-related uncertainties stem from technology and resources (Moenaert and Souder 1990; Green et al. 1995).

6.2.3.3 Impact of Project Management Approach

Project teams constitute the typical second-level organisational form in innovation projects. The characteristics of project organisations are subject to different demands based on the prime organisation of the respective company. Depending on the degree of involvement of the individual project participant in the project team and the make-up of the project leadership, different kinds of interfaces appear between project team and prime organisation, for instance in terms of accessing common resources (Salomo et al. 2003; Donnellon 1993).

The generally positive influence of team work on project success has been stated in various empirical studies (Gemünden and Högl 1998; Lechler 1996). A study by Clark and Fujimoto (1991) proves that project success defined by the factors time, cost, and quality of product development projects is mainly influenced by the above-mentioned structural characteristics of the project team. Clark and Fujimoto further prove that, in comparison to functional teams, weighted teams correlate with a more positive outcome in terms of the success factors of time, cost, and quality. The question of how the choice of a certain team approach influences individual success factors at a company level remains unanswered.

Ancona and Caldwell (1992) demonstrate in their study that the amount as well as the kind of external interaction positively influences the productivity of a team in terms of quality, cost, and time. By placing employees from all departments into one new product development project team, the interaction of the team in its organisational context can be enhanced (Ancona and Caldwell 1992).

Pinto et al. (1993), in their research regarding multiple-department collaboration, identify five factors for which they postulate a direct influence on the interdepartmental collaboration of a team. For three of these factors (super-ordinate goals, physical proximity, and project team rules) they find a significant positive influence (Pinto et al. 1993).

In another empirical study concerning the consumer goods industry, Sethi (2000) proves that the quality of newly developed products positively correlates with the integration of information within multiple-department project teams. Functional diversity, the measure of the different departments of the team members, has no influence on the quality of newly developed products. Sethi comes to the conclusion that in addition to functional diversity, the way in which different departments integrate their information and carry out the integration and coordination process is an important determining factor (Sethi 2000). Karlsson and Ahlström (1996) arrive at a similar conclusion in their study of the influence of cross-functional project teams on production-oriented product design.

While these empirical studies reveal clear results regarding the influence of cross-functional team work on innovation project success, a specific analysis of the demand for coordination between project teams and their organisational environment is still missing. Clark and Fujimoto (1991) describe the effects on success factors like quality, time, and cost for weighted teams versus functional structures. Nevertheless, they leave the question of how to respond to the demand for coordination mostly unanswered. In his research on the reorganisation of product development at Toyota, Nobeoka (1995), in addition to finding positive effects in a weighted project team structure compared to cross-functional coordination, identifies a substantially higher demand for tools to coordinate specific project activities.

While distinct results have been found on the moderating effect of the respective phase of an innovation project and of technical- and market-related uncertainty on the cross-functional need for coordination, no empirical connection between the basic types of project management and the need for coordination in innovation projects has been proven as yet. A possible reason for it is the fact that, in the literature, project management has been considered an instrument of interface management or a coordination mechanism without testing different kinds of project management as to their coordination effect (Brockhoff and Hauschild 1993).

Nevertheless, depending on the kind of project management, different interface patterns originate from the various roles, responsibilities, and duties of the actors involved in an innovation project. In this chapter, the basic types of project organisation (functional, lightweight, heavyweight, autonomous) are thus considered moderating factors in terms of the need for coordination in innovation projects.

6.2.4 The Time Dimension: The Foundation Process

The questions of 'when' and 'how' (the tools or mechanisms used) should be considered core issues of multiple-function coordination in innovation projects, since these touch on the central design parameters of coordination (Benkenstein 1987a). The coordination of an innovation team with its organisational surroundings can be carried out during the different phases of an innovation project. For this reason, the question of coordination timing must be answered phase-specifically. The project phase has already been identified as a moderating factor with regard to the demand for coordination in innovation projects. To further understand the division of a project into phases, it is useful to investigate the models for describing innovation processes.

Related specifically to the time dimension, coordination can be further distinguished as active, which foresees coordination, or reactive, which reacts to turbulence

(Reger 1997). Kieser and Kubicek (1992) refer to these two types as *pre-coordination* and *feedback coordination.* In the case of pre-coordination, the main goals are translated one by one into executable, well-coordinated activities. Feedback coordination is carried out as a reaction to turbulence that occurs as a consequence of uncertainty in the execution of a company's processes. Inasmuch as future developments can be grasped and foreseen with satisfying results, pre-coordination for the coordination of partial systems is generally sufficient. Especially in case of the realisation of innovation projects, the environment is usually described as being dynamic and uncertain due to a multitude of possible disturbances. For this reason, feedback coordination is considered important in innovation projects (Reger 1997; Benkenstein 1987b).

While various models that divide the innovation process into different phases can be found in the literature, a common standard has not yet been established. This results from the fact that innovation processes – which depend, among other things, on industry, company, and complexity – happen to be very diverse.

A widely used model is Thom's three-phase model, which consists of idea creation, idea acceptance, and idea realisation (Thom and Etienne 2000). Other authors arrive at four phases (Buggie 2001), five phases (Ulrich and Eppinger 1995; Sommerlatte 2000; Herstatt and Verworn 2003) or even up to eight phases (Trott 1998; Winkler 1999) when describing the innovation process.

Although these different models vary in terms of number of phases, their central assumptions are similar (Hauschildt and Petersen 1987):

- Different kinds of activities appear during the innovation process (assumption of existence);
- Comparable jobs are combined into task bundles (assumption of bundles), and
- An optimum order of jobs or job bundles can be determined (assumption of order).

To summarise, it is more important for the present study to focus less on specific phases and instead more on the identification of critical points in given innovation projects, keeping in mind the analyses of the effects of certain tools of coordination (form of coordination). It similarly appears that certain decisions or determinations (for instance, project team building, freezing of specifications) are to be made during the project life cycle (Hauschildt 1991) regardless of the division of phases or the degree of parallelism of tasks during the project. With the review of such critical points, the problem of overlapping project phases in terms of the empirical analysis of individual innovation projects is eliminated (Bailetti et al. 1998; Schmidt and Calantone 2002).

6.2.5 The Instrument Dimension: Coordination Instruments

The scientific literature presents a variety of suggestions for options for achieving the required demand for coordination and decreasing barriers at the interfaces in project development processes. Some actions are categorised under integration as well as coordination mechanisms. While in the German literature, terms such as *tools of interface management* (e.g. Brockhoff 1996; Specht 2000) and *coordination mechanisms* (Kieser and Kubicek 1992; Schneider and Müller 1993) dominate, the English literature uses mostly the term *integration mechanisms* (Leenders and Wierenga 2002). In this study, the terms *tool* and *mechanism* are used interchangeably.

Coordination mechanisms appear to have been systematised in the literature according to differing points of view. A view of the tools takes place at a similar level as the understanding of coordination. First of all, a number of authors can be found in the literature (Euringer 1995; Griffin and Hauser 1992; Maltz 2000; Maltz et al. 2001; Moenaert et al. 1994; Rueckert and Walker 1987), who, under the topic of tools, offer mainly approaches to improving the flow of information between the departments involved in a product development process. The scope of effects of the tools mentioned thus encompasses the structural design of the information flow, which is influenced by information and communication channels and information storage.

A high number of tools explicitly focus on these issues, in order to conquer socio-emotional barriers and harmonise interfaces (Lawrence and Lorsch 1986; Kahn 1996; Pinto et al. 1993). Such tools consist mostly of approaches to shaping company culture and multiple-department activities, tasks, and action frames.

In addition, authors exist who do not consider the construct of interaction and collaboration in the categorisation of tools (Brockhoff and Hauschildt 1993; Schmidt 1996; Specht 2000; Thom 1980; Wolfrum 1994). They therefore analyse tools not by their effects, but according to their organisational point of action. This categorisation focuses mainly on the structural freedom of management to design interfaces, and restores the perspective to the structural characteristics of an interface.

This perspective is found especially in the works of Schmidt (1996) and Thom (1980). Both of them describe tools that categorise the structural characteristics of an organisation. According to Schmidt (1996), the design of the organisational action parameters of organisation structure, leadership, interface controlling, and employee leadership positively influences the human factor. The coordination tools that Schmidt termed *action parameters* thus become options for influencing the probability of innovation project success. With active configuration of these parameters it is possible to influence the behaviour of an individual employee in an innovation project.

In analogy to Schmidt, Thom (1980) mentions organisational structure, temporal project, and macro structure, as well as formal and informal decision rules, as coordination-enhancing action parameters. A comprehensive overview and discussion of the different approaches to the systematisation of coordination tools as well as an overview of the empirical findings regarding the effects of the tools of coordination can be found in Lühring (2006). Review of this text indicates that the same coordination tools have sometimes been categorised according to different criteria. The overall picture regarding the possible mechanisms that can be used to satisfy the need for innovation project coordination still appears confusing.

Additionally, it is remarkable that the interface between R&D and marketing still seems to be the main focus of all papers.

In terms of a project-specific view, only a few mechanisms can be applied directly. Some basic conditions under which the coordination is facilitated are described (for instance, company culture, open-door policies, vision of the company, networking, competence centre, matrix organisation). These suggestions rightfully indicate that the creation of innovation-promoting organisational structures depends on the frame for given projects, which has a substantial influence on successful outcome. No absolute rules or measures can be formulated with regard to which actions should be carried out to boost coordination in a specific project. A division into project- and organisation-related tools is done only in exceptions (see Wolfrum 1994). In this sense, these suggestions remain at the abstract level of general structural models for the creation of innovative organisational structures.

Some of the above-mentioned mechanisms have resulted in proposals that in and of themselves contain a large bundle of individual actions (for instance, Simultaneous Engineering, Stage-Gate-Processes). Depending on project management approach, project management also stipulates differing demands for coordination activities.

Existing empirical studies concerning the employment of different tools of coordination provide useful results only with regard to individual mechanisms, for instance for the employment of promoters or the physical closeness of teams or departments in innovation projects (Folkerts and Hauschild 2002; Allen 1977; Leenders and Wierenga 2002; McDonough et al. 2001). Regarding the formalisation of project activities (use of coordination methods) and decentralisation (Moenaert et al. 1994), empirical findings offer little guidance on how specific the formalisation should be, how decision-making structures should be designed, or with which mechanisms the required external effects can be accomplished.

In researching specific projects, then, the focus must be on the creation of a framework to identify the specific coordination tools used to manage interfaces. With regard to the effects of these tools, research should focus on determining which mechanisms can be used to overcome the empirically determined barriers to coordination.

6.3 Research Questions and Conceptual Design

In preparation for this empirical research, five research questions were formulated:

1. What tools are employed to coordinate innovation projects?
   This question aims at filling the previously identified gap in the literature. The aim is to identify tools that can be used to coordinate innovation projects in practice (Olsen et al. 2001: 270). Furthermore, as previous works have strongly focussed on the interface between marketing and R&D (Song et al. 1997: 36;

Song et al. 1998: 291), the need for coordination in innovation projects has been captured only partially. Closely related to the first question is the second:

2. How are coordination tools specifically applied in practice?
   When tools for interface management are employed, the following questions arise: Which departments appear in this context? Are these initiators? How much lead time is given during initiation of this process, and which other departments are involved in the initiation process?
   In this context, it is also interesting to know whether tool use varies between departments or groups of departments or whether all departments use the same tools in the same way. This leads to the next research question:
3. How do the coordination tools used affect the successful performance of coordination functions between organisation- and project-level structures?
   This question is based on the assumption that coordination between a company's innovation teams and departments has no purpose in and of itself, but is conducted to perform specific coordination functions. The questions that have to be answered here are whether certain tools perform the coordination function more or less effectively than others, and whether there is a major difference between such tools concerning their effect on coordination.
4. How does the application of different coordination tools influence the success of individual project phases or the project as a whole?
   The overall goal of every innovation project is to achieve success. In terms of this study, it is of specific interest to determine which cause-and-effect relationships exist between project success and the application of specific interface management tools. For this purpose, we need to determine different measures of success or points within the various phases of an innovation project, which then enables us to analyse the use of tools throughout the innovation process. In addition, the question concerning the cause-effect relationship between the interface management tools and overall project success must also be answered.
5. Which situation- or context-related factors determine the application of coordination tools in innovation projects?
   Current research has shown that project phase and degree of innovative work function as context-related factors. In cases where demands for coordination lead to the application of interface management tools, it might be supposed that their application is also influenced by project phase and degree of innovation. This aspect will be further analysed through this fifth and final research question.

Figure 6.2 shows our research framework, including the various dimensions of coordination. Section 6.2.3 of this chapter discussed the three **moderating factors** on coordination demand that appear during the phases of the innovation process: (1) critical points or gates in the project; (2) project team layout (basic types of project management), and (3) project uncertainty. These input variables are treated as moderating factors in the conceptual model.

The central points of this study are the interface management **coordination tools** that are employed during innovation projects. Due to the difficult systematisation of coordination tools and the number of details involved in their application in various

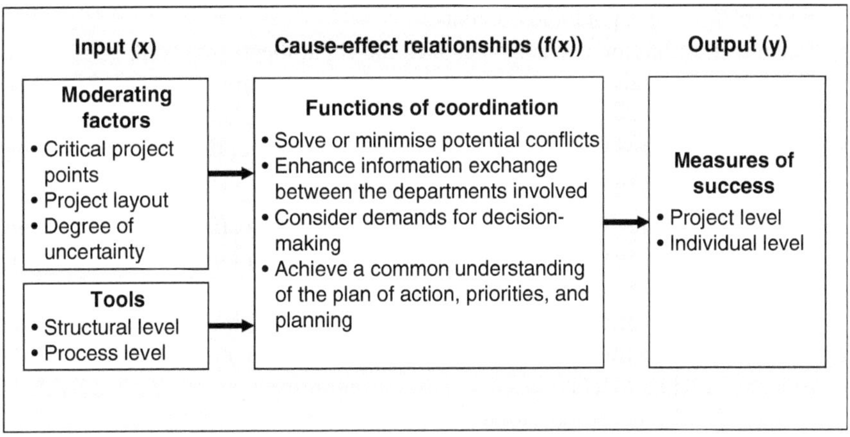

Fig. 6.2 Conceptual design as basis for empirical investigation; the question of the total demand for coordination between the departments (see Lühring 2003)

innovation projects, it was first necessary to develop a system for grouping coordination tools, which was then used in this research.

The interface problem involved leads back to the fundamental design characteristics of organisations: the principles of structural and process organisation. An interface-dependent problem can, therefore, only have a coordinating effect at the levels of process and structure during the innovation process.

The structure level includes the tools (Picot et al. 2001) with which – at least for a limited period of time for one or more phases – the structure of the project organisation is determined. **Structural-level characteristics** are:

- Formalisation (Thom 1980),
- Centralisation (Manns 1992; Staehle 1999),
- The system of responsibilities,
- Communication channels (Shannon and Weaver 1949; Euringer 1995), and
- The closeness of the places of work (Allen 1977).

The tools with which a punctual influence on the project process is achieved belong to the **process level** (Specht 2000; Knickel 1997; Köhler 1992). This includes process characteristics such as

- Work sharing and
- Determination of task order, as well as
- Correlation (Küpper 1981).

By analysing which tasks and contents are coordinated between what departments and with the support of which tools, the task and content dimension was also taken into consideration.

Section 6.2.4 addressed the various causes for disturbances between the different parties involved in an innovation project, thus underlining the need for coordination. Employment of individual coordination tools is efficient and effective when those tools are able to:

- Solve or minimise potential conflicts,
- Enhance information exchange between the departments involved,
- Consider demands for decision-making (support the process of making decisions), and
- Achieve a common understanding of the plan of action, priorities, and planning among all the participants involved.

For this reason, the conceptual model is based on the **functions of coordination**, which enable an evaluation of the coordination tools with regard to their effect in specific project situations during empirical study.

Finally, analysis of the effects of specific coordination tools on interface management (or their ability to address the functions of coordination outlined above) is considered with regard to the success of the various project phases. Both individual- and project-level criteria were used.

6.4 Methods

The employment of coordination tools for the management of organisational interfaces during the innovation process depends greatly on the specific innovation approach. This is why a comparative case-study approach was selected as a research method to answer the research questions presented here. The case studies were used to generate a better understanding of the problems involved and to develop a set of hypotheses. In the light of this study, issues concerning the dimensions of 'how' and 'what' as well as 'why' were taken into consideration. In terms of approach, the study follows the structure of explorative-explanatory case study.

6.4.1 Data Generation

Three in-depth case studies were developed based on practical case work. One recently completed large-scale innovation project was chosen from each of three different companies. Two of the three companies are suppliers in the automotive and/or rail industry. The third company is a major electronics-industry company. Two projects were customer-initiated; the third was an internally initiated innovation project.

Qualitative, partially standardised interviews were used as the main data source. Interviews were carried out with project leaders, team members, and decision makers from the various departments involved in each of the three projects. The number of interviews per project was between 15 and 20. Moreover, project documents, such as minutes, change protocols, timetables, and task lists, were used. In a number of workshops, participant feedback on project results was discussed and the results were completed, revised, or corrected accordingly.

In two out of the three companies, ongoing innovation work was observed to get a feel for the way these companies manage their innovation work and coordination patterns across the entire project life cycle.

The content of the first project was the delivery of a prime spring module for the wheelset of a railcar bogie. The prime spring module consists of 16 parts, 3 of which were produced by the company; the rest were ordered externally. The company is a globally recognised technology partner in the railcar industry. This customer-oriented development project was characterised by moderate technology- and market-related uncertainty. The evaluation of technological and market-related uncertainty was carried out through the interviews. In this context, customer orientation means that the project was carried out for a certain customer on the basis of a contractual specification made by said customer. This customer-oriented project was led by two managers, one from development and one from sales. Nevertheless, there were no fixed contacts in the different departments for the project. The project lasted for about 1 year.

The second project was conducted at a leading automotive supplier that specialises on vibration technology. It was directed at the development of a new rear axle spring for a new generation of cars. The entire system was composed of 14 parts, combined into 5 bushes. The project also had a customer orientation and was characterised by a low-to-medium technology- and market-related uncertainty, which varied during the project. Due to changing customer demands, the project management approach had to be changed several times. This caused several coordination problems. The project lasted for about 3 years.

The scope of the third project was the development of a new generation of coin changers for the US and European markets. The company is one of several global leaders in electronic checking and cashless payment systems. The uncertainty of this project, which did not involve a customer, was analysed as low to medium with regard to technology and market issues. The remarkable aspect of this project was that – unlike in the other projects – from the first project activities, a fully functional project team was set up. The coordination pattern during the project was positively influenced by this. The project lasted for about 18 months.

6.4.2 *Data Analyses*

Data analysis was qualitative, focusing on content analysis. The assembled data (minutes from interviews and workshops, data from project documents, observations from project monitoring) served as an indicator for certain issues or facts of the case or project stage in question. In light of this study, these issues or facts described events or situations, for which a system of categorisation was developed in close relation to the research questions.

To develop separate categories, a combination of theoretically and empirically oriented approaches was selected (Früh 1998). The theory-oriented approach served as the foundation for the design of the evaluation categories. These were derived on a theoretical basis from existing knowledge. The coordination dimensions (see Fig. 6.1) developed in Sect. 6.2 and the conceptual model (see Fig. 6.2) served as a basis for the evaluation design. The data and knowledge generated were further analysed to develop new evaluation categories.

The categorical systems developed were used for analytic evaluation of the interviews as well as for analysis of project-related data. For this purpose, the evaluation dimensions that matched the dimensions of the research model had to be determined (see Fig. 6.2). For each evaluation dimension, categories were defined and partial subcategories assigned. With regard to the model illustrated in Fig. 6.2, for example, 'exchange of information' is a category of the dimension 'functions of coordination'. The categories and subcategories were specified with indicators, which served as a basis for the interpretation of the data sample. Indicators were formulated activities with which the categories or subcategories could be clearly identified. The content of the data material was extracted when one of these pre-defined indicators was found for a given category. The extracted contents were then taken into consideration for the case studies. In order to achieve additional transparency, so-called 'anchor examples' (Hugl 1995) were used. These are examples that fall into a certain category and can serve as illustrative examples for the category in question.

6.5 Results and Generation of Hypotheses

Based on the findings of the case studies, a set of hypotheses was developed. The formulation of the hypotheses is intended to determine dependence between at least two facts. The frame of validity for the developed hypotheses is shown in Fig. 6.3. This Figure is based on the study of the model discussed in Sect. 6.3.

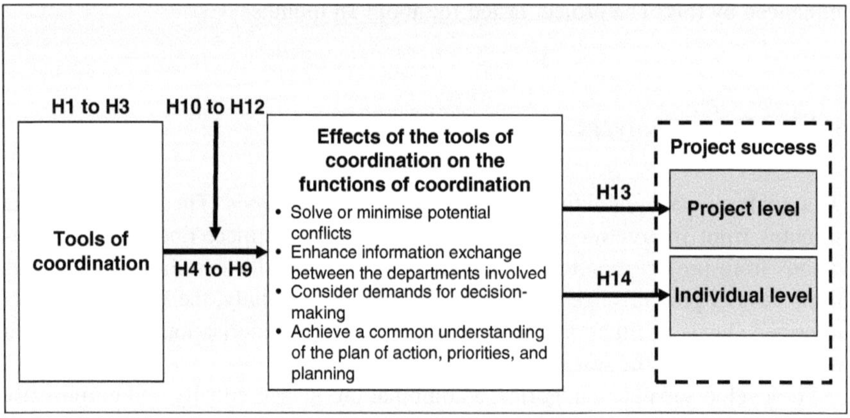

Fig. 6.3 Model for the formulation of hypotheses

Hypotheses 1–3 relate to the interdependency effects between coordination tools (H1, H2) as well as to initiating tool employment (H3). Hypotheses 4–9 encapsulate the effect of coordination tools in relation to the functions of coordination (H4 to H9). Hypothesis 10–12 target the moderating factors related to the employment of coordination tools (H10 to H12). Hypotheses 13 and 14 capture the influence of the employed coordination tools on project success. In this relationship, a distinction is made between the project level (H13) and the individual level (H14).

6.5.1 Hypotheses Regarding the Employment and Effects of Coordination Tools

Based on empirical results regarding the interdependencies of specific coordination tools, the following hypotheses can be formulated:

H1: The more clearly the role of process and expert promoter is performed by the project leader, the lower the significance of the use of formal coordination tools.

H2: The higher the congruency between project task and responsibility, the lower the demand for coordination tool use.

Hypothesis 1 focuses on the diverse roles that project leaders have taken in two case studies on the one hand, and (in the first phases), in one case study on the other. The comparison of the case studies demonstrates that the role of project leader as a promoter of processes and expertise can compensate for a deficiency in formal coordination. Formal tools of coordination are tools whose employment is planned and executed in advance. The empirical basis for Hypothesis 2 is one case study in which the formulation of development models and the assignment of fixed responsibilities in terms of product modules in the project team resulted in better transparency within the project, thus leading to a lower demand for additional coordination. With regard to initiating the employment of coordination tools, the following hypothesis can be formulated:

H3: The more deeply the departments of the prime organisation are embedded in the project, the higher their motivation to employ coordination tools.

This hypothesis is based on empirical findings from one case study that demonstrated that the embedding of functional units (departments) of the prime organisation increases the willingness and motivation to employ coordination tools. An explanation for this was found in the fact that the early involvement of different departments clarifies coordination demands at an early stage.

Regarding the effects of coordination tools on the successful performance of coordination functions, the following hypotheses can be formulated:

H4: The more specifically a company's documented procedural instructions are adapted to project structure, the higher their contribution to successful performance of coordination functions.

H5: The greater the number of coordination instruments employed between departments, the higher their contribution to successful performance of coordination functions.

H6: The more interdepartmental development methods are employed on a workshop basis, the higher their contribution to the coordination of innovation projects.

H7: The more spontaneous bilateral communication that complements the formal communication structure is employed, the higher the degree of successful performance of coordination functions.

H8: The more clearly project tasks and results are compiled in a central controlling document, the higher the degree of successful performance of coordination functions.

H9: The use of formal decision instances that involve management increases the binding character of partial results during an innovation project.

Hypothesis 4 targets the empirical results of the three case studies, which postulate that while the decision rules form the frame for the execution of given innovation projects, they need to be adapted to specific project demands to cover the complete 'project reality'. One case study indicated that after project-specific adoption of the process model ('a quality-planning process') during the second phase of the project, the necessary degree of coordination could be achieved.

Hypotheses 5 and 6 refer to the findings from all three case studies indicating that it is important to reach a common understanding of all individuals involved regarding the challenges, interdependencies, and goals of the project. This goal is reached especially when coordination tools are applied across departments.

Hypothesis 7 is also based on empirical findings from all three case studies pointing out that spontaneous bilateral communication is one of the most relevant success factors in the coordination of innovation projects. This communication is necessary to be able to react in the short run to complex, rapidly changing tasks. Bilateral communication is especially necessary if project tasks involve more than one department and coordination is needed to ensure project progress. For this coordination to be efficient and goal-oriented, it has to be regarded as complementary to the formal communication structure, which expands overall understanding among the project team. In this manner, various tasks and their respective priorities can be managed.

Hypothesis 8 refers to the empirical findings from one case study that show that the use of project duty lists and respective milestones can significantly increase coordination in innovation projects.

Hypothesis 9 refers to the same case study, in which the decisions of a formal steering committee composed of general management and heads of departments led to a high commitment to milestones. This was particularly focused on the approval of project needs (project duty lists) and the definition of specifications (requirement list) at defined points of the project cycle.

6.5.2 Hypotheses Regarding Moderating Factors in the Use of Coordination Tools

Regarding the moderating factors on the employment of coordination tools, the following hypotheses can be formulated:

H10: High uncertainty in an innovation project, especially as recognised by the staff involved in the project, leads to a strong embedding of the departments and, accordingly, to better employment of coordination tools between the individuals involved.

H11: In the case of dynamic uncertainty during the project cycle, the absence of formal coordination tools will lead to conflicts between the individuals involved in the project.

The empirical basis for Hypothesis 10 is a comparison between two of the three case studies. In the first study, due to obvious and recognised high uncertainty, intense coordination efforts were launched before the start of the project; this effort was even more intense than in other projects at the same company. In addition to official channels, departments had been involved to analyse the basic feasibility of the project to be undertaken. In comparison, the other case study initially showed only limited coordination effort, due to a general perception that the project could be solved with existing product components.

Hypothesis 11 is derived from the results of one study, which proved that a dynamically changing degree of uncertainty during the project cycle was compensated for early on by the employment of appropriate coordination tools. This dynamically changing degree of uncertainty was a major cause for conflicts in phases 2 and 3.

H12: The earlier formal tools of coordination are employed in innovation projects, the stronger their effect with regard to the creation of a common understanding of the project among all individuals involved.

Hypothesis 12 focuses on results from the case study in the electronics industry. This study showed that the early formal involvement of all departments involved in the entire project delivered a high degree of common understanding of project goals, task interdependencies, and challenges during the project.

6.5.3 Hypotheses Regarding the Effects of Coordination Tool Employment on Success

---

H13: The more strongly a common understanding of project goals, task interdependencies, and challenges during the project is supported by applying the tools of coordination, the higher the success of each separate project phase.

The effects of coordination tools on success at the project and individual level were analysed separately. The empirical findings allow the formulation of following hypotheses on project- (H13) and individual-level (H14) success:

Hypothesis 13 is founded on the case studies in the vehicle supplier industry on the one hand and the electronics industry case study on the other. In the electronics project, formal tools of coordination were used from the beginning, furthering the common understanding expressed in Hypothesis 13. In this manner, a high contribution to the success of each project phase and the project as a whole was made.

This common understanding was achieved only during the late phases in the other projects, resulting in negative effects on the dimensions of the project success.

H14: The earlier the departments from the prime organisation are involved in an innovation project, the higher the personal satisfaction regarding the point in time of involvement and the project coordination.

The individual ratings of each project involved in all three cases demonstrated a dependency between a formal involvement in the project and individual satisfaction with project coordination and the point in time with respect to individual involvement in the project.

6.6 Summary and Outlook

The studies conducted reinforce the significance of coordination tools for innovation projects. It is apparent that companies are faced with the problem of defining a formal frame to enable successful innovation project execution on the one hand while leaving this structure flexible enough to meet the varying coordination demands of individual projects on the other.

It was found that coordination tools that enable the creation of a common understanding among the departments involved an innovation project are of significance for project execution. In this instance, a common understanding refers to project goals, interdependencies, and task priorities as well as to special challenges of certain other departments. This common understanding is achieved through a formal project frame complemented by spontaneous, informal, and usually bilateral communication. In addition to the definition of the structure, the definition of processes is also part of the formal frame, for instance in form of scheduled project review meetings.

Furthermore, the studies reinforced the essential role played by the project leader in the coordination of innovation projects. The formal coordination of tasks is only one function of the project leader. Depending on the uncertainty of the project, the project leader also needs social competence and expertise that enable him to support the departments in their project-related work.

The case studies underlined the high importance of the employment of development methods such as Quality Function Deployment (QFD). Only inclusive, cross-departmental application of the methods, optimally in workshop form, can achieve the needed coordination effect on innovation projects.

As a moderating factor in the employment of tools of coordination, uncertainty was further investigated in the case studies. The comparison of the empirical findings from each study underlines the importance of coordination efforts in reducing uncertainty. One interesting result of the analysis was that while estimated technological and market-oriented absolute uncertainty was lower at the beginning of the project, the dynamics of the degree of uncertainty changed during the course of innovation projects. This represents a challenge for the use of coordination instruments.

In terms of the use of coordination tools in practice-based innovation, the integration of users presents itself as another challenge. From this, other interfaces that need to be coordinated in the course of an innovation project arise beyond the functional borders of a company. The coordination barriers indicated in Sect. 6.2.2 thus take on additional significance. Linguistic problems, different understandings of concepts, missing structural connections, or unclear responsibilities are only some examples of potential barriers to be considered when integrating users. The application of coordination tools must counteract these potential barriers. Other special challenges in the coordination of practice-based innovation lie in the following areas:

- Overcoming spatial distance between the different actors (company and users).
- Processing the generated information and data.
- Ensuring the simplicity of using individual coordination-advancing methods, such as Quality Function Deployment.

In terms of the execution of future research in the field of coordination of innovation projects, the results of this study lead to following recommendations:

- Investigate industry-specific differences regarding the use of coordination instruments in innovation projects.
- Investigate the need for and outcome of certain coordination tools in practice-based innovation projects.
- Investigate the use of coordination instruments in radical innovation projects. Above all, examination of the hypotheses regarding the role of the formal project framework would be of interest.
- Detailed investigations of sub-processes in innovation projects, in particular in the early phases before actual project release.
- Execution of differentiated, quantitative studies on the basis of the generated hypotheses.

To sum up, the results of the study prove the great importance of the application of coordination tools in the realisation of innovation projects. It has demonstrated that, above all, these special tools generate a large benefit, which promotes a common understanding among partners regarding processing of project tasks. The form of use plays an important role in addition to the basic use of individual coordination tools.

References

Allen, T. J. (1977). *Managing the flow of technology*. Cambridge, MA: The MIT Press.

Ancona, D. G., & Caldwell, D. F. (1992). Bridging the boundary: External activity and performance in organizational teams. *Administrative Science Quarterly, 37*, 634–665.

Bailetti, A. J., Callahan, J. R., & McCluskey, S. (1998). Coordination at different stages of the product design process. *R&D Management, 28*(4), 237–248.

Benkenstein, M. (1987a). *F&E und Marketing – Eine Untersuchung zur Leistungsfähigkeit von Koordinationskonzeptionen bei Innovationsentscheidungen*. Wiesbaden: Gabler. (In German.)

Benkenstein, M. (1987b). Koordination von Forschung & Entwicklung und Marketing. *Marketing – Zeitschrift für Forschung und Praxis, 2*, 123–132.

Brockhoff, K. (1996). Schnittstellenprobleme: Sechs Fälle aus der Wirtschaftspresse. In K. Brockhoff (Ed.), *Management von Innovationen* (pp. 203–206). Wiesbaden: Gabler. (In German.)

Brockhoff, K., & Hauschild, J. (1993). Schnittstellen-Management – Koordination ohne Hierarchie. *Zeitschrift für Führung und Organisation, 62*(6), 396–403. (In German.)

Buggie, F. D. (2001). The four phases of innovation. *Journal of Business Strategy, 22*(5), 36–42.

Bullinger, H. J., Kugel, R., Ohlhausen, P., Stanke, A., et al. (1995). *Integrierte Produktentwicklung – zehn erfolgreiche Praxisbeispiele*. Wiesbaden: Gabler. (In German.)

Clark, K. B., & Fujimoto, T. (1991). *Product development performance*. Boston: Harvard Business School Press.

Donnellon, A. (1993). Crossfunctional teams in product development: Accommodation the structure to the process. *Journal of Product Innovation Management, 10*, 377–392.

Ehrlenspiel, K. (1995). *Integrierte Produktentwicklung: Methoden für die Prozess-organisation, Produkterstellung und Konstruktion*. München/Wien: Hanser. (In German.)

Euringer, C. (1995). *Marktorientierte Produktentwicklung: Die Interaktion zwischen F&E und Marketing*. Wiesbaden: Gabler. (In German.)

Folkerts, L., & Hauschild, J. (2002). Personelle Dynamik in Innovationsprozessen – Neue Fragen und Befunde zum Promotorenmodell. *Die Betriebswirtschaft, 62*(1), 7–23. (In German.)

Frese, E. (1988). *Grundlagen der Organisation*, 4. durchgesehene Auflage. Wiesbaden: Gabler. (In German.)

Früh, W. (1998). *Inhaltsanalyse: Theorie und Praxis*, 4. überarbeitete Auflage. Konstanz: UVK Medien. (In German.)

Gemünden, H. G., & Högl, M. (1998). Teamarbeit in innovativen Projekten: Eine kritische Bestandsaufnahme der empirischen Forschung. *Zeitschrift für Personalforschung, 12*(3), 277–301. (In German.)

Gerpott, H. (1991). *F&E und Produktion, Theoretische und empirische Analysen zu Schnittstellenproblemen im Innovationsprozess unter besonderer Berücksichtigung personalwirtschaftlicher Aspekte*. München: VVF. (In German.)

Green, S. G., Gavin, M. B., & Aiman-Smith, L. (1995). Assessing a multidimensional measure of radical technological innovation. *IEEE Transactions on Engineering Management, 42*(3), 203–214.

Griffin, A., & Hauser, J. R. (1992). Patterns of communication among marketing, engineering and manufacturing: A comparison between two new product teams. *Management Science, 38*(3), 361–373.

Griffin, A., & Hauser, J. R. (1996). Integrating R&D and marketing: A review and analysis of the literature. *Journal of Product Innovation Management, 13*, 191–215.

Gupta, A. K., & Wilemon, D. (1988). Why R&D resists using marketing information. *Research-Technology Management, 31*(Nov./Dec.), 36–41.

Gupta, A. K., Raj, S. P., & Wilemon, D. (1986). A model for studying R&D-Marketing interface in the product innovation process. *Journal of Marketing, 50*, 7–17.

Hauschildt, J. (1991). Zur Messung des Innovationserfolges. *Zeitschrift für Betriebswirtschaft, 61* (4). (In German.).

Hauschildt, J., & Petersen, K. (1987). Phasen-Theorem und Organisation komplexer Entscheidungsverläufe – weiterführende Untersuchungen. *Zeitschrift für Betriebswirtschaftliche Forschung, 39*, 1043–1062. (In German.)

Herstatt, C., & Verworn, B. (2003). Bedeutung und Charakteristika der frühen Phasen des Innovationsprozesses. In C. Herstatt & B. Verworn (Eds.), *Management der frühen Innovationsphasen* (pp. 3–15). Wiesbaden: Gabler. (In German.)

Hugl, U. (1995). *Qualitative Inhaltsanalyse und Mindmapping: Ein neuer Ansatz für Datenauswertung und Organisationsdiagnose*. Wiesbaden: Gabler. (In German.)

Kahn, K. B. (1996). Interdepartmental Integration: A definition with implications for product development performance. *Journal of Product Innovation Management, 13*, 137–151.

Kahn, K. B. (2001). Market orientation, interdepartmental integration and product development performance. *Journal of Product Innovation Management, 18*, 314–323.

Karlsson, C., & Ahlström, P. (1996). The difficult path to lean product development. *Journal of Product Innovation Management, 13*, 283–295.

Kieser, A., & Kubicek, H. (1992). *Organisation*, 3. neu bearbeitete Auflage. Berlin/New York: de Gruyter. (In German.)

Knickel, V. (1997). Gestaltung von Kommunikationsprozessen an Schnittstellen in der Produktentwicklung. In G. Warnecke (Ed.), *Produktionstechnische Berichte. Band 26.* Kaiserslautern: Universität Kaiserslautem. (In German.)

Köhler, R. (1992). Absatzorganisation. In E. Frese (Ed.), *Enzyklopädie der Betriebswirtschaftslehre, Bd. 2: Handwörterbuch der Organisation*, 3. Aufl. (pp. 36–56). Stuttgart: Schäffer-Poeschel. (In German.)

Küpper, H. C. (1981). Ablauforganisation. In F. X. Bea, E. Dichtl, & M. Schweitzer (Eds.), *Grundwissen der Ökonomik*. Stuttgart/New York: Fischer. (In German.)

Lawrence, P. R., & Lorsch, J. W. (1986). *Organization and environment: managing differentiation and integration*. Boston: Harvard Business School Press.

Lechler, T. (1996). *Erfolgsfaktoren des projektmanagements*. Peter Lang: Frankfurt am Main. (In German.)

Leenders, M. A. A. M., & Wierenga, B. (2002). The effectiveness of different mechanisms for integrating marketing & R&D. *Journal of Product Innovation Management, 19*, 305–317.

Lucas, G. H., & Bush, A. J. (1988). The marketing-R&D interface: Do personality factors have an impact? *Journal of Product Innovation Management, 5*, 257–268.

Lühring, N. (2003). Innovationsfördernde Organisationsstrukturen unter Berücksichtigung früher Innovationsphasen. In C. Herstatt, & B. Verworn (Eds.), *Management der frühen Innovationsphasen* (pp. 117–144). Wiesbaden: Gabler. (In German.)

Lühring, N. (2006). *Koordination von Innovationsprojekten*. Wiesbaden: DUV. (In German.)

Maltz, E. (2000). Is all communication created equal? An investigation into the effects of communication mode on perceived information quality. *Journal of Product Innovation Management, 17*, 110–127.

Maltz, E., Souder, W. E., & Kumar, A. (2001). Influencing R&D/Marketing integration and the use of market information by R&D managers: intended and unintended effects of managerial actions. *Journal of Business Research, 52*, 69–82.

Manns, J. R. (1992). *Produktinnovation als Ergebnis der Koordination von F&E und Marketing. Ursachen, Auswirkungen und Lösungsmöglichkeiten, dargestellt am Beispiel eines Unternehmens der mobilen Kommunikation*. Ludwigsburg/Berlin: Verlag Wissenschaft & Praxis. (In German.)

McDonough, E. F., III, Kahn, K. B., & Barczak, G. (2001). An investigation of the use of global, virtual, and colocated new product development teams. *Journal of Product Innovation Management, 18*, 110–120.

Moenaert, R. K., & Souder, W. E. (1990). An information transfer model for integrating marketing and R&D personnel in new product development projects. *Journal of Product Innovation Management*, *7*(2), 91–107.

Moenaert, R. K., Souder, W. E., DeMeyer, A., & Deschoolmeester, D. (1994). R&D marketing integration mechanisms, communication flows and innovation success. *Journal of Product Innovation Management, 11*, 31–45.

Nijssen, E. J., Biemans, W. G., & de Kort, J. F. (2002). Involving purchasing in new product development. *R&D Management, 32*(4), 281–289.

Nobeoka, K. (1995). Reorganizing for Multi-Project management: Toyota's new structure of product development centers. In *Kobe Economic & Business Review, 40th Annual Report*. Kobe University: Research Institute for Economics and Business Administration.

Olsen, E. M., Walker, O. C., Ruekert, R. W., & Bonner, J. M. (2001). Patterns of cooperation during new product development among marketing, operations and R&D: Implications for project performance. *Journal of Product Innovation Management, 18*, 258–271.

Osterloh, M., & Frost, J. (2000). Von der Strukturgestaltung zur Prozessorganisation. Überlegungen zur Koordinations-, Orientierungs- und Motivationsdimension der Organisation. In M. K. Welge, A. Al-Laham, & P. Kajüter (Eds.), *Praxis des strategischen Managements: Konzepte – Erfahrungen – Perspektiven* (pp. 279–294). Wiesbaden: Gabler. (In German.)

Picot, A., Reichwald, R., & Wigand, R. T. (2001). *Die grenzenlose Unternehmung*, 4. Auflage. Wiesbaden: Gabler. (In German.)

Pinto, M. B., Pinto, J. K., & Prescott, J. E. (1993). Antecedents and consequences of project cross-functional cooperation. *Management Science, 10*, 1281–1297.

Reger, G. (1997). *Koordination und strategisches Management internationaler Innovations-prozesse*. Bamberg: Difo-Druck. (In German.)

Rueckert, R. W., & Walker, O. C. (1987). Marketing's interaction with other functional units: A conceptual framework and empirical evidence. *Journal of Marketing, 51*(Jan), 1–19.

Rühli, E. (1992). Koordination. In E. Frese (Ed.), *Handwörterbuch der Organisation,* 3. Auflage (pp. 1164–1175). Stuttgart: Poeschel. (In German.)

Salomo, S., Gemünden, H. G., & Billing, F. (2003). Dynamisches Schnittstellenmanagement radikaler Innovationsvorhaben. In C. Herstatt, & B. Verworn (Eds.), *Management der frühen Innovationsphasen* (pp. 161–194). Wiesbaden: Gabler. (In German.)

Schmidt, R. (1996). Der Interface-Check als Analyse der organisatorischen Schnittstelle "F&E und Marketing". *JFB, 1*, 4–22. (In German.)

Schmidt, J. B., & Calantone, R. J. (2002). Escalation of commitment during new product development. *Journal of the Academy of Marketing Science, 30*(2), 103–118.

Schneider, D. J. G., & Müller, M. E. (1993). Durch Koordination zu marktgerechten Innovationen. Schnittstellengestaltung zwischen F&E und Marketing. *Marktforschung & Management, 1*, 6–13. (In German.)

Seidel, M. (1996). *Zur Steigerung der Marktorientierung der Produktentwicklung – Analyse der Interaktion zwischen F&E und Marketing im Innovationsprozess*. Dissertation. St. Gallen: Universität St. Gallen. (In German.)

Sethi, R. (2000). New product quality and product development teams. *Journal of Marketing, 64* (April), 1–14.

Shannon, C. E., & Weaver, W. (1949). *The mathematical theory of communication*. Chicago/ London: University of Illinois Press.

Shanteau, J., & Rohrbaugh, C. (2000). Social/psychological barriers to successful management of technological innovation. In B. Green, R. Cressy, & F. Delmar (Eds.), *Risk behaviour and risk management in business life*. Dordrecht: Kluwer.

Shaw, V., & Shaw, C. T. (1998). Conflict between engineers and marketers. *Industrial Marketing Management, 27*, 279–291.

Sherman, J. D., Souder, W. E., & Jenssen, S. A. (2000). Differential effects of the primary forms of cross functional integration on product development cycle time. *Journal of Product Innovation Management, 17*, 257–267.

Silber, H. (1986). Innovation durch Querschnittskoordination. *Zeitschrift für Führung und Organisation, 4*, 243–251. (In German.)

Sommerlatte, T. (1988). Visionen des Wandels. In A. D. Little (Ed.), *Management des geordneten Wandels* (pp. 19–36). Wiesbaden: Gabler. (In German.)

Sommerlatte, T. (2000). Längerfristige Zukunftssicherung durch Innovationsmanagement. In K. Steinle et al. (Eds.), *Vitalisierung: Das Management der neuen Lebendigkeit* (pp. 192–206). FAZ: Frankfurt am Main. (In German.)

Song, X. M., & Parry, M. E. (1997). Teamwork barriers in Japanese high technology firms: The socio-cultural difference between R&D and marketing managers. *Journal of Product Innovation Management, 14*(5), 356–367.

Song, X. M., Montoya-Weiss, M. M., & Schmidt, J. B. (1997). Antecedents and consequences of cross-functional cooperation: A comparison of R&D, manufacturing and marketing perspectives. *Journal of Product Innovation Management, 14*, 35–47.

Song, X. M., Thieme, R. J., & Xie, J. (1998). The impact of cross-functional joint involvement across product development stages: An exploratory study. *Journal of Product Innovation Management, 15*, 289–303.

Souder, W. E. (1988). Managing relations between R&D and marketing in new product development projects. *Journal of Product Innovation Management, 5*, 6–19.

Specht, G. (2000). Schnittstellenmanagement: Marketing und Forschung & Entwicklung. In A. Hermann et al. (Eds.), *Kundenorientierte Produktgestaltung* (pp. 267–285). München: Vahlen.

Staehle, W. H. (1999). *Management*, 8. Auflage. München: Vahlen. (In German.)

Staudt, E., Mühlemeyer, P., & Kriegesmann, B. (1991). Schnittstelle = Bruchstelle? *Absatzwirtschaft, 11*, 108–114. (In German.)

Steven, M. (2001). Die Koordination im Unternehmen. *WISU, 7*, 965–970. (In German.)

Stuffer, R. (1994). *Planung und Steuerung der Integrierten Produktentwicklung*. München/Wien: Hanser. (In German.)

Thom, N. (1980). *Grundlagen des betrieblichen Innovationsmanagements*, 2. völlig neu bearbeitete Auflage. Hanstein: Königstein/Ts. (In German.)

Thom, N., & Etienne, M. (2000). Organisatorische und personelle Aspekte für ein erfolgreiches Innovationsmanagement. In A. Clermont, W. Schmeisser, & D. Krimphove (Eds.), *Personalführung und Organisation* (pp. 283–294). München: Vahlen. (In German.)

Trott, P. (1998). *Innovation management and new product development*. London: Financial Times Pitman Publishing.

Ulrich, K. T., & Eppinger, S. D. (1995). *Product design and development*. New York: McGraw-Hill.

Wiebecke, G., & Tschirky, H. (1987). Interface zwischen Forschung + Entwicklung und Marketing. *io Management Zeitschrift, 56*(1), 23–26. (In German.)

Wijinstra, F. J. Y., & van Stekelenborg, R. H. A. (1996). *The role of purchasing in new product development*. Eindhoven: University of Technology.

Winkler, T. (1999). *Entwicklung eines Instrumentariums zur Erfolgsanalyse von Innovationsprojekten*. Frankfurt am Main: Lang. (In German.)

Wolfrum, B. (1994). Schnittstellenprobleme zwischen F&E und Marketing im Innovationsmanagement. *WISU, 12*, 1016–1022. (In German.)

Chapter 7
Measuring the Impact of Innovation Intermediaries: A Case Study of Tekes

Margaret Dalziel and Satu Parjanen

Abstract Innovation intermediaries are believed to have a beneficial influence on innovation processes, but as there are no universally accepted metrics of intermediary performance it is difficult for these organisations to provide the evidence of their contributions. We present a general-purpose methodology for measuring the impact of innovation intermediaries that applies across all types of intermediaries. We demonstrate the methodology by assessing the impact of the Global Access Program (GAP), which is made available to Finnish firms through Tekes, an intermediary organisation whose mission is to enhance Finnish industry through technology and innovation. Our findings show that the GAP program has had an impact on the performance of participating firms in terms of revenue growth, exports, new international customers, and employment growth. Consistent with our expectation that impacts on firm performance are a consequence of earlier impacts on firm resources and capabilities, we find a statistically significant relationship between the immediate impact of strategic information and advice, and information and advice on new markets, and longer term impact on firm performance.

7.1 Introduction

According to Ståhle et al. (2004) intermediary organisations can be defined in two ways. A narrow definition describes intermediaries as information transmitters. The broader one covers both the information transmitting related to the substance and the direct and indirect influence that the intermediaries have on the structure and dynamics of their environment. Similarly, Dalziel (2010) defines innovation

M. Dalziel (✉)
Telfer School of Management, University of Ottawa, Ottawa, Canada
e-mail: dalziel@telfer.uottawa.ca

S. Parjanen
Lahti School of Innovation, Lappeenranta University of Technology, Lahti, Finland

H. Melkas and V. Harmaakorpi (eds.),
Practice-Based Innovation: Insights, Applications and Policy Implications,
DOI 10.1007/978-3-642-21723-4_7,

intermediaries as organisations or groups within organisations that work to enable innovation, either directly by enabling the innovativeness of one or more firms, or indirectly by enhancing national, regional, or sectoral innovative capacity. Examples include research and technology development institutes, consortia, networks, and programs; science, technology, and business parks and incubators; university technology transfer offices; and industry and trade associations, economic development agencies, and chambers of commerce.

Intermediary organisations perform three innovation system functions: funding support, networking and collaboration, and other supportive functions (Howells 2006; Inkinen and Suorsa 2010). In the Finnish case, intermediary organisations are predominantly understood to be publicity owned or semi-public regional companies, non-profit organisations (Koskenlinna et al. 2005) or privately owned firms (Suvinen et al. 2010). It has been estimated that in Finland there are approximately 80 regional development companies, 22 technology centres, 70 incubator organisations and 40 other innovation-supporting organisations (Koskenlinna et al. 2005). Despite the growing importance of intermediary organisations in the Finnish innovation system, such organisations have received little attention in the literature. One of the reasons for this might be that the impact of intermediary organisations is difficult to measure.

While there are general-purpose indicators for measuring the performance of firms and universities, there are no such indicators for measuring the performance of innovation intermediaries. As a consequence, past studies of intermediary performance have used a range of performance indicators that vary according to intermediary objectives, the nature of participating firms, and the availability of data. For example, researchers have measured the performance, or effect of the activities of:

- The US Advanced Technology Program using patents (Sakakibara and Branstetter 2003) and the likelihood of raising additional financing (Feldman and Kelley 2006)
- Chinese science parks using park-level revenues and exports (Yu and Heshmati 2007), and Swedish science parks using firm-level revenues, employment, and profits (Löfsten and Lindelöf 2002)
- US regional institutes using pollution prevention and quality management capabilities (McEvily and Zaheer 1999)
- Canadian industry associations using frequency of citation as a collaborator or as a source of ideas (Dalziel 2006)
- The Responsible Care program of the US Chemical Manufacturer's Association using environmental performance (King and Lenox 2000)
- US technology transfer offices using the innovativeness of new products (Kaufmann and Tödtling 2001), the number of startups (Di Gregorio and Shane 2003; Markham et al. 2005), and licensing agreements and revenues (Siegel et al. 2003)
- The US Small Business Innovation Research program using revenues (Audretsch et al. 2002).

In this chapter we present a general-purpose methodology for measuring the impact of innovation intermediaries that applies across all types of innovation intermediaries. The methodology was developed by the lead author and Brian Barge of The Evidence Network, and is being used for research purposes by the former and commercialised by The Evidence Network[1]. We demonstrate the methodology on a program made available to Finnish firms through Tekes, one of the leading technology agencies in Finland. The Global Access Program (GAP) is offered to non-US firms by the Anderson School of Management at the University of California at Los Angeles. Through Tekes, over 100 Finnish firms have participated in the GAP program since 1999.

In the next section we present the theory that underlies the measurement methodology. We then describe our sample, measures, and results. Our results show that the GAP program has had an impact both on the resources and capabilities of participating firms and on their performance in the market in terms of increased revenues, exports, new international customers, and employment. Furthermore, we are able to show a statistically significant relationship between impact on firm resources and capabilities, and firm performance in the market. These findings give testimony to the merits of both the Global Access Program and the impact assessment methodology.

7.2 Theory

The basic premise is that innovation intermediaries, despite their diversity, can be described as an overarching class of organisations whose members share common goals. Despite their diversity, innovation intermediaries, ranging from small economic development organisations to large and sophisticated research institutes, seek to enhance the innovative capacity of the regions, nations, or sectors of which they are members (Furman et al. 2002). While innovation intermediaries may impact a variety of actors, ranging from firms to university researchers, hospitals, and government policy makers, their primary target is typically firms. Accordingly, we focus on the efforts of innovation intermediaries to enhance the innovativeness of their client or member firms, in the interests of facilitating increases in their viability, growth, or profitability. We use a logic model approach to link intermediary purpose to intermediary outputs and impacts on participating firms.

First used in the 1970s, logic models are narrative or graphical depictions of processes in real life that communicate the assumptions that underlie the expectation that an activity will lead to a specific result. Logic models illustrate a sequence

[1] www.theevidencenetwork.com

of cause-and-effect relationships – a systems approach to communicating the path toward a desired result (Millar et al. 2001; Renger and Titcomb 2002). Logic models serve numerous functions, one of which is the identification of a set of performance indicators that may be used to monitor and summarise performance for funders and decision makers (McLaughlin and Jordan 1999; Rogers 2005).

Table 7.1, below, is a general logic model for innovation intermediaries that focuses on their impacts on the firms that are their clients or members. It illustrates how innovation intermediaries work to fulfil their missions, and how their impact can be measured. As shown in the leftmost column, innovation intermediaries express their purpose in terms such as national competitiveness, regional economic development, industry strength, or viable new ventures. Using inputs including knowledgeable people and relationships with external individuals and organisations, they conduct activities that result in outputs such as business, scientific, and technological knowledge, equipment, and facilities; design, testing, prototyping, scale-up, intellectual property management, and licensing services; and events, conferences, seminars, and meetings.

They conduct activities in the interests of achieving immediate and intermediate impacts on the firms that are their members or clients, and long-term impacts in the form of socio-economic benefits. The immediate impacts of innovation intermediaries are improvements in the resources or capabilities of client or member firms, intermediate impacts are improvements in the performance of client or member firms, and long-term impacts affect communities, industries, economies, societies, and the environment.

Working backwards, from right to left, shows how different types of impact are achieved. The achievement of long-term impact depends on the achievement of intermediate impact, which in turn depends on the achievement of immediate impact. So, for example, an innovation intermediary that seeks to create economic growth in a region (its purpose and desired long-term impact) does so by facilitating improvements in the performance of local firms (its desired intermediate impact), either by facilitating firm growth or the creation of new ventures. It facilitates firm growth and the creation of new ventures by facilitating improvements in the resources and capabilities of local firms (its desired immediate impact). The fundamental insight is that innovation intermediaries achieve their desired intermediate and long-term impacts indirectly by affecting the resources and capabilities of the firms with which they work.

Innovation intermediaries hope to have long-term impacts that correspond to their missions. But the measurement of long-term impact is difficult because changes in the economy, the environment, or society are brought about by the collective action of many players. So it is difficult to attribute such changes to the activities of a single organisation. But as long-term impact is facilitated by the achievement of intermediate impact, evidence of intermediate impact is suggestive of possible long-term effects.

Table 7.1 General logic model for innovation intermediaries

Purposes	Inputs	Outputs	Impacts		
			Immediate impacts	Intermediate impacts	Ultimate impacts
			Firms with increased resources or capabilities	Firms with improved performance	Socio-economic benefits
• National competitiveness	• Knowledgeable people	• Business, scientific and technological knowledge, relationships, equipment, facilities	• Information, advice	• Increased revenues	• Sustainable wealth and jobs
			• Opportunities for promotion, influence	• Increased employment	• Community, regional, national economic, social, and environmental benefits
• Community, regional economic development	• Relationships		• Business linkages	• New products, services	
• Industry strength	• Equipment, facilities				
	• Funding				
• Viable new ventures		• Design, testing, prototyping, intellectual property management, licensing services	• Research linkages	• Faster time to market	
			• Technology services	• Increased market share	
		• Plans, proposals, projects	• Access to financing	• Reduced environmental impact	
		• Events, conferences, seminars, meetings	• Complementary business services	• Increased valuation	
		• Websites, blogs, reports, directories, newsletters		• Increased investment	
		• Access to financing			

Source: The Evidence Network

7.3 Tekes and the Global Access Program

7.3.1 Tekes: the Finnish Funding Agency for Technology and Innovation

Tekes is the most important publicly funded expert organisation for financing research, development and innovation in Finland (Inkinen and Suorsa 2010). Tekes was founded primarily to assist Finland in the economic recession of the 1970s. In 1984, Tekes began to operate both nationally and internationally. Tekes experienced a dramatic growth in the late 1990s as a result of a marked increase in public technology funding. In turn, Tekes was granted the responsibility for finding the optimum ways to invest these funds. Tekes now supports wide-ranging innovation activities in research communities, industry and service sectors. Every year, Tekes finances some 1,500 business research and development projects, and almost 600 public research projects at universities, research institutes and polytechnics. For example in year 2009 Tekes funded 2,329 enterprises (Tekes Annual Review 2009). Research, development and innovation funding is targeted to projects that create the greatest long-term benefits for the economy and society.

7.3.2 The Global Access Program (GAP)

The Global Access Program (GAP) at the University of California at Los Angeles (UCLA) Anderson School of Management is designed to provide low cost management consulting services to non-US companies seeking strategic advice, particularly concerning operating in and selling to international markets. The GAP program is an educational program that matches a team of students from the Fully Employed MBA program with existing international technology companies to develop a business strategy that enables the companies to move to the next stage of their corporate development. The students work with key members of the organisation to address strategic business issues and to develop a comprehensive and independent business plan. The plan typically involves launching a new product into an existing market, launching an existing product into a new geographical market, offering an existing product into a new market segment, developing a new product for a specific market segment, or leveraging existing intellectual property into new products and new markets.

The GAP program works with foreign organisations throughout the world to assist their technology companies in developing a global strategy. The partners are GAP-sponsoring organisations located abroad with the mission of enhancing their country's economic development and technological innovation. Tekes entered the Global Access Program in 1999. Since then over 100 Finnish companies have participated in the GAP program with the help of Tekes funding. The GAP program is targeted to Tekes clients that are independent businesses or a small business unit within a larger corporation. To be able to take part in the GAP program the company should have business based on an innovative product idea, adequate technological knowhow, paying clientele, a sustainable competitive advantage, adequate funding or positive cash flow for two years ahead, and a web site in English.

According to the UCLA Anderson School of Management the purpose of the GAP program is accelerated business growth. As shown in Table 7.2, the outputs of participation in the GAP program consist of a comprehensive, independent assessment and evaluation of company's current business plans, corporate growth strategies and market opportunities, extensive primary and secondary market research and analysis, an investment-quality, professional business plan with full financial projections, application of state-of-the-art management frameworks and analysis techniques, specific recommendations to accelerate the company's growth, final plan review and constructive feedback from outside executives including attorneys, accountants, technology experts, angel investors and venture capital professionals, and access to UCLA Anderson's extensive business network of

Table 7.2 GAP logic model

Purposes	Inputs	Outputs	Impacts		
			Immediate impacts	Intermediate impacts	Ultimate impacts
			Firms with increased resources or capabilities	Firms with improved performance	Socio-economic benefits
• Accelerated business growth	• Participation of Finnish company executives	• Primary research (100+ interviews)	• Information and advice: - Strategic information and advice - Feedback on products, services	• Increased revenues - Change in revenues - Export sales	• Sustainable wealth and jobs • Community, regional, national economic, social, and environmental benefits
	• UCLA MBA students	• Interpretation and analysis of research findings	- On selling in new markets - On operating in new markets - On raising capital	• Increased employment - Change in employment • Increased market share	
	• UCLA faculty and staff	• Executive education	• Business linkages - With service providers • Complementary business services	- New international customers • Increased investment - Financing	
	• Tekes personnel and funding	• Business and consultancy contacts	- Business planning services - Executive education services		

students, alumni, faculty and consultants. The immediate impacts of the GAP program are increased resources or capabilities of participating firms which can be categorised as information and advice, business linkages and complementary business services. Intermediate impacts of the GAP program include increased revenues, employment, market share and investment. Socio-economic benefits of the GAP program may include sustainable wealth and jobs, environmental and health care improvements and increased social wellness.

7.4 Measuring the Impact of the Global Access Program

7.4.1 Sample

A total of 60 Finnish firms participated in the GAP program between 2004 and 2009 inclusive. Fifty-three of these firms were asked to respond to the survey, the remaining firms had either closed, been acquired, or no longer employed the executive who had engaged with the GAP program. After two email reminders and telephone calls to non-respondents, 33 firms responded to the survey for a response rate of 62%.

The multiple-choice questionnaire consisted of 29 questions as follows:

- Four questions regarding the importance of GAP offerings (outputs)
- Eight questions regarding immediate impact
- Five question pairs regarding firm outcomes and intermediate impacts on firm outcomes
- Four questions regarding firm attributes
- Three questions regarding the year of participation and satisfaction with GAP.

While measuring the immediate impact of intermediary activities on firm resources and capabilities is straightforward, measuring the intermediate impact of intermediary activities on the performance of participating firms requires isolating the impact of intermediary activities from the many other factors that may impact firm performance. Studies that rely on secondary for assessing intermediary impact typically rely on multiple years of data to control for other factors such as firm effects and effects of the business environment, and to distinguish selection effects from treatment effects (Sakakibara and Branstetter 2003).

We use primary data and so are able to rely on survey respondents to judge the impact of intermediary activities on firm performance. Our approach minimises bias in two ways. First, we ask firms to access the impact of intermediary activities on specific dimensions of the performance (e.g. revenues, market share, employment), rather than asking a more general question. Measures that are specific, such as impact on revenues, are less likely to suffer from biased responses than measures that are general, such as impact on overall firm performance (Cote and Buckley 1987). Second, we precede each question on the impact of intermediary activities on specific dimensions of firm performance with a question on the level or change in performance on that dimension. This focuses the mind of the respondent on the specific dimension of firm performance and allows us to gather data on firm performance. For example, to determine the impact of an intermediary on the revenues of participating firms, we may first ask about changes in revenues and then ask the respondent to gage the degree to which those changes are attributable to activities of the intermediary in question.

7.4.2 Measures

7.4.2.1 Firm Description

We control for five firm attributes that might impact the firm's assessment of the impact of the GAP program on their performance in the market (intermediate impact). Age indicates the firm's age in years when it participated in the GAP program and has an average value of 11.18 years. Respondents were asked to indicate the number of employees and the amount of annual revenues, both on four-point scales. Responses to these questions were multiplied to get an indicator of firm size that ranged from a low of one (less than 20 employees and less than 500 K Euros in revenues) to a high of 16 (more than 100 employees and more than 2 million Euros in revenues). IT industry and health industry are dummy variables that are 1 if the firm is in the respective industry and 0 otherwise. Nineteen respondents indicated that their firm belonged to the information and communication technologies industry, six respondents indicated that their firm belonged to the biotechnology, health, or medical industry, and eight respondents indicated that their firm belonged to other industries. Finally, we control for firm growth as firms that are growing may be inclined to be more generous in their assessment of intermediary impact. Respondents were asked to indicate the change in employment since their firm's participation in GAP on a five point scale that ranged from a low of 'decreased employment' to a high of 'increased by more than 100%', and to indicate the change in revenues since their participation in GAP on seven point scale that ranged from a low of 'decreased by more than 100%' to a high of 'increased by more than 100%'. Responses to these questions were multiplied to get an indicator of firm growth that ranged from a low of three to a high of 35.

7.4.2.2 Measures of the Importance of Offerings and Impact

Factor analysis was used to consolidate measures of the importance of GAP offerings and GAP impact. As shown in Table 7.3, the four measures of the importance of GAP offerings were reduced to two factors (custom offerings, general offerings), the seven measures of immediate impact were reduced to three factors (strategic information and advice, new markets, general assistance) and the

Table 7.3 Factor analysis

Type of measures	Measures	Factors
Importance of offerings	Primary research	Custom offerings
	Interpretation and analysis	(Cronbach's Alpha = .85)
	Executive education	General offerings
	Business and consultancy contacts	(Cronbach's Alpha = .89)
Immediate impact	Strategic information and advice	Strategic information and advice
	Feedback on products, services	(Cronbach's Alpha = .90)
	Business planning services	
	Selling new markets	New markets
	Operating new markets	(Cronbach's Alpha = .81)
	Raising capital	General assistance
	Service provider linkages	(Cronbach's Alpha = .81)
	Executive education	
Intermediate impact	Change in revenues	Intermediate impact
	Change in employment	(Cronbach's Alpha = .93)
	New international customers	
	Export sales	
	Financing	

Table 7.4 Descriptive statistics and correlations

	1.	2.	3.	4.	5.	6.	7.	8.	9.	10.	11.
1. Age											
2. Size	−.17										
3. IT industry	.17	.09									
4. Health industry	−.33	−.11	−.55**								
5. Growth	−.32	.37*	.02	−.15							
6. Custom offerings	−.33	.11	−.12	.10	−.10						
7. General offerings	−.39	.12	.02	.14	−.11	.00					
8. Strategic info, advice	.10	−.02	.01	.15	−.13	.25	.43*				
9. New markets	−.02	.08	−.37*	.24	−.02	.20	.38*	.00			
10. General assistance	−.02	−.16	.34	−.21	.08	.03	.43*	.00	.00		
11. Intermediate impact	−.10	.27	−.07	−.03	.44*	.28	.25	.35*	.49**	.10	
Mean	11.18	7.45	.58	.18	15.45	0.00	0.00	0.00	0.00	0.00	0.00
Standard deviation	7.12	4.86	.50	.39	8.42	1.00	1.00	1.00	1.00	1.00	1.00
Minimum	2	1	0	0	3	−2.28	−1.88	−2.58	−1.88	−1.63	−1.17
Maximum	27	16	1	1	35	1.10	2.56	1.88	2.46	2.51	3.58

* = $p < .05$; ** = $p < .01$

five measures of intermediate impact were reduced to a single intermediate impact factor. All composite measures of the importance of offerings and impact are reliable as indicated by the Cronbach alphas. Descriptive statistics and correlations are reported in Table 7.4.

7.4.3 Results

We first present data on the importance of the GAP program offerings and the immediate and intermediate impact of the GAP program on participating firms. Figure 7.1 shows the average importance of the GAP program offerings as assessed by participating firms. Four types of offerings were assessed: primary research (student teams conduct over 100 interviews on behalf of their client firm), interpretation analysis of interview data and data provided by the client firm, executive education, and the provision of business and consulting contacts. As shown in Fig. 7.1, the firm-specific primary research and interpretation of analysis activities were judged to be very important while the executive education and the provision of contacts were judged to be somewhat important.

Figure 7.2 provides data on the immediate impact of the GAP program on the resources and capabilities of participating firms. Three dimensions of immediate impact were considered: information and advice (strategic information and advice, feedback on products and services, and information and advice on selling in new markets, operating in new markets, and on raising capital), business linkages (linkages with service providers), and business services (business planning services

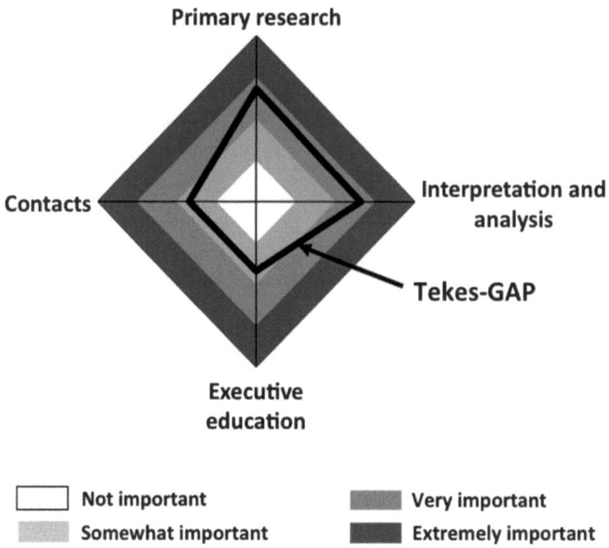

Fig. 7.1 Average importance of GAP offerings

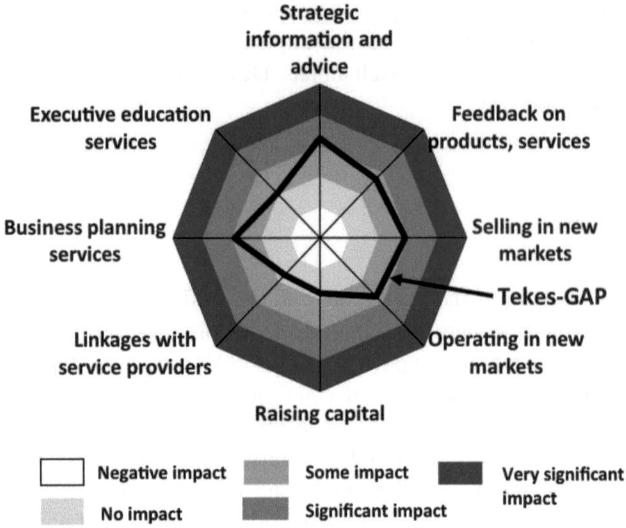

Fig. 7.2 Average assessment of immediate impact

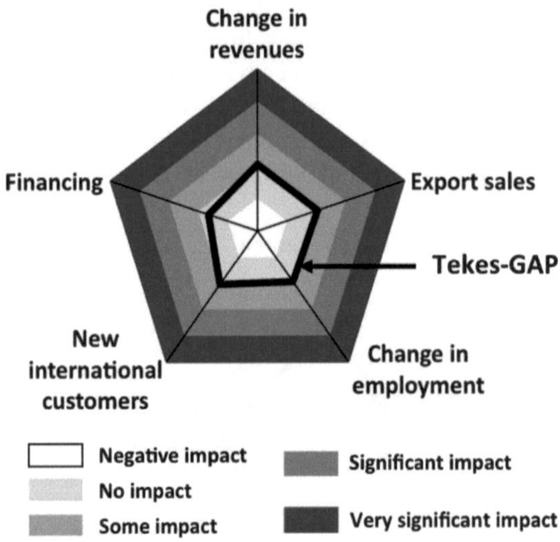

Fig. 7.3 Average assessment of intermediate impact

and executive education). As shown in Fig. 7.2, on average, strategic information and advice, information and advice on selling in new markets, and business planning services had the greatest immediate impact on participating firms. Information and advice on raising capital and the provision of linkages with service providers had the least impact.

Figure 7.3 provides data on the intermediate impact of the GAP program on the performance participating firms in the market. Four dimensions of immediate impact were considered: revenues (changes in revenues, export sales), employment (change in employment), market share (new international customers), and investment (financing). As shown in Fig. 7.3, on average, firms report that participation in the GAP program had an impact on their revenues, export sales, and new international customers.

As indicated by our logic model, our theory is that intermediate impact is predicted by immediate impact. Linear regression was used to test for a significant relationship between immediate impact and intermediate impact where our measure of intermediate impact is the composite measure of intermediate impact that was created by using factor analysis to combine respondents' assessments of the five measures of intermediate impact: changes in revenues, export sales, employment, new international customers, export sales, and financing.

As shown in Table 7.5, we test three models, Model 1, which includes only control variables, Model 2, which includes control variables and the importance of offerings variables, and Model 3, which includes control variables and the immediate impact variables. As expected, Model 1 explains a negligible amount of the variance in intermediate impact ($R^2 = .07$), although the growth variable is significant suggesting that growing firms are indeed more generous in their assessment of intermediary impact than firms that are not growing, as reported by previous studies (Inkinen and Suorsa 2010).

Model 2, which tests for a relationship between respondents' assessments of the importance of program offerings and their assessment of intermediate impact on firm performance, does not show a statistically significant relationship, but does explain a greater amount of the variance in intermediate impact than does Model 1 ($R^2 = .23$).

Table 7.5 Linear regression on intermediate impact

	Model 1 control variables	Model 2 program offerings	Model 3 immediate impact
Constant	−1.37	−1.29	−1.48
Age	0.36	0.35	−0.16
Size	0.72	0.32	0.39
IT industry	−0.54	−0.62	0.19
Health industry	0.08	−0.61	−0.81
Growth	2.05*	2.62*	2.78*
Output: Custom offerings		1.80	
Output: General offerings		1.93	
Immediate impact: Strategic information and advice			3.25**
Immediate impact: New markets			3.56**
Immediate impact: General assistance			0.13
Model characteristics			
n	32	29	32
F	1.44 (5 dof)	2.21 (7 dof)	4.65** (8 dof)
Adjusted R^2	.07	.23	.48

dof, Degrees of freedom
* = $p < .05$; ** = $p < .01$

Model 3 tests for a relationship between respondents' assessments of immediate impacts on firm resources and capabilities, and their assessment of intermediate impact on firm performance. As per our theory, Model 3 explains a significant amount of the variance in respondents' assessments of intermediate impact ($R^2 = .48$) and shows that immediate impacts of strategic information and advice, and information on new markets, are both significant predictors of intermediate impact on firm performance ($p<.01$ in both cases).

7.5 Summary and Conclusion

Supporting intermediaries may be the most effective way in which governments and regional authorities can increase the innovativeness of firms and regions. By conducting and supporting innovation activities intermediaries may increase the efficiency of innovation systems by allowing business people and researchers to specialise, focusing on the activities they do best. But the performance of innovation intermediaries is difficult to measure and attempts to measure their performance using inappropriate indicators risk diverting them from their most important work.

Using a novel methodology for measuring the impact of innovation intermediaries on firms, we have shown that the GAP program is having an impact on the performance of participating firms and explain how this impact on firm performance is being achieved. Participating firms reported that the GAP program had an impact on their performance in terms of revenue growth, exports, new customers, employment growth, and financing. The GAP program achieved this impact on firm outcomes by impacting firm resources and capabilities by providing strategic information and advice, information and advice on new markets, and general business assistance in terms of executive education and information and advice on financing and business services. Statistical analysis of the factors that are associated with impact on firm performance showed a statistically significant relationship between strategic information and advice and information and advice on new markets, and firm performance. Immediate impact achieved through the provision of general business assistance was not a statistically significant predictor of intermediate impact on firm performance.

Overall, our findings on the impacts being achieved by the GAP are encouraging. Not only do most respondents assess core program activities to be 'very important' or 'extremely important', but overall impact on the resources and capabilities of companies is being manifested in improved company performance in the marketplace. These results are impressive as the companies that participated in the GAP are established and successful prior to participation, requiring targeted and effective interventions to achieve performance improvements. These findings will facilitate discussions among Tekes and the Global Access Program management, client companies, and other stakeholders to ensure the GAP offers a set of continually improving services, and that those services continue to positively impact the GAP's client companies.

This assessment of the impact of the GAP program on participating firms demonstrates the ability of the general-purpose impact assessment methodology to provide a theoretically-grounded assessment of a specific program. Our logic model expresses the a priori expectation that immediate impacts on firm resources and capabilities will lead to subsequent impacts on firm performance, an expectation that holds across all types of innovation intermediaries. While other researchers have shown relationships between engagement with intermediaries and firm resources and capabilities (Human and Provan 1997; McEvily and Zaheer 1999) or performance outcomes (Feldman and Kelly 2006; Sakakibara and Branstetter 2003) their approaches are specific to the types of intermediaries being studied and are not generalisable across intermediary types.

Acknowledgments The authors would like to acknowledge the assistance of Minh Lam and Raine Hermans of Tekes, Bob Foster and Elwin Svenson of the UCLA Global Access Program, and Brian Barge and Andree Boisvert of The Evidence Network.

References

Audretsch, D. B., Link, A. N., & Scott, J. T. (2002). Public/private technology partnerships: Evaluating SBIR-supported research. *Research Policy, 31*, 145–158.

Cote, J. A., & Buckley, M. R. (1987). Estimating trait, method, and error variance: Generalizing across 70 construct validation studies. *Journal of Marketing Research, 24*(3), 315–318.

Dalziel, M. (2006). The impact of industry associations. *Innovation: Management, Policy & Practice, 8*, 296–306.

Dalziel, M. (2010). Why do innovation intermediaries exist? 2010 DRUID Conference, London, UK. http://www2.druid.dk/conferences/viewabstract.php?id=500976&cf=43.

Di Gregorio, D., & Shane, S. (2003). Why do some universities generate more start-ups than others? *Research Policy, 32*, 209–227.

Feldman, M. P., & Kelley, M. R. (2006). The ex-ante assessment of knowledge spillovers: Government R&D Policy, economic incentives and private firm behavior. *Research Policy, 35*, 1509–1521.

Furman, J., Porter, M., & Stern, S. (2002). The determinants of national innovative capacity. *Research Policy, 31*, 899–933.

Howells, J. (2006). Intermediation and the role of intermediaries in innovation. *Research Policy, 35*, 715–728.

Human, S. E., & Provan, K. G. (1997). An emergent theory of structure and outcomes in small-firm strategic manufacturing networks. *Academy of Management Journal, 40*, 368–403.

Inkinen, T., & Suorsa, K. (2010). Intermediaries in regional innovation systems: High-technology enterprise survey from Northern Finland. *European Planning Studies, 18*(2), 169–186.

Kaufmann, A., & Tödtling, F. (2001). Science-industry interaction in the process of innovation: The importance of boundary-crossing between systems. *Research Policy, 30*, 791–804.

King, A. A., & Lenox, M. J. (2000). Industry self-regulation without sanctions: The chemical industry's responsible care program. *Academy of Management Journal, 43*(4), 698–716.

Koskenlinna, M., Smedlund, A., Ståhle, P., Köppä, L., Niinikoski, M.-L., Valovirta, V., Halme, K., Saapunki, J., & Leskinen, J. (2005). *Välittäjäorganisaatiot – moniottelijat innovaatioita edistämässä* [Intermediary organisations – allrounders promoting innovations]. Technology Review 168/2005. Helsinki: Tekes. (In Finnish.)

Löfsten, H., & Lindelöf, P. (2002). Science parks and the growth of new technology-based firms – academic-industry links, innovation and markets. *Research Policy, 31*, 859–876.
Markman, G. D., Phan, P. H., Balkin, D. B., & Gianiodis, P. T. (2005). Entrepreneurship and university-based technology transfer. *Journal of Business Venturing, 20*, 241–263.
McEvily, B., & Zaheer, A. (1999). Bridging ties: A source of firm heterogeneity in competitive capabilities. *Strategic Management Journal, 20*, 1133–1156.
McLaughlin, J. A., & Jordan, G. B. (1999). Logic models: a tool for telling your program's performance story. *Evaluation and Planning, 22*, 65–72.
Millar, A., Simeone, R. S., & Carnevale, J. T. (2001). Logic models: a systems tool for performance management. *Evaluation and Program Planning, 24*, 73–81.
Renger, R., & Titcomb, A. (2002). A three-step approach to teaching logic models. *American Journal of Evaluation, 23*(4), 493–503.
Rogers, P. J. (2005). Logic model. In S. Mathison (Ed.), *Encyclopedia of evaluation* (pp. 232–235). Thousand Oaks, CA: Sage.
Sakakibara, M., & Branstetter, L. G. (2003). Measuring the impact of US research consortia. *Managerial and Decision Economics, 24*, 51–69.
Siegel, D. S., Waldman, D., & Link, A. (2003). Assessing the impact of organizational practices on the relative productivity of university technology transfer offices: An exploratory study. *Research Policy, 32*, 27–48.
Ståhle, P., Smedlund, A., & Köppä, L. (2004). Välittäjäorganisaatioiden rakenteelliset ja dynaamiset haasteet: osaselvitys innovatiivisesta johtamisesta monen toimijan verkostossa [Structural and dynamic challenges of intermediary organisations: A partial investigation of innovative management in a multi-actor network]. Finland's Ministry of Trade and Industry. http://ktm.elinar.fi/ktm_jur/ktmjur.nsf/All/82C22418EEFA2836C2256F26003D29AA/$file/jaettu_johtajuus_final.pdf. (In Finnish.)
Suvinen, N., Konttinen, J., & Nieminen, M. (2010). How necessary are intermediary organizations in the commercialization of research? *European Planning Studies, 18*, 1365–1389.
Tekes Annual Review 2009. http://www.tekes.fi/en/community/Annual%20review/341/Annual%20review/1289.
Yu, Z., & Heshmati, A. (2007). Growth and performance of science parks in China. In A. Heshmati (Ed.), *Recent developments in the Chinese economy* (pp. 55–82). New York: Nova Science Publishers.

Part II
Micro-, Meso- and Regional Level Applications

Chapter 8
Developing a Framework for Innovation and Learning in the Workplace

Lotte Darsø and Steen Høyrup

Abstract In this chapter an analytical framework is developed for conceptualisation and analysis of the interplay between innovation and workplace learning. By introducing the concepts of preject and project as arenas for the innovation process and by pointing out some important differences in these arenas, this chapter mainly directs attention to the less developed and discussed concept of preject. Learning theory is examined from the perspective of the individual in his or her social setting, and the concepts of innovative learning, adaptive learning, and reproductive learning are introduced and discussed. By relating these concepts to a problem matrix developed by Darsø (2001), we are able to, firstly, differentiate between preject, problem solving, and project, and, secondly, associate innovative learning, adaptive learning, and reproductive learning with the above-mentioned corresponding arenas. The chapter concludes with a figure displaying the framework.

8.1 Introduction

The intent of this chapter is to build a framework as a potential analytical tool for research as well as a tool for practitioners to develop workplaces as fruitful arenas of innovation and learning.

After defining innovation, we examine the innovation process and selected innovation process models. We use the Minnesota Innovation Research Program (MIRP) framework as our point of departure to provide the overall picture and to serve as our figure-ground. Our focus is on human activities that may lead to innovation. As our figure, we have chosen the early chaotic phase of innovation

L. Darsø (✉) • S. Høyrup
Department of Learning, Danish School of Education, University of Aarhus/Copenhagen Campus, Copenhagen, Denmark
e-mail: LDA@dpu.dk

H. Melkas and V. Harmaakorpi (eds.),
Practice-Based Innovation: Insights, Applications and Policy Implications,
DOI 10.1007/978-3-642-21723-4_8,

and as our ground the social and cultural organisational environment. We begin to build our framework by introducing two concepts of working with innovation: preject versus project. Preject was coined (Darsø 2001) to differentiate the early divergent process of innovation from project, which per definition is goal-directed and convergent. The next step is to examine learning and the learning environment or context. Here we try to understand why recent literature on learning in chaotic environments often portrays learning as adaptive and reproductive. When examining a case study on learning in an emergency department (Goldman et al. 2009), we find important similarities to problem-solving, e.g. as conceptualised by Isaksen (1988). By applying a problem map developed by Darsø (2001), it becomes apparent that the innovation process in the preject can be distinguished from problem-solving. We argue that the preject should be categorised as an explorative search, because here the problem is yet to be identified. We then introduce Billett's understanding of learning by bringing into focus the interaction between the affordances and constraints of the social setting (i.e. the workplace), on the one hand, and the agency and biography of the individual participant, on the other (Billett 2004). The notion of 'self-directed' information search in the above mentioned case study by Goldman et al. (2009) has inspired us to suggest a new notion of 'inquiry-directed' search. We then introduce Ellström's concepts of developmental learning versus reproductive learning (Ellström 2010). Returning to the preject versus the project modes of working as our guiding perspectives, we are now able to identify important organisational conditions for enhancing and hindering learning and innovation in the workplace. Finally, inspired by Weick and Westley (1996) we juxtapose these conditions in order to understand how learning and innovation can analytically inform, enrich, and challenge each other.

8.2 The Innovation Process

The term innovation was coined by Joseph Schumpeter in the 1930s and defined as a novelty that creates economic value (Schumpeter 1934). This entails that ideas, creativity, or inventions can only be called innovation when they have been successfully launched or implemented into the market. Thus, the innovation process involves the entire progression from the emergence of the first vague idea to the finished 'product' being applied, bought and taken in by the users, customers, and clients. The most comprehensive innovation process model was developed by Van de Ven and colleagues in 1989 in the Minnesota Innovation Research Program, a longitudinal multiple study (Van de Ven et al. 1989). In comparison with other simpler innovation process models (e.g. the typical stage-gate model), this framework illustrates the richness and vast complexity involved in innovation processes.

As seen in Fig. 8.1, the model involves 12 steps divided into three main phases (Van de Ven et al. 1999): initiation, development, and implementation/termination. Initiation consists of steps 1–3: gestation, shock, and plans. Van de Ven et al. describe *gestation* as an extended period of 'setting the stage' for innovation.

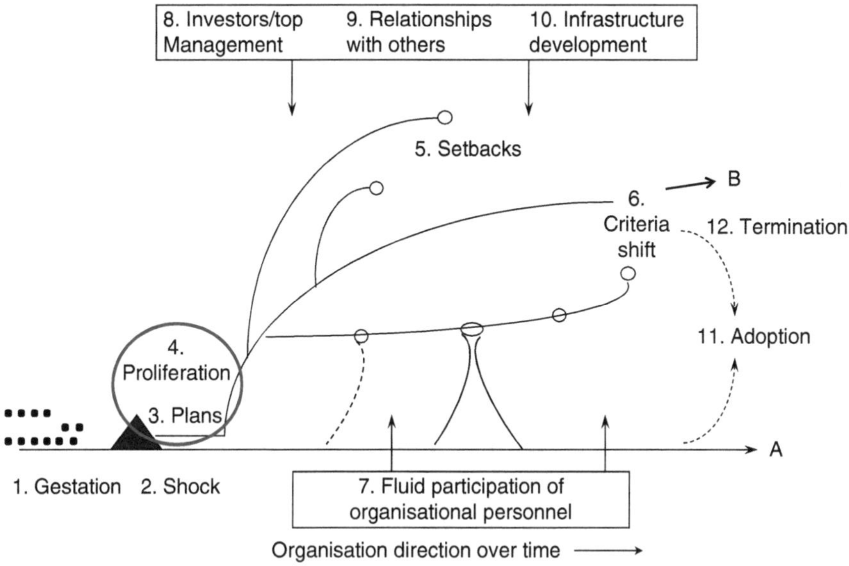

Fig. 8.1 Key components of the innovation journey (Van de Ven et al. 1999: 25)

Changes in customer needs, new technology, legislation, etc. slowly but surely make the organisation receptive to innovation so that when the *shock*, e.g. a critical incident, occurs people in charge will react and begin to focus on innovation as a way to grow and overcome the predicament. Ideas and suggestions flow and *plans* are made for investing resources and time in an innovation project.

Development consists of steps 4–10: proliferation, setbacks, criteria shift, fluid participation of organisational personnel, investors/top management, relationships with others, and infrastructure development. *Proliferation* is a divergent process of many parallel paths of ideas and activities. *Setbacks* happen when unexpected developments occur, when the environment changes, or because of surprising information and knowledge. *Criteria* for success and failure change with time and with the development of the project, often involving dynamic power struggles in the organisation. *Fluid participation of organisational personnel* influences the innovation process in various ways. History and continuity can be problematic as people walk in and out of projects and organisations. Projects undergo ups and downs to such a degree that Van de Ven et al. talk about changing human emotions as 'gut-wrenching' experiences (1999: 24). *Investors and top management* perform different roles during the innovation process, which can have a significant influence on both process and outcome. *Relationships with others* concern networks, collaborations, and partnerships with external stakeholders. This can involve economic investments or marketing relationships with distributors, etc. and can consequently be quite risky. *Infrastructure development* is often needed for commercialising the innovation, such as changing institutional norms and financing agreements or other types of resources.

Implementation/termination consists of the last two steps, 11–12: adoption and termination. In Fig. 8.1, the letters A and B indicate ongoing operations in the general direction and an innovation proceeding in a new direction, respectively. *Adoption* involves introduction, application, and diffusion of the innovation. Adopters often modify the innovation in order to make it fit their specific needs as well as their local context. A classical organisational challenge is to link the new with the old. *Termination* is not as easy as it may sound. For some innovation teams it can be difficult to realise that the project must come to an end. Evaluation, feedback, and learning from the process become complicated and tricky, because so many people have been involved, which tends to entail diverse perspectives, attributions, and opinions.

Evidently, challenges and opportunities involved in innovation are manifold and thus impossible to cover in a single chapter. The above-mentioned framework offers a valuable, research-based foundation for selecting and discussing some important conditions for successful innovation in organisations.

We are mostly interested in the human aspects of innovation and learning in the workplace, which we consider a form of *social innovation*. The term has been used by Drucker regarding renewal based on social needs, e.g. in the public sector (Drucker 1985). Yet, social innovation can also be used regarding renewal of social interaction, social relations, and social behaviour, e.g. in the workplace. Consequently, our focus is not on technology, product development, strategy, research and development, production, marketing, and sales. Our argument is that innovation always involves people, and while research on management of innovation is abundant, the employee and workplace perspective has been somewhat neglected. This chapter will focus on the early innovation process, corresponding to steps 3–4, plans and proliferation, in the MIRP model. We introduce a specific framework, *The Diamond of Innovation* (Darsø 2003), based on empirical research, which conceptualises the innovation process as preject versus project (Darsø 2001).

8.3 The Stage-Gate Model

In most companies the innovation process is represented by the classical stage-gate model (e.g. Cooper 1986), which works more or less like organisational management models in providing an organisational overview. Stage-Gate models are descriptive, e.g. by indicating in which department the innovation project physically takes place. A stage-gate model can also focus on activities, such as idea generation, screening, development, marketing, etc. or it can be based on gates related to criteria that needs to be met and decisions of go/no-go. The main problem with stage-gate models is the so-called 'over-the-wall' problem. This has been discussed by Takeuchi and Nonaka (1986), who came up with three expressive metaphors: relay race, rugby model, and sashimi.

However useful the stage-gate model is for organisational overviews, it reveals little about what is actually going on during the innovation process. Another way to understand the innovation process is to examine when the process opens up or diverges and when it closes down or converges. How are ideas and concepts developed? When and how are the different elements of the innovative idea or concept defined? In traditional product development certain features must be defined at specific milestones, after which there is no way back – except starting all over. This is the case in the car industry, for instance. In the 1990s, however, it was found that Toyota had developed a different model, which was much more open. In fact, the idea was to keep everything open for as long as possible in order to adjust measurements of all the parts towards standards and prices (Ward et al. 1995). The newest innovation process models are even more open, such as the 'never-finished' model (Austin and Darsø 2009) found in IT companies that keep adding and updating software in new versions. It should be emphasised that all the mentioned models concern the innovation process conceptualised from an internal organisational perspective and should be distinguished from the concept of 'open innovation' coined by Henry Chesbrough (2003). The latter concept is related to strategy and business models and designates open interaction between an organisation and external networks and partnerships.

As business develops from producing products towards producing and selling services, and as products and services become more oriented towards involving experience (Pine and Gilmore 1996), the traditional innovation process models become more and more obsolete. New types of models have been suggested, such as the Ibbotson theatre model, the innovation journey, and an art-making model (Austin and Darsø 2009).

8.4 Divergence – Convergence

The innovation process consists of several periods of opening up and closing down, i.e. of divergence and convergence. Divergence is aimed at exploring, finding out, asking questions and discovering new possibilities, whereas convergence is aimed at reaching the set goal, making decisions, limiting possibilities and controlling the results. Evidently, both are needed in the innovation process. It is also evident, however, that the balance has been extremely biased towards the convergent side. In the second half of the twentieth century, innovation and project management were closely linked (Cooper 1986), which can be seen in the massive literature published on project management. Here a project is defined as a task that is goal directed, involving a group of people (usually from different departments) with a budget and a specific time limit. Projects are per definition convergent. How do we articulate the corresponding divergent counterpart? The term preject has been suggested by Darsø: "...the preject is characterised as nonlinear, divergent and process driven..." (2001: 196). The Diamond of Innovation, developed through action research, illustrates the concept of preject (Darsø 2003); see Fig. 8.2.

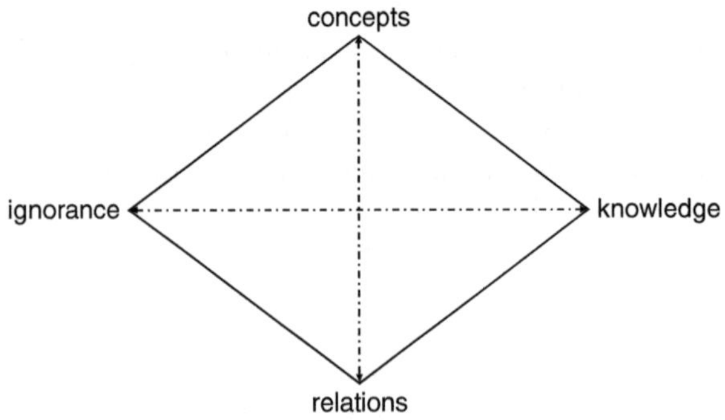

Fig. 8.2 The diamond of innovation (Darsø 2001: 336)

8.5 The Diamond of Innovation

The Diamond of Innovation has been developed in a 3-year research project that aimed at mapping the early phases of innovation in groups working in a pharmaceutical company (Darsø 2001). The research questions were: How is innovation born? What encourages and impedes innovation in the early phases? Focus was on human communication and interaction, and the model is based on an empirical study of how innovation happens in practice. The model is conceptual and is intended to illustrate dynamic interaction. The innovation process is constituted by two simultaneous and interconnected dynamics; a *knowledge dynamic* between knowledge and ignorance and a *communication dynamic* between relations and concepts. The parameters can be seen as opposites (analytically), though it should be emphasised that in real life situations they co-exist as complex and dynamic innovation processes. In the following we will describe each of the four parameters: knowledge, ignorance, relations, and concepts.

8.5.1 Knowledge

It comes as no surprise that *knowledge* is necessary for innovation. The surprising part is that knowledge can also curb innovation. How is that explained? It appears that knowledge is often accompanied by a subjective ingredient of opinion or attitude, which is not conscious to the knower. Most knowledge is to some degree influenced by underlying assumptions. Groups working with innovation are usually put together by criteria of diversity in order to represent different departments, professions and markets. This way each group member is regarded an 'expert' in his/her area, and statements are consequently not questioned. The data (i.e. taped conversations from meetings) from the study demonstrated how 'experts' would restrain possibilities by saying things like: 'That's impossible', 'can't be done in X market', etc. The dilemma is that, on the one hand, knowledge is the foundation for innovation (Simon 1986) and, on the other hand, innovation is often successful because people do not know it cannot be done. Knowledge then must be critically scrutinised in order for groups to distinguish between data, facts, opinions, attitudes, experiences, suppositions, and beliefs.

8.5.2 Ignorance

The opposite of knowledge is *ignorance*; that which we do not know. When applying the Diamond of Innovation as the starting point, the theme or subject should be represented in the middle where the axes cross. Moving from the middle to the left there are different degrees of ignorance. Close to the middle is what we know that we do not know. Further to the left is what we do not know that we do not know. All the way towards ignorance is what we cannot fathom could be known. Why is ignorance needed for innovation? An important finding from the study was that innovation was triggered by open questions, i.e. ignorance. The problem is, however, that most people shy away from ignorance, because it makes them look 'stupid', whereas contributing from the arena of knowledge makes people feel competent. Asking ignorant questions not only makes people feel incompetent, it also frightens them to be working from an arena of uncertainty, shaky ground and chaos. In fact, it takes courage to navigate in chaos, but how can that be accomplished? This is where the other dynamic comes in.

8.5.3 Relations

Another important finding was the significance of relations. *Relations* are the invisible and tacit ties between people; the way people bond with each other. Relations are about how open or closed we are when meeting new people. Relations are about sympathy and incomprehension, about attraction and repulsion, about inclusion and exclusion and about trust and distrust. In the study, the quality of the relations in the group was the decisive factor whether the group would venture into ignorance or not. It showed that relations characterised by trust and respect constituted a constructive foundation for innovative crystallisation. Relations are typically formed as a tacit, subconscious process whenever people meet. Yet, they can also be shaped or created explicitly. For that to happen we need the fourth dimension, concepts.

8.5.4 Concepts

Concepts are forms, structures or descriptions that can help conceptualise ideas and themes. While relations are tacit, concepts aim at turning speech and conversations explicit. Concepts are important in innovation and a successful innovation process in the initial stages often concludes with a new concept, which will, naturally, have to be tested at a later stage. This is crystallisation (Darsø 2001: 173): "...innovative crystallisation is defined as the outcome of a process involving collective transformation of accumulated and integrated ideas into a new conceptualisation or prototype". Conceptualising is an ongoing process of describing, illustrating and clarifying ideas, thoughts, knowledge, and ignorance. Communication is central to innovation which many groups fail to acknowledge. They talk for hours and not much comes out of it. Ways of dealing with this challenge is for instance to use metaphors, associations, examples, images, or drawings to share what lies behind the words; the more facets, nuances and depth, the richer the concept.

8.6 Dynamics of the Preject

The dynamic of communication can be understood as an iterative process moving between relations and concepts. Both parameters will be at play in communication whether the focus is on relations or on concepts. If a group decides to work on developing relations of trust and respect, it will evidently involve concepts, and, vice versa, when the group develops concepts, it will naturally involve relations. Likewise, the dynamic of knowledge-ignorance is an iterative process moving between knowledge and ignorance. Talking about what the group knows will reveal gaps of not knowing and asking open questions will lead to some answers and to seeking information and data. The dynamics occur simultaneously in groups working on innovation and are inherently complex. In order to complete the first building block of the framework, we will now return to the concept of project.

8.7 Project Management

As mentioned earlier, the project format has become the most applied system for working with innovation. The concept of project has its roots in scientific management, i.e. in a functionalist paradigm. A project is defined as a temporary endeavour with a specific beginning, a precise goal, a budget, an assigned group of people from different departments, and a definite date for delivery of the result. Project management was implemented primarily in engineering, construction, and the military around the 1950s as an alternative to industrialism and standardised bureaucracies for evoking renewal and change. Later, project management also became established in other types of organisations, particularly with the introduction of the matrix organisation. Projects have defined milestones, and it is the responsibility of the project manager to plan tasks, time, and resources well, for instance in a Gantt chart.[1] For many years, project management followed the

[1] Named after Henry Gantt (1861–1919), who invented it.

functionalist paradigm of a stage-gate model. Most prominent became Robert G. Cooper, who in 1986 published the book 'Winning at New Products', in which innovation was portrayed and understood as product development. His proposed stage-gate process had 5 stages: scoping, build business case, development, testing and validation, as well as launch. This model has greatly influenced how companies organise and work with innovation in practice. However, in the 1990s, a new paradigm emerged contemplating the project as a temporary organisation. This indicated a shift from an instrumental product approach towards an increased focus on the process and on human interaction, which led to a focus on expectations, human actions, and learning (Lundin and Söderholm 1995; Packendorff 1995).

In spite of these attempts to renew the rationale behind project management, most efforts are still directed towards improving the existing toolbox of project management. Here, two statements, both deriving from Cicmil et al. (2009: 83–84), are relevant for critical reflection:

> "Given the above characterisation of projects as a progressive and alternative way of working, it appears on one level paradoxical that the normative toolbox of Project Management originates from the very same conceptual and ideological foundations as Fordist mass manufacturing." "A growing body of literature, as well as a growing body of empirical evidence and the voices of numerous practitioners indicate that accepting and applying these widely promoted project management 'good practice' standards does not eliminate project failures, nor does it guarantee project success (Williams 2004). On the contrary, a number of studies within the field of project management suggest that it is the use of project management, or a certain conception of project management, as a methodology for organisational innovation and change which is at the heart of project failures (Currie 1994; Thomas 2000; Maylor 2001; Geraldi et al. 2008)." (Cicmil et al. 2009)

From these critical statements it could be concluded that project management has gone awry. The problem is, however, not project management per se. The real problem is that there is no alternative, that project management is considered the only correct approach. Project management works well when the goal has been established and specific tasks can be defined and delegated. Yet, whenever conditions change, when critical incidents happen and when unexpected outcomes are obtained, a much more constructive approach to deal with these would be through the preject. We contend that all projects ought to be preceded by prejects, but what is even more imperative is that both forms should be existing and available all the way through innovation projects; see Fig. 8.3.

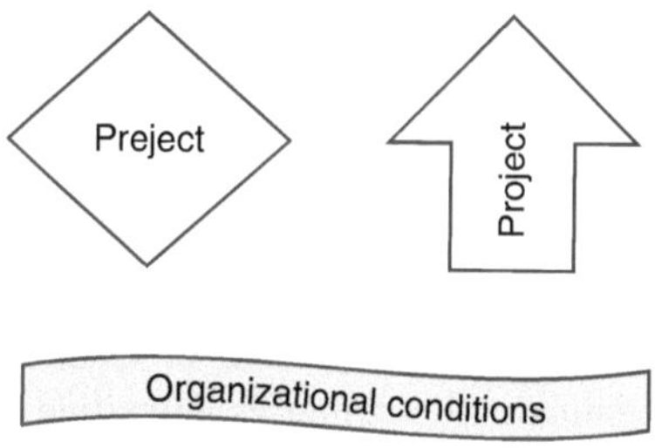

Fig. 8.3 Preject – project (Darsø 2001: 196)

At the beginning of this chapter we stated that we intend to discuss innovation and learning. We will therefore leave the theme of innovation for a short while to continue with learning.

8.8 Linking Learning and Innovation

Only little research has been conducted in the field of linking learning and innovation, especially concerning the micro level of learning (group/team/individual) (Crossan and Apaydin 2010), or – in other terms – with focus on creating a learning perspective on innovation. This may be due to the many different approaches to learning and the apprehension of learning as highly diverse processes occurring at different levels: the individual level, group level, organisational level, network level, and the interconnection of these levels. Most research in the field relates innovation to the organisational level and tends to conclude that organisational learning is an antecedent for innovation and that organisational learning influences innovation in different positive ways. Yet, this approach to learning does not help us gain insight into the micro processes of learning and innovating; it does not enlighten us about what is actually going on.

As mentioned earlier, we are mostly interested in the human aspects of innovation and learning in the workplace. Consequently, our chosen learning approach emphasises the following perspective: the individual in his or her social setting. Experience and what people do with their experience lies at the core of our conception of learning (Jarvis 2006: 6–7). Actions take place in social settings and experiences occur at the intersection of the individual and the social world. Learning often occurs at this point of interaction. A constructive approach to link learning and innovation may therefore be to see the social processes and the social system, constituted by the preject and project logic, as the social context of the learning process. In the following, we will apply the preject – project logic as a means to enquire how innovative processes and learning processes interact and create preconditions and implications for each other.

8.8.1 Disjuncture

In spite of the many divergent views on learning, it seems that researchers to a high extent agree on what initiates the learning process, i.e. the question: What are the triggers of learning? Peter Jarvis emphasises that the driving force for learning often is a certain tension – a disjuncture – between the individual (inner self) and the outer world (Jarvis 2006: 7). Disjuncture is a kind of disharmony between what the individual brings to the situation (experiences, cognitive structure, etc.) and what the situation demands of the individual. It is thus the desire of the individual to overcome this disharmony and return to a state of harmony, which constitutes an important motivation for learning. This idea of disharmony or imbalance – as the trigger for learning – is shared by many learning theorists. Dewey, for instance, states that 'habits do not work' as the starting point of reflective thinking. He adds (Dewey 1997: 74):

> "In cases of striking novelty or unusual perplexity, the difficulty, however, is likely to present itself at first as a shock, as emotional disturbance, as a more or less vague feeling of the unexpected, of something queer, strange, funny, or disconcerting."

Based on Kurt Lewin's thinking, Schein states that uncertainty, in terms of lack of confirmation or disconfirming cues (in relation to existing cognitive system, attitudes, etc.) from the social setting, is the starting point for learning (Schein and Bennis 1965). At the same time, this uncertainty creates a mild or strong anxiety in the individual. This anxiety may foster or be a barrier to learning. Too strong anxiety activates defence mechanisms that hinder learning.

Ellström also adheres to this thinking of disharmony in describing the driving forces for learning and innovation processes. Basic to his thinking is "that tensions and contradictions between work processes as officially prescribed (the explicit dimension) and as perceived and performed in practice (the implicit dimension) create potentials for learning and practice-based innovations in an organisation" (Ellström 2010: 32).

Comparing steps 1–3 in the model of Van de Ven with the above mentioned conceptions of learning triggers, it seems evident that it is basically the same conditions that initiate innovative and learning processes; gestation may be out of reach for most employees, but may form the starting point for management. Shock – the critical incident – corresponds very well to the disjuncture and disharmony as triggers for learning. Plans may be the actions that have the potential to create new experiences and reconstruction of experiences, i.e. learning. The next phase might take the form of the preject; a phase where divergent and convergent processes alternate, although the divergent processes are more prominent. Looking at the preject as an arena of learning, how does the preject relate to learning in terms of fostering different kinds of learning processes? Here the question is: What kind of learning processes are spurred by the social processes predominant in the preject imbedded in the innovation processes?

8.8.2 The Preject in a Chaos Perspective: Learning in Chaos

Chaos is commonly understood as a condition of disarray and confusion, where order is absent (Goldman et al. 2009; Fitzgerald and Van Eijnatten 2002). Referring to Van Eijnatten and Putnik chaos may be understood as "a condition of disarray, discord, confusion, upheaval, bedlam, and utter mess arising from the complete absence of order" (Van Eijnatten and Putnik 2004: 419). Yet, according to chaos theory and complexity theory, chaos and order in organisations are not to be seen as opposites, but as complementary. This is underlined in the term *chaord*, coined by

Dee Hock, combining the words 'order' and 'chaos' into the single term 'cha-ord' to indicate the simultaneous presence of both. A chaordic system is a holon with behaviour that is both unpredictable and patterned at the same time (Van Eijnatten and Putnik 2004).

The preject may thus be characterised as a chaordic system. It is remarkable that environmental dynamics such as disruption, confusion and chaos historically have been seen as negative factors, often indicating the impending demise of an organisation (Goldman et al. 2009: 556). Looking at the potential of the chaordic system to foster or impede different kinds of learning, it is important to note that there are high levels of autonomy for individuals working within successful complex systems (Goldman et al. 2009: 560). Autonomy is often stated as an important supporting factor in workplace learning (Ellström 2010).

In a thorough literature review addressing learning at work accomplished by Fenwick (2008), the author only finds a few articles that make use of complexity theory. It seems that complexity theory is a rarely used frame of reference in the field of workplace learning. Nevertheless, it appears to be fruitful because it treats learning as an inventive/adaptive activity produced continuously through action and relations of complex systems, particularly triggered by disturbances (Fenwick 2008: 236). This is quite in line with our approach. Yet, it is remarkable that much learning in chaos is described as adaptive and reproductive learning. Survival in the system is important (adaption), and in the case study of Goldman et al. most of the learning seems to be reproductive (learning tips and tricks) (Goldman et al. 2009). Learning as an adaptive/reproductive activity implies that the learning product is not new; it often includes some kind of socialisation, implying that the individual complies with the demands at hand, which is contrary to innovation.

What about *innovative* learning, which means that what individuals learn – for instance by co-creating knowledge – is new to the organisation? Fenwick states that the concept of 'context', i.e. the context of learning, seems to be important for the categories of learning we identify: adaptive learning basically implies conceptualising context as a container in which the individual moves about (Fenwick 2008: 237). If, however, we conceptualise context as a web of relations, it becomes possible to identify learning – workplace learning – as exploration, which is innovative learning. This means that learning involves new behaviour and creation of new knowledge that may change interaction, relationships, and knowledge in the organisation, i.e. create process or organisational innovation.

8.8.3 Adaptive and Innovative Learning in the Preject: What Do the Learning Processes Look Like?

When applying the term adaptive learning we should call attention to what Ellström calls the inbuilt duality of the concept: "In one sense adaptive learning is about learning to handle certain tasks or to master the norms, practices and routines in an organisation. In another sense adaptive learning is about the learning and reproduction of a prescribed order..." (Ellström 2010: 33).

In literature, we find several examples of adaptive learning in settings of chaos that are comparable to the conditions of the preject. It seems possible to identify:

- Different types of learning processes
- Context factors supporting the learning processes

An example of adaptive learning in a chaotic environment is presented by Goldman et al. (2009). The concrete working and learning environments – the emergency departments – are chaotic by nature and characterised as busy environments, and those working in them are subject to frequent interruptions. The workload is unpredictable and uncontrollable (Goldman et al. 2009: 559). Goldman et al. identified four different types of what they call learning episodes, activities, or events during which learning did occur:

- *Participation in the environment* concerns the day-to-day work processes that involve seeing and managing patients, observing and talking with others and participating in formal education. These activities facilitate the participants' general knowledge and understanding of emergency medicine (Goldman et al. 2009: 563–564).
- *Focused learning moments* involve short, focused learning moments when something very specific was learned. This is about 'learning tips' related to particular procedures or ways of managing patients. This learning always involves another source such as an attending physician, colleagues, consultants or published material.
- *Repetitive cycles* involve repetition; repeated occurrences of the same patient symptoms or situations. This may imply development of skills concerning patient management, multitasking or specific procedures (Goldman et al. 2009: 565).
- *Intense experiences* include a high level of interpersonal exchange between the participant and other healthcare professionals involved in patient management. These experiences are identified as sources of substantial learning.

All four types of learning involve adaptive and reproductive learning, and it is remarkable that different aspects of the chaotic learning environment clearly support this learning. We can understand this by turning to Billett (2004), who argues for a view of workplace learning as the interaction between the opportunities or constraints of the environment and individual agency. The degree of relatedness between what the workplace affords and what the individual engages in affects the nature of learning (Goldman et al. 2009: 557). What then characterises the chaotic learning environment in this perspective? In relation to individual agency, the author underlines self-direction, i.e. self-directed information search. It seems that the chaotic system produces conditions of great autonomy and freedom for the participants, which support learning.

The characteristics of the chaotic learning environment are *role expectations* within the system that promotes *participation* and implies possibilities to observe, act and talk with others (peers and experts); expectations initiated by the individual, and a great volume of cases to observe and act upon. The great volume ensures diversity in cases; difficult cases (problems) as well as unusual cases. Moreover, the corresponding experiences also ensure repetition of situations and problems to frame and act upon. Repetition pushes the individual to act and reflect and it implies facilitation of learning including practice and *reflection,* which often lead to additional self-directed learning, i.e. possibilities to ask questions and elicit feedback, when wanted. This may include self-directed activities in terms of asking people to share knowledge and possibilities to follow-up on self-directed activities. Intense experiences typically involve high levels of interpersonal exchange between professionals.

The chaotic environment can be characterised as a forceful stimulating learning environment, because chaos, on the one hand, ensures a vast amount of affordances and cases that include great diversity, severe cases and problems, and in that way directs access to real life 'learning-material', including repetitions and pressure to act, frame and solve problems. On the other hand, *social relations* are vital in the chaotic context: We apprehend a social system as a system that supports action and problem-solving, a system that supports participation in a social unit in which observation of peers and experts and sharing of knowledge are prominent features. Mistakes may be unavoidable but they are conceptualised in reflection activities. Furthermore, the system provides freedom and autonomy for the individual and paves therefore the way for self-directedness. In this way, it is the individual or the agency that creates the basic relation between the two sides of the system that makes it a potentially strong learning system. Goldman et al. state that one of the most striking findings about learning in the chaotic emergency department environment was how much self-direction was involved (Goldman et al. 2009: 568).

The chaotic context of an emergency unit resembles the preject regarding the important role of social relations. As pointed out earlier, the quality of relations concerning trust and respect constitutes the foundation for successful innovation processes. In that respect we find significant similarities. But there are, however, also some important differences that become apparent when we try to categorise these two contexts in a problem map (Darsø 2001: 146); see Fig. 8.4. The problem map is a matrix with one axis referring to the problem (as known or not known), and one axis referring to the solution (as known or not known). Obviously, when both problem and solution are known, work is in the area of *routines*. In the emergency unit there are situations where routines can be applied, but the majority of situations are characterised by at least one unknown feature. It can be a known problem with an unknown solution, i.e. unknown to the doctor or nurse, until a colleague with more experience can provide it, i.e. *problem-solving*. Or the problem can be more complex (not known), which means that several solutions are available, i.e.

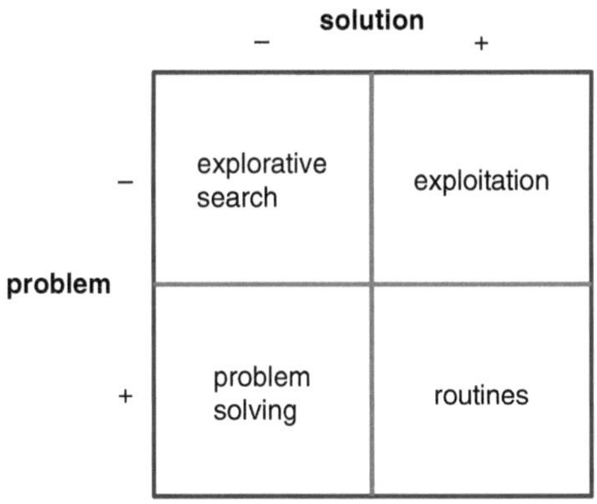

Fig. 8.4 Problem map[2] (Darsø 2001: 146)

exploitation. Whatever the case in the emergency room, it will rarely be characterised as *explorative search* – fortunately for the patients.

In comparison, the preject is distinguished by the fact that the participants neither know the problem nor the solution. Thus, the preject is per definition *explorative search*. The central purpose of the preject is to explore what the problem could be, i.e. to determine the problem or rather to identify the 'threshold' question (Austin and Darsø 2009). The idea of identifying a threshold question, which can set the direction for the search, is underlined by a quote attributed to Einstein, in which he states that if he had one hour to solve a significant problem he would spend 55 minutes on identifying the question and only 5 minutes on the solution (Vogt et al. 2003). Why? Because finding the right question implicitly holds the solution. Furthermore, the preject explorative search is guided by insistent and inexorable enquiry. We therefore argue that while individual agency takes place as *self-directed information search* in chaotic learning contexts, such as emergency units, individual and team agency take place as *enquiry-directed search* in the learning context of the preject and innovation processes.

8.9 Innovative Learning: What Are the Preconditions of Learning?

Innovative learning seems to be an underdeveloped field of research. In relatively new books by Jarvis (2006) and Illeris (2006) they only deal with the subject in a few lines, and both refer to Botkin et al. (1979), who make a distinction between *maintenance learning* and *innovative learning*. *Maintenance learning* is the

[2] In Figure 8.4 minus indicates: not knowing (the problem or solution). Plus indicates: knowing (the problem or the solution).

acquisition of fixed outlooks, methods, and rules for dealing with known situations and recurring situations. It enhances our problem-solving ability for problems that are given. It is the type of learning designed to maintain an existing system or an established way of life. Maintenance learning tends to take for granted those values inherent in the status quo and to disregard all other values.

Innovative learning – according to Botkin et al. – occurs in times of turbulence, change, or discontinuity. These conditions seem to match the essence of chaos. Innovative learning is the type of learning that can bring change, renewal, restructuring, and problem reformulation. Using the term *developmental learning* (synonymous to innovative learning) implies a strong emphasis on the subjects' capacity for self-management and their preparedness to question, reflect on, and, if necessary, transform established practices in the organisation into new solutions or ways of working (Ellström 2010: 34). In conceptualising *practice-based innovation* Ellström prefers to use the term developmental learning, synonymous to innovative learning (Ellström 2010).

Characteristics of learning contexts that foster innovative learning are thus new problems and unknown situations. Unknown situations may arise when individuals, groups or organisations face situations they have never met before, including complex situations, turbulence, change and discontinuity. This is to a certain extent in line with the characteristics of the preject.

What are then the decisive factors that foster innovative learning? Ellström states that "innovation may be viewed as the result of learning and knowledge creation through which new problems are defined and new knowledge is developed to solve them" (Ellström 2010: 36).

Using Billett's concept of *affordances of the workplace* as our point of departure, three factors seem to be prominent for distinguishing adaptive and innovative learning: (1) traits in the culture and management of the organisation, (2) the character of the tasks/problems that challenge the individual at work, and (3) structural conditions of the workplace to allow employees to share and re-combine knowledge in new ways (Hargadon 2002).

Culture and management should value and support that the individual questions the taken-for-granted in offering possibilities to: challenge routines and possibly the security of the organisation, reflect on basic premises and conditions, behave in an explorative mode, be given the possibilities to experiment or make mistakes, behave in a self-directed manner and make efforts to create new ideas and knowledge.

In order to understand the impact of the *task- or problem-structure* that challenges the individual at work, we turn to a conceptualisation elaborated by Herbst (1977). In broad terms, the problems we meet and can learn from may diverge in nature, due to three parameters: (1) How strict, precise or specified is the problem situation at the outset? (2) The procedure to solve the problem: Is there a predetermined procedure to reach the solution or are there many possibilities? (3) The solution to the problem, the end result: Is there a predetermined 'correct' result or can a broad range of results be accepted as satisfying solutions?

By combining the degree of specification/predetermination with these three parameters, it is possible to distinguish between the most open problems (low specification on all three parameters), the closed problems (specification on all three parameters) and various types of semi-open/closed problems. Evidently, open problems are conducive to innovative learning. This could be the reason why the case on learning in the emergency unit chaos does not produce a high degree of innovative learning. In the practice of emergency, medicine professionals can challenge problems that are open in the outset, but procedures and product (diagnosis and treatment) are rather predetermined elements in the problem situation.

Regarding *structural conditions in the workplace*, we would like to draw on Hargadon's thinking (2002). The basic question he enlightens is: *How existing knowledge becomes the raw material from which individuals in organisations construct innovative solutions?* Hargadon constructs a complicated model linking learning and innovation in organisations. Only some essential ideas should be presented here. A basic idea is that innovation is created by moving ideas and knowledge in the organisation from where they are known to where they are not known. In this process, new combinations of existing ideas and knowledge are created. These new combinations are *the essence of innovation*, whereas the process of knowledge sharing and recombination of existing knowledge represents *the learning process*. This is not an easy and ongoing process in organisations, due to the fragmented nature of the larger social structures predominant in many workplaces. The fragmented social structure is created to promote specialist development and professional knowledge, but it does not support innovative processes. One important means to create innovative capacity is consequently to create possibilities to bridge multiple domains and move ideas from where they are known to where they are not (Hargadon 2002: 44). The essence of learning is knowledge-moving and sharing as well as recombination of existing knowledge, whereby innovative learning and innovations are created in the organisation.

8.10 Summary and Conclusion

We started by introducing the concepts of preject and project as two different modes of working with innovation in the workplace. In a time perspective, this would mean that the preject precedes the project. Yet as mentioned, in real life they alternate as projects often meet unforeseen challenges that, in a way, turn them into prejects for a while, until the challenges have been dealt with. The preject is divergent and has its focus on identifying the goal, while the project is convergent and begins when the goal has been identified. Chaos therefore prevails in the preject compared to a dominance of order in the project.

The next step was to link innovation with learning focusing on the individual in his or her social setting. Here we found that disjuncture is a common trigger for both innovation and learning. By examining literature from the field of chaos theory applied in organisations, we found – to our surprise – that chaotic environments apparently support adaptive and reproductive learning, as demonstrated in the case study by Goldman et al. (2009) from an emergency department. To understand this we applied a problem matrix by Darsø (2001), which helped us differentiate chaos in an emergency unit from chaos in a preject. While the former generally concerns problem-solving, the latter concerns explorative search. In relation to the matrix, projects can be categorised as routines.

Weick and Westley juxtapose order and disorder in their conception of learning (1996: 456):

> "The act of repunctuating continuous experience is what we mean by learning. What people learn are intersubjective meanings embedded in culture. To make repunctuation even a possibility, organisation must be reduced and doubt and curiosity must be cultivated. These changes, which mix together order and disorder, juxtapose sufficient order to sustain a learning entity and sufficient disorder to mobilise forgotten material and new alternatives. This juxtaposition is dynamic and represents a transient window of opportunity."

Under this heading our basic perspective on the learning – innovation relation is that innovation may be apprehended as the result of learning. Problem-solving is an important part of learning and in this process problems may be framed in new ways and the solutions may create new insight and knowledge. Conceptualised as arenas of learning, preject and project support adaptive and reproductive learning. Moreover, both types of learning arenas provide many learning conditions that generally support learning.

The preject, however, also seems to support innovative learning, as the need for self-direction and autonomy is often greater here. Cultural/managerial factors are also prevailing, such as support for questioning the taken-for-granted, challenging routines, reflection on basic premises and conditions, explorative behaviour, possibilities to experiment, etc. Ultimately, the open character of problems in the preject and the possibility for people to build bridges between domains of

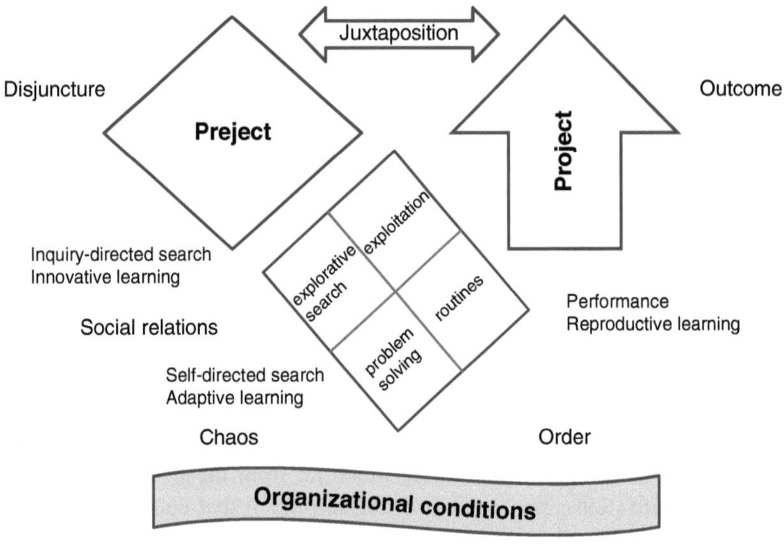

Fig. 8.5 The framework for innovation and learning in the workplace

specialised knowledge to recombine existing knowledge in new ways are dominant features of the preject. The preject thus constitutes a strong learning environment that supports the production of creative ideas, new questions, new knowledge and new solutions to new as well as known problems. These important resources can then feed into the project. The framework we elaborated is shown in Fig. 8.5. The core and the integrating element of the framework is the juxtaposition of the preject and the project in terms of innovation and learning. The preject and the project inform, enrich, and challenge each other through the interplay of innovation and learning.

References

Austin, R., & Darsø, L. (2009). Innovation processes and closure. In N. Koivunen & A. Rehn (Eds.), *Creativity and the contemporary economy* (pp. 55–81). Copenhagen, Denmark: Copenhagen Business School Press.

Billett, S. (2004). Workplace participatory practices. Conceptualizing workplaces as learning environments. *Journal of Workplace Learning, 16*(6), 312–324.

Botkin, J. W., Elmandjra, M., & Malitza, M. (1979). *No Limits to Learning: Bridging the Human Gap, A Report to the Club of Rome*. Exeter, UK: A. Wheaton and Co. Ltd.

Chesbrough, H. W. (2003). *Open innovation. The new imperative for creating and profiting from technology*. Boston, MA: Harvard Business School Press.

Cicmil, S., Hodgson, D., Lindgren, M., & Packendorff, J. (2009). Project management behind the facade. *Ephemera, 9*(2), 78–92.

Cooper, R. G. (1986). *Winning at new products*. Reading, MA: Addison-Wesley.

Crossan, M. M., & Apaydin, M. (2010). A multi-dimensional framework of organizational innovation: a systematic review of the literature. *Journal of Management Studies, 47*(6), 1154–1191.

Currie, W. (1994). The strategic management of a large scale IT project in the financial service sector. *New Technology, Work and Employment, 9*(1), 19–29.

Darsø, L. (2001). *Innovation in the making*. Frederiksberg, Denmark: Samfundslitteratur.

Darsø, L. (2003). Findes der en formel for innovation? [Is there a Formula for Innovation?]. *Børsen Ledelseshåndbøger: Innovations- og Forandringsledelse*. Copenhagen: Børsen Forum A/S. (In Danish.)

Dewey, J. (1997). *How we think*. Mineola, New York: Dover.

Drucker, P. F. (1985). *Innovation and entrepreneurship: practice and principles*. London: Pan Books.

Ellström, P.-E. (2010). Practice-based innovation: a learning perspective. *Journal of Workplace Learning, 22*(1/2), 27–40.

Fenwick, T. (2008). Understanding relations of individual-collective learning at work: A review of research. *Management Learning, 39*(3), 227–243.

Fitzgerald, L., & Van Eijnatten, F. (2002). Chaos speak: A glossary of chaordic terms and phrases. *Journal of Organizational Change Management, 15*(4), 412–423.

Geraldi, J. G., Turner, J. R., Maylor, H., Söderholm, A., Hobday, M., & Brady, T. (2008). Innovation in project management: voices of researcher. *International Journal of Project Management*, *26*(5), 586–589.

Goldman, E., Plack, M., Roche, C., Smith, J., & Turley, C. (2009). Learning in a chaotic environment. *Journal of Workplace Learning, 21*(7), 555–574.

Hargadon, A. B. (2002). Brokering knowledge: Linking learning and innovation. *Research in Organizational Behavior, 24*, 41–85.

Herbst, P. G. (1977). *Alternativ til hierarkisk organisasjon [An alternative to hierarchical organisation]*. Oslo: Tanjum-Norli. (In Norwegian.)

Illeris, K. (2006). *How we learn. Learning and non-learning in school and beyond*. London and New York: Routledge.

Isaksen, S. G. (1988). Innovative problem solving in groups. New methods and research opportunities. In Y. Ijiri, & R. L. Kuhn (Eds.), *New directions in creative and innovative management. Bridging theory and practice*. Cambridge, MA: Ballinger Publishing Company.

Jarvis, P. (2006). *Towards a comprehensive theory of human learning. Lifelong learning and the learning society* (Vol. 1). London and New York: Routledge.

Lundin, R. A., & Söderholm, A. (1995). A theory of the temporary organization. *Scandinavian Journal of Management, 11*(4), 437–455.

Maylor, H. (2001). Beyond the Gantt chart: project management moving on. *European Management Journal, 19*(1), 92–100.

Packendorff, J. (1995). Inquiring into the temporary organization: New directions for project management research. *Scandinavian Journal of Management, 11*(4), 319–333.

Pine, B. J. I., & Gilmore, J. H. (1996). *The experience economy. Work is theatre & every business a stage*. Boston: Harvard Business School Press.

Schein, E. H., & Bennis, W. G. (1965). *Personal and organizational change through group methods*. New York: Wiley.

Schumpeter, J. A. (1934). *The theory of economic development*. Cambridge, MA: Harvard University Press.

Simon, H. A. (1986). What we know about the creative process. In R. L. Kuhn (Ed.), *Frontiers in creative and innovative management*. Cambridge, MA: Ballinger Publishing Company.

Takeuchi, H., & Nonaka, I. (1986). The new product development game. *Harvard Business Review, 64*, 137–146.

Thomas, J. (2000). Making sense of project management. In R. A. Lundin, & F. Hartman (Eds.), *Projects as Business Constituents and Guiding Motives*. Boston: Kluwer Academic Press.

Van de Ven, A. H., Angle, H. L., & Poole, M. S. (Eds.). (1989). *Research on the management of innovation: The Minnesota studies*. New York: Harper & Row Publishers, Ballinger Division.

Van de Ven, A. H., Polley, D. E., Garud, R., & Venkataraman, S. (1999). *The innovation journey*. Oxford and New York: Oxford University Press.

Van Eijnatten, F., & Putnik, G. D. (2004). Chaos, complexity, learning, and the learning organization. Towards a chaordic enterprise. *The Learning Organization, 11*(6), 418–129.

Vogt, E. E., Brown, J., & Isaacs, D. (2003). *The art of powerful questions; catalyzing insight, innovation, and action*. Mill Valley, CA: Whole Systems Associates.

Ward, A., Liker, J. K., Cristiano, J. J., & Sobek, D. K., II. (1995). The second Toyota paradox: How delaying decisions can make better cars faster. *Sloan Management Review, 36*, 43–61 (Spring).

Weick, K. E., & Westley, F. (1996). Organizational learning: Affirming an oxymoron. In S. Clegg, C. Hardy, & W. R. Nord (Eds.), *Handbook of organization studies* (pp. 440–458). London: Sage.

Williams, T. (2004). *Assessing and building on the underlying theory of project management in the light of badly over-run projects*. Paper presented at PMI Research Conference, London, UK.

Chapter 9
Fostering Practice-Based Innovation Through Reflection at Work

Per Nilsen and Per-Erik Ellström

Abstract Practice-based knowledge built up through experience is needed to obtain expertise and skills in occupations and work tasks, but this form of knowledge is not conducive to change or renewal. The development of practice-based innovations can be fostered by creative learning made possible through the deliberate use of research-based knowledge to challenge well-established attitudes, beliefs, and practices. Reflection in the workplace can be used as an instrument to integrate research-based knowledge with practice-based knowledge to scrutinise and possibly revise prevailing thought and action patterns, thus facilitating creative learning. Reflection is often not allotted formal priority on the management agenda, yet the potential of reflection cannot be fully realised without formalising this as an expected, legitimised activity in the workplace. Thus, decisions and planned mechanisms, structures, and procedures are needed to facilitate and support reflection in the workplace.

9.1 Introduction

Innovation has traditionally been viewed as the result of technological progress and planned investment in research and development. However, it is now widely recognised that innovations can also be a function of the informal learning and ongoing knowledge creation that takes place in the production of goods and services in organisations (e.g. Edquist 2005; Lam 2005; Lorenz and Lundvall 2006; Lundvall and Nielsen 1999). This perspective on innovations is not new, however,

P. Nilsen (✉)
Department of Medical and Health Sciences, Linköping University, Linköping, Sweden
e-mail: per.nilsen@liu.se

P.-E. Ellström
Helix Vinn Excellence Centre, Department of Behavioural Sciences and Learning,
Linköping University, Linköping, Sweden

H. Melkas and V. Harmaakorpi (eds.),
Practice-Based Innovation: Insights, Applications and Policy Implications,
DOI 10.1007/978-3-642-21723-4_9, © Springer-Verlag Berlin Heidelberg 2012

as Arrow (1962) and other researchers noted in the 1960s that innovations could be the result of experience-based learning at work. An important implication of this broader view of innovations is the need to consider the workplace as an important site for learning that creates scope for innovation.

Practice-based innovations refer to the employees' or management's renewal of their own operations in some respect, for example, by developing and implementing new working methods, routines, products or services, where this renewal is based on informal learning through work processes within the operations concerned (Ellström 2010). Thus, this definition focuses on workplace learning as a fundamental mechanism behind practice-based innovation processes. This endogenous bottom-up approach to innovation contrasts with the traditional approach, which assumes that innovation is primarily the result of top-down decision-making and implementation of exogenously developed solutions.

Workplace learning can be seen as everyday learning processes that arise within the daily work routine; the challenges of work itself, the organisation of work, and social interactions at work are important sources of this learning (Höyrup and Elkjaer 2006). Workplace learning has attracted greater interest, as workers in the knowledge society must become flexible life-long learners with the ability to reflectively analyse and solve problems, make decisions, possess interpersonal skills, and have a contextual understanding. Employees today need to go beyond formal training to learn a range of vocational, interpersonal, and organisational skills that were not part of previous job demands (Cressey et al. 2006).

This chapter discusses workplace learning that can result in process innovation in working methods and production processes. The emphasis is on intra-organisational innovation processes in the workplace, rather than inter-organisational processes. This focus may be seen as a limitation given the recognised importance of networks and learning in interplay with other organisations, which often form part of an innovation system approach (Edquist 1997; Lundvall 1992). However, process innovations stem from intra-organisational learning processes to a greater degree than product innovations relating to goods and/or services (Hommen 2000).

The purpose of this chapter is to describe how the development of practice-based innovations can be fostered by creative learning made possible through the deliberate use of research-based knowledge to challenge well-established attitudes, beliefs, and practices. Practice-based knowledge built up through experience is needed to obtain expertise and skills in occupations and work tasks, but this knowledge is not conducive to change or renewal. A central argument of this chapter is that reflection in the workplace can be used as an instrument to integrate research-based knowledge with practice-based knowledge in order to facilitate creative learning involving scrutiny and possible revision of prevailing thought and action patterns.

The chapter begins with an explanation of the concepts of research-based and practice-based knowledge (Sect. 9.2). The next section (9.3) discusses limitations of practice-based knowledge in achieving learning that leads to development of new solutions and ideas. The concepts of adaptive and creative learning are introduced and explored. This is followed by a discussion of the potential of using research-based knowledge to achieve creative learning and a description of the process of integration of research-based and practice-based knowledge (Sect. 9.4). Section 9.5 elaborates on the concept of reflection and how it can yield creative learning. The concluding section addresses some of the challenges involved in achieving reflection in the workplace. Some of the aspects of the chapter are also discussed in light of the evidence-based agenda, which has attracted widespread attention in recent years.

9.2 Research-Based and Practice-Based Knowledge

The concepts of research-based knowledge and practice-based knowledge are important to the discussion in this chapter. Key characteristics of research-based and practice-based knowledge are summarised in Table 9.1. The table presents a simplified dichotomisation of the two knowledge concepts (Nilsen 2010).

Research-based knowledge is derived from empirical research as well as concepts, theories, models, and frameworks, which can be seen as analytical tools for understanding, explaining, and making predictions about a given subject matter. This knowledge is scientifically grounded, and is generated in a highly structured and meticulous process that typically begins with a thorough analysis of the problem under study. Practice-based knowledge is gradually built up from practitioners' experience, and is manifested in their craft expertise and skills.

The source of practice-based knowledge can be a specific problem that requires a solution. Different types of problems can also be the starting point for developing research-based knowledge, but this process typically involves a great deal of problematising of the issue under study before research questions are formulated and the issue is investigated. Thus, research-based knowledge cannot provide quick

Table 9.1 Key characteristics of research-based and practice-based knowledge

Characteristics	Research-based knowledge	Practice-based knowledge
Rationale for knowledge development	Obtaining improved understanding or explanation of problems	Finding solutions to problems
Desirable knowledge attributes	Generalisability and objectivity	Hands-on applicability in concrete everyday situations
Generalisability of the knowledge	Usually high ambitions for generalisability	Context-specific, with limited generalisability
Articulation of the knowledge	Explicit codified knowledge, primarily expressed in writing	Usually tacit knowledge expressed in action
Ease or difficulty with which the knowledge is shared	Easy to share	Difficult to share
Accessibility and availability of the knowledge	Public, available in certain channels, e.g., journals and reports	Unique, personal, and difficult to access

solutions to problems in the way that practice-based knowledge can. Practice-based knowledge predominantly serves to solve everyday problems that occur in the daily flow of events, which means that the knowledge is connected to specific situations or environments. The subjective and context-bound nature of practice-based knowledge limits its generalisability. In contrast, research-based knowledge strives to be objective and generalisable.

Research-based knowledge is explicit, primarily articulated in different types of text, which facilitates communication and knowledge exchange. A broader definition of research-based knowledge also includes other forms of codified knowledge, which may not have been subjected to such a rigorous quality control process, e.g. different types of reports and results. Practice-based knowledge, on the other hand, tends to be tacit, expressed through action rather than in words. This type of knowledge can be difficult to access and communicate to others.

In practice, aspects of research-based and practice-based knowledge are intertwined. It is rarely an either/or choice for practitioners; more often a question of making sense of many different types and sources of knowledge, some of which may be research-based and others that are practice-based. It is neither possible nor desirable to draw a strict line of demarcation between these two types of knowledge. Practitioners may acquire research-based knowledge as part of their professional education and by keeping up to date with research advances by participating in conferences and continuing professional education; practice-based knowledge is acquired through experiences in everyday work and discussions with colleagues and customers.

Both research-based and practice-based knowledge are required to achieve a high level of competence, which is an individual's ability to act knowledgeably, effectively, deliberately, strategically, and reflectively in a situation (Svensson et al. 2004). Either knowledge aspect may dominate under different circumstances and may depend on factors such as education, profession, or type of activity. For instance, it has been shown that psychiatrists and psychologists use research-based knowledge to a far greater extent than social workers (Bergmark and Lundström 2002). There can also be considerable variation in the balance of research-based and practice-based knowledge among individuals within the same profession (Hannan et al. 2000; Hemsley-Brown and Sharp 2003; Rosén 1994).

9.3 Learning and Innovation

Practice-based knowledge is indispensable to perform many work tasks well and solve different types of problems that are encountered in the course of daily work. Thus, people perform tasks, solve problems, and cooperate with others, based on knowledge and skills resulting from previous learning. Although such actions are clearly deliberate and reflective under certain conditions, under other conditions they are best described as automatic or routinised. Based on theory and research within the field of cognitive action theory, Ellström (2006a) conceptualised informal learning at work as interplay between four levels of knowledge and action involved in carrying out a task:

1. *Skill-based action*, i.e., routinised (automatised) actions that are typically performed smoothly on the basis of implicit (tacit) knowledge and with little subjective effort, conscious attention, or control;
2. *Rule-based action*, i.e. actions that are performed on the basis of a rule or procedure that is based on experience (learning) from similar previous occasions, instructions, or communication with others;
3. *Knowledge-based action*, i.e. actions that are consciously controlled, generated, and selected on the basis of analyses of tasks and goals in relation to factual knowledge and/or more general theoretical knowledge that, in principle, can be articulated and codified; and
4. *Reflective action*, i.e. actions based on evaluations and reflections concerning not only the performance and consequences of actions, but also reflections concerning the task and the goals at hand (cf. the notion of reflection-on-action as proposed by Schön [1983] and the notion of critical reflection as discussed by Mezirow [1991]).

A basic assumption behind this model is that reflection and action at one level do not exclude parallel or integrated activity at other levels. In practice, the optimal handling of a certain task may require performance at different levels, in sequence or in parallel (Olsen and Rasmussen 1989). Higher-level cognitive functions are assumed to monitor ongoing actions and to anticipate upcoming problems and demands in the environment. Consistent with this idea, Giddens (1984) refers to the 'reflexive monitoring of action' to indicate the active character of routine actions. In this kind of reflective process, logical inconsistencies as well as self-reflection become of focal concern, which also makes a reliance on meta-cognitive knowledge necessary, i.e. knowledge about oneself and one's own knowledge, its scope and limits, strengths and weaknesses (e.g. Eraut 2004; Flavell 1979).

In the remainder of this section, the levels of action presented above are used as a starting point and a framework for discussing learning and innovation in organisations. Specifically, learning and innovation are conceptualised as *dialectic* or *dual* processes depending on interplay between the establishment of effective routines and habits, on the one hand, and on the capacity to unlearn and change routines and habits that are no longer effective, on the other.

9.3.1 Establishing Routines Through Adaptive Learning

The consolidation of thought and action patterns as routines or habits may be seen as a way to successfully cope with the daily flow of events, problem situations, and contradictory demands, while maintaining a sense of security and stability in life (Giddens 1984). This can also be illustrated by our tendency to develop cognitive shortcuts (heuristics or rules of thumb) concerning how something works and what is possible and not possible, how things should be seen, and what outcomes are likely (Gigerenzer et al. 1999; Gladwell 2005; Lehrer 2009). There are also shortcuts in the form of habits that are more or less automatic responses that develop through learning as behaviour is repeated in the same context (Verplanken and Wood 2006).

Cognitive shortcuts and habits can be very efficient for dealing with new situations by treating them like something that has been encountered before. They preserve crucial time, energy, and mental capacity, thus freeing mental resources for other purposes, including creative work and innovations. Cognitive shortcuts essentially leap frog over analysis to reach conclusions without going through the intermediary steps required for questioning and, if necessary, modifying perceptions, thoughts, feelings, and behaviours (Rushmer and Davies 2004). This is also in line with theories on ego defence and perceptual set, which explain why people are less likely to act on new information and may even avoid information that challenges present attitudes, beliefs, and behaviours (Cramer 1991; Nevin 2006). Meanwhile, cognitive dissonance theories posit that people bring attitudes, beliefs, and behaviours into line to reduce unpleasant feelings of dissonance (Festinger 1957). The human tendency to avoid the unknown in favour of the familiar, certain, and controlled can also function as a barrier to change. Inertia to change can also be seen at group and organisational levels. Colleagues can convey strong group norms and role expectations that constrain change in a group (Conner and Norman 2005), whereas the dominant logic of an organisation (e.g., the cultural norms and beliefs concerning where an organisation comes from and what it stands for) can function as a filter to screen out information that does not conform, thus reducing the organisation's perceived need to change (Schein 1985).

The establishment of routines or habits for handling problems and demands in daily life occur through mechanisms of learning, specifically, what has been called adaptive learning (Ellström 2001, 2006a). This mode of learning can be conceptualised as a process in which the learning subject manages to handle a certain task or situation at a skill-based level of action, that is, in a routinised way. Instances of adaptive learning are easy to find in organisations, including, for example, the mastering of new tasks or situations, or learning to follow certain instructions. Adaptive learning is also central to the socialisation of individuals to an organisation. The focus then is to what extent the people in an organisation are acquiring the code, and thus learning how one should or must think and act in different situations. Adaptive learning yields efficient, effective, and reliable task performances that are stable over time. Variations in performance within and between individuals in an organisation are reduced, promoting homogeneity (Ellström 2010). In line with this, the decisive criterion for successful adaptive learning is that the task concerned can be performed rapidly and with a low percentage of error (Argote 1999; Argyris and Schön 1978; March 1991). In practice, measures to reduce variation in the performance of a task may include formalisation through written rules and instructions, limited autonomy, and the formulation of clearly specified tasks and goals.

Adaptive learning (like other forms of learning) does not take place easily, however. The literature is full of cases where the implementation of prescribed working methods, procedures, or policies has failed (implementation failures). At the same time, what on the surface appears to be a successful implementation may, on closer examination, prove to be a pro forma adaptation to stipulated work processes; that is, learning has taken place in the sense that people have picked up the prescribed way of perceiving and speaking (rhetorical learning), but they have not appropriated the prescribed work process in practice (Wertsch 1998).

However, and this is the other side of the coin, once established, routinised actions are very difficult to change and unlearn, especially if one relies on intellectual and verbal forms of education and training (Frese and Zapf 1994). There is ample theoretical and empirical evidence to suggest that there are strong mechanisms at work to preserve established attitudes, beliefs, and behaviours at the individual, group, and organisation levels. People function well at a routinised level of action until problems or surprises of one kind or another arise (Gersick and Hackman 1990). This is clearly in line with Dewey's (1933) view of reflection as a process that begins with a perceived difficulty or disturbance in a routinised task or situation. However, as mentioned above, routinisation may also impede our detection and management of contextual changes or disturbances (Gersick and Hackman 1990). In consequence, in many situations we tend to ignore or misinterpret changes in our surroundings in order to maintain existing structures and patterns of thought and action. Thus, learning that is based only on practice-based experiential knowledge is not sufficient to challenge the prevailing status quo of well-learned and routinised thought and action patterns.

9.3.2 Unlearning of Established Routines Through Creative Learning

Adaptive learning can be contrasted with creative (or developmental) learning, which is associated with questioning and possibly revising established perspectives, routines, and practices. Creative learning is also broadly similar to concepts such as Argyris and Schön's (1978) 'double-loop learning', Engeström's (1988) 'expansive learning', and Mezirow's (1991) 'transformative learning'. Creative learning is assumed to occur when individuals or groups in an organisation question established thought and action patterns and break out from routinised actions to develop new ways of handling the duties and often complex problems involved in a job. The focus is on promoting and exploring variation and diversity in thought and action, which means that creative learning calls for risk taking, acceptance of failures, and a scope for experimenting with and testing alternative ways of acting in different situations (Ellström 2006b, 2010).

Creative learning entails an upward movement from the level of skill-based (routinised) action to the levels of knowledge-based or reflective action. Thus, in accordance with Dewey (1933: 29), it could be argued that creative learning requires the cultivation (through previous learning) of habits of reflection, that is, 'habits of critical examination and inquiry'. In this view, conflicts and ambiguity are not potential threats to learning or to efficient performance, but rather potentials for triggering creative learning processes. Creative learning could be expected if an individual (or group), while working at a routinised level of action, encounters an unfamiliar problem or a new situation for which there is no ready-made rule or solution available (cf. Gersick and Hackman 1990). In such a situation, the alternative to seeking a solution through trial and error is to engage in a process of problem solving at a knowledge-based or reflective level of action. In this way, creative learning can be a driving force for change and innovation in organisations (Ellström 2010). One important mechanism for this to happen is the planned or unplanned variations that always exist in the performance of work. This variation may lead to discoveries, ideas, and new actions that, under certain conditions, can transform established routines and practices in an organisation (Miner et al. 2001).

Opportunities for creative learning in an organisation depend on a complex interplay between a range of factors, including task characteristics (complexity, variety, and control), opportunities for feedback and evaluation, the type and degree of formalisation of work processes, organisational arrangements for employee participation in problem handling and developmental activities, and learning resources in terms of time and possibilities for knowledge exchange (Ellström 2006b, 2001). Creative learning requires sufficient scope and resources for experimenting with and testing alternative ways of acting in different situations, that is, a high degree of autonomy (McGrath 2001). However, the degree of autonomy in performing a task appears to be a necessary, but not sufficient precondition for creative learning. In addition, as discussed extensively elsewhere (e.g. Ellström 2001), individuals or groups must have the subjective capacity required to make use of the autonomy afforded by their jobs. This in turn is related to, for example, previous experience with similar tasks, the individual's knowledge and understanding of the task at hand, self-confidence, and occupational identity. Many of the characteristics and traits associated with creativity are also beneficial to creative learning; for example, openness to experience, imagination, drive, ambition, independence, and nonconformity (Feist 1999).

9.3.3 Innovation as a Balancing Act Between Adaptive and Creative Learning

Creative learning is often assumed to be beneficial, in contrast with adaptive learning, which is implicitly understood to be less valuable. However, both adaptive and creative learning have important functions and should be seen as complementary modes of learning in relation to innovation processes. An overly strong emphasis on adaptive learning means that issues of routinisation and efficiency are likely to dominate over idea development and change. In contrast, too strong emphasis on creative learning entails a risk that issues of creativity, innovation, and renewal drive out concerns related to the efficiency and stability of current practices (Ellström 2006b). Elmholdt and Brinkmann (2006: 179) caution that an exclusive focus on creative learning may disrupt existing well-functioning structures in workplaces. They argue that creative learning must rest on a "persistent reproduction of stable forms of practice in the workplace".

This is also in line with well-established models of learning and innovation (e.g. Dewey 1933; Engeström 1988; Weick and Quinn 1999). According to these models, as well as the ideas presented above, the learning and innovation process begins with questioning, a disturbance or the emergence of a problematic situation in the conduct of a task, or in the interplay with other people. This leads to unlearning of routinised patterns of thought and action, and to a search for new ways of dealing with the disturbance or the problematic situation at hand. As argued by Lam (2005), innovations may be viewed as the result of learning and knowledge creation through which new problems are defined and new knowledge is developed to solve them. In a following phase, this new knowledge and new ways of handling the problems at hand need to be effectively implemented and mastered through processes of adaptive learning in an ongoing cyclical process of adaptation and transformation (for further development of this view, see Ellström 2010). Thus, it is not an issue of either/or, adaptive, or creative learning, but of a question of both/and where the two types of learning can be seen as extremes on a continuum of learning. Both adaptive and creative learning are needed in most organisations, as there is often a requirement to deal alternately with well-known tasks and handle new problematic situations. The challenge is one of providing opportunities for creative learning in organisations without sacrificing the necessary adaptive learning, or vice versa.

9.4 Integration of Research-Based and Practice-Based Knowledge

It has increasingly been recognised that creative learning can be facilitated by purposively using research-based knowledge to challenge established ways of seeing and understanding the world (e.g. Fitzgerald and Dopson 2005; Hörberg 1996; Ohlsson 2002; Rosell 2002). For instance, Svensson (2002: 178) argues that research-based knowledge makes it possible to move beyond the specific 'here-and-now' circumstances of current work practices. He believes that research-based knowledge can "broaden the views, provide unexpected perspectives, and show new opportunities for action". Bilson's (2006) studies on research use in health care have demonstrated that research can be used by practitioners to consider their work practice in a new light, and to reflect on the underlying taken-for-granted assumptions that shape many aspects of their work. Using various forms of research can thus provide a means to release creativity, according to Bilson (2006).

The process of integrating research-based and practice-based knowledge has been understood from a constructivist perspective, that is, individuals actively construct new personal knowledge by combining different forms of knowledge. Research-based knowledge typically goes through processes of interpretation, negotiation, debate, and social influence before it is accepted and applied in practice. For instance, it has been shown that trust in respected colleagues is an important determinant of acceptance of new evidence in health care, emphasising the socially constructed nature of evidence and other forms of research-based knowledge (Fitzgerald and Dopson 2005). Thus, before research-based knowledge leads to changes in thinking and behaviour, the newer knowledge has to be actively related to what individuals already know.

How research-based and practice-based knowledge are integrated and the outcomes of this process have been described somewhat differently by different researchers. There appears to be consensus that this process can create *new* knowledge and foster creative learning that may contribute to creating new innovative solutions. Fitzgerald and Dopson (2005) view the relationship between research-based and practice-based knowledge as 'circular', stating that the two knowledge aspects reinforce each other as they become woven together. Desforges (2000) describes this integration as a 'transformation' process, in which 'old' knowledge in a particular problem context is merged with 'new' knowledge. Nutley et al. (2007: 43) depict the process as 'complex, varied, and unique' and believe the end result represents "an interaction between the ideas and findings contained in the research and the existing knowledge, experiences, and attitudes of the practitioners themselves".

Explicit research-based knowledge can be embodied as tacit practice-based knowledge through learning by doing and different forms of practice and training. An example may be a health care practitioner who participates in a course on cognitive behavioural therapy (CBT), which results in the practitioner's use of certain CBT components as part of their own patient counselling. This process is usually referred to as internalisation. The integration of research-based and practice-based knowledge can also result in externalisation of tacit practice-based knowledge, that is, tacit knowledge is articulated as explicit knowledge, perhaps being triggered by dialogue with colleagues (Nonaka et al. 2001). An example may be a health care practitioner who takes part in a course on Stages of Change (Prochaska and DiClemente 1983), a theory on how interventions can be matched to individuals' motivation to change a behaviour; this can yield improved understanding and provide concepts that make it easier for practitioners to describe their own way of working and discuss this with colleagues, potentially leading to improvements in their practice.

When practitioners make their practice-based knowledge explicit, it can be shared by others and become the basis for new knowledge and learning. However, externalising tacit practice-based knowledge can be difficult. Tacit knowledge has traditionally been viewed as less important than explicit knowledge. Methods that are perceived as well known, rational, and logical tend to be favoured over tacit knowledge in the form of rules of thumb or intuition. Another difficulty stems from the fact that people are not always aware of the full extent of their tacit knowledge. Time raises further difficulties for sharing tacit knowledge, as the internalisation of this form of knowledge may take considerable time. Distance is yet another restrictive factor because face-to-face interaction is usually seen as a prerequisite for sharing tacit knowledge (Hislop 2005; Nonaka et al. 2001).

Knowledge transfer processes and issues concerning how knowledge is shared in organisations are studied in knowledge management. Research in this field has progressed from an objectivist perspective that focused on the codification and infrastructure to allow knowledge to be shared, to a more practice-based perspective that is more concerned with how interpersonal knowledge sharing can be achieved through diverse forms of interaction and communication. The objectivist perspective assumes that knowledge can exist in a fully explicit and codified form, independent of human beings. This view is challenged by the practice-based perspective, which posits that knowledge is personal and that it is impossible to totally disembody knowledge from people into a fully explicit form (Hislop 2005). In contrast to codification approaches to knowledge management, more recent personalisation strategies focus on ways to improve the face-to-face sharing of tacit knowledge between practitioners (Hansen et al. 1999; Hunter et al. 2002).

9.5 Reflection to Foster Creative Learning

Reflection provides a mechanism by which to use research-based knowledge, thus providing opportunities for creative learning. The concept of reflection is generally understood as a means of translating experience into learning, by examining one's responses, beliefs, and actions, to draw conclusions to enable better choices or actions in the future. The major outcome of reflection is learning, as the understanding gained is integrated into one's experience.

John Dewey's writings in the early part of the twentieth century constitute the foundation for the concept of reflection. Dewey saw the act of reflection as central to human learning and conceptualised it as "active, persistent, and careful consideration of any belief or supposed form of knowledge in the light of the grounds that support it and the further conclusion to which it tends" (Dewey 1933: 9). Reflection, according to Dewey, allows individuals to expose and scrutinise their habitual taken-for-granted responses to everyday life experiences. He proposed that every learning episode should be considered as one that involves a "reconstruction or reorganisation of experience which adds to the meaning of experience, and which increases ability to direct the course of subsequent experience" (Dewey 1933: 89–90).

Although Dewey is widely considered to be the founding father of reflection, it was Donald A. Schön who popularised the concept in 1983 with the publication of his seminal book *The reflective practitioner* (Schön 1983). Schön's work was important in bridging "the chronically difficult vacuum that exists between theory and practice" (Redmond 2006: 31). Since then, many articles and books on the topic of reflection have appeared and many theorists have contributed to the development of the concept of reflection in diverse ways.

Many authors have emphasised the importance of reflection as a means of identifying and scrutinising the assumptions that underlie one's thoughts and actions. For instance, Boud et al. (1985) believe that reflection prepares the individual for new experiences and leads to new skills, ideas, mental models, and mind sets. Similarly, Mezirow (1991) argues that the process leads to a new interpretation involving a change in the individual's meaning schemes or a transformation of meaning perspectives. Seibert and Daudelin (1999) also imply that a change in thinking occurs as part of the reflective process. Brookfield (1990) emphasises reflection as a means to identify and question actions, habitual behaviours, and assumptions that sustain attitudes and beliefs.

Reflection is typically viewed as a cognitive process or activity, but some writers (e.g. Boud et al. 1985) have also pointed to the relevance of the emotions of the individual in the reflective process. Although usually described as an individual process, the concept of reflection has also been expanded to encompass more group-centred reflection in the workplace. People tend to encounter complex problems and work on projects that require contributions from and involvement by many people, which makes more contextual and social relations perspectives on reflection highly relevant from a workplace perspective (Schenkel 2006). The term productive reflection was introduced by Boud et al. (2006) to reflect the importance of more collective forms of reflection at work.

The type of learning that is generated in reflection processes depends on the depth or level of reflection. Moon (1999) describes five stages of reflection that reflect different levels of complexity of learning. 'Noticing' is the acquisition of sensory data. 'Making sense' involves seeking a coherency in the material that is perceived. 'Making meaning' is where the new material of learning is related to that which is already known. 'Working with meaning' occurs when the learner thinks things over until they generate new meaning, which Moon considers fundamental to deep learning. 'Transformative learning' is the result of persistent work towards understanding, although Moon also notes that a sudden transformation of understanding is possible.

Somewhat similarly, Strampel and Oliver (2007) describe a model that distinguishes between different levels of reflection depending on the input or source of the reflective action. The lowest level of reflection in their model, 'stimulated reflection', merely involves noticing that something does not quite fit. Reflection at the next level, 'descriptive reflection', means that learners recall what has taken place and their reactions to the situation. 'Dialogic reflection' is a re-evaluation of an experience, which involves relating and connecting new knowledge with prior knowledge, testing for internal consistency between new ideas and existing knowledge, and making the knowledge one's own. Dialogic reflection is a process of searching for meaning, coming to an understanding, and applying new knowledge. 'Critical reflection' is the highest level of reflection (other theorists have used the term critical reflection to denote the examination of social and political assumptions). This level involves application of new knowledge acquired through dialogic reflection, which may include doing things differently, clarifying issues, or developing new skills.

9.6 Challenges of Reflection in the Workplace

There are a number of conditions to achieving reflection in the workplace. Reflection requires the individual's active engagement. Loughran (1996) contends that reflection is dependent on open-mindedness, wholeheartedness, and responsibility on the part of the learner. Langer (1989) similarly argues that reflection can occur only when the individual makes a conscious choice to be mindful. This view is shared by both Boud et al. (1985) and by Mezirow (1991) who state that conscious awareness and deliberate choice are prerequisites to reflection.

Workplace reflection also depends on resources and certain structural and procedural conditions. There is a need for forums that provide legitimacy for reflection and the formal opportunity to meet and discuss matters of relevance for reflection. Time is an important resource for reflection because, as Ellström (2011: 115) points out, it takes "time to observe, time to think, and time to exchange ideas with others". However, although time is often emphasised as a primary factor in reflection, Fenwick (2003) argues that the subjective awareness of the learning opportunities encountered at work may be more important than the allocation of 'objective' time for learning at work. These various conditions, in turn, are dependent on the extent to which the organisational culture fosters reflection and learning at work (Docherty et al. 2006).

The tension between time for production and time for activities that involve reflection, learning, and innovation has long attracted the attention of researchers who have studied learning in organisations. The challenge is to find a suitable balance between adaptive learning and measures that bring about stability, efficiency, and short-term results, on the one hand, and activities that promote creative learning, innovation, and long-term development and survival, on the other hand. As argued by March (1991), an obvious difficulty in establishing these priorities is that the benefits of learning activities in terms of increased innovativeness and long-term survival are both more remote in time and less reliable than the more easily calculated value of measures to boost day-to-day production. Furthermore, the decisions made when it comes to allocating resources for workplace learning are highly related to prevailing views concerning management strategies and leading actors' perceptions of the possible, desirable and, for the time being, appropriate way of managing a production system (Ellström and Kock 2009).

Management interest in the creation and maintenance of learning mechanisms has been found to vary considerably, from total unawareness of the possible existence of learning and reflection in the workplace, to ranking such activities as primary means of competition (De Geus 1992; Garvin 2000; Shani and Docherty 2003). Many researchers have emphasised the importance of viewing reflection as a matter of design, not of evolution, meaning that conscious, active decisions and planned mechanisms, structures, and procedures are needed to facilitate and support reflection because this is often not formally allotted clear priority on the management agenda in many organisations. At the same time, it has been suggested that the dynamic, open, and unpredictable nature of reflection processes can make formalisation difficult. For instance, Cressey et al. (2006: 23) warn that reflection can only be "tamed and domesticated at the risk of destroying what it can offer". However, it is questionable whether the potential of reflection can be fully realised without formalising this as an expected legitimised activity in the workplace.

9.7 Summary and Conclusion

This chapter describes the key characteristics of practice-based knowledge and research-based knowledge and how these types of knowledge can be integrated to achieve creative learning that can be a source for practice-based innovation. Reflection in the workplace is proposed as a mechanism by which this integration can take place. Although the literature on reflection provides strong arguments in favour of this activity, empirical research has described many challenges to reflection in the workplace. Reflection does not automatically yield deep learning. Indeed, the 'reflection level' models of Moon (1999) and Strampel and Oliver (2007) suggest that creative learning is only possible if reflection is sufficiently advanced.

This chapter suggests that creative learning can be facilitated by deliberately drawing on concepts, theories, models, frameworks, and empirical findings of research to achieve deeper, more advanced reflection. However, the notion of using research-based knowledge to improve practice is far from novel. The importance of research-based knowledge is central to the evidence-based practice agenda, which assumes that the rational and systematic application of research findings can yield more effective and accountable practice. Evidence-based practice is seen as a response to growing demands in many fields that professional practice should be underpinned by scientific knowledge and evidence as to the effectiveness of various practices (Trinder 2006). Evidence-based practice has been described in terms of prescribing methods and routines to reduce variations in different practice contexts, thus making practice more standardised, codified, and transparent (Dopson et al. 2005). Evidence-based practice implies that professionals engage in predominantly adaptive learning. The focus is very much on the implementation of and adherence to a particular best practice. In contrast, this chapter emphasises the potential importance of using research-based knowledge to *challenge* established patterns of thought and action. Creative learning involves experimentation and testing of alternative ways of acting in different situations, activities that seem far removed from the ideals of evidence-based practice.

This chapter points to many research challenges: investigation into the extent to which reflection that specifically draws on research-based knowledge occurs in various practice contexts; how such reflection is organised; barriers and facilitators for this reflection; the extent to which reflection is part of the management agenda; and how resources and other conditions that foster this reflection are prioritised in different organisations. The increased emphasis on evidence-based practice in many practice contexts underscores the importance of conducting research into questions of the degree to which an evidence-based practice provides opportunities for creative learning and practice-based innovation. The issues are manifold and complex, but they need to be addressed for better understanding of how the potential of integrating practice-based and research-based knowledge through reflection in the workplace can be realised.

References

Argote, L. (1999). *Organizational learning: Creating, retaining, and transferring knowledge.* Boston, MA: Kluwer.

Argyris, C., & Schön, D. (1978). *Organizational learning: A theory of action perspective.* Reading, MA: Addison-Wesley.

Arrow, K. J. (1962). The economic implications of learning by doing. *The Review of Economic Studies, 29*(3), 155–173.

Bergmark, Å., & Lundström, T. (2002). Education, practice and research. Knowledge and attitudes to knowledge of Swedish social workers. *Social Work Education, 21*(3), 359–373.

Bilson, A. (2006). Rationalities, reflection and research. In S. White, J. Fook, & F. Gardner (Eds.), *Critical reflection in health and social care* (pp. 135–143). New York: Open University Press.

Boud, D., Keogh, R., & Walker, D. (Eds.). (1985). *Reflection: Turning experience into learning.* London: Kogan Page.

Boud, D., Cressey, P., & Docherty, P. (2006). Setting the scene for productive reflection. In D. Boud, P. Cressey, & P. Docherty (Eds.), *Productive reflection at work* (pp. 3–10). Abingdon, UK: Routledge.

Brookfield, S. (1990). Using critical incidents to explore learners' assumptions. In J. Mezirow & Associates (Ed.), *Fostering critical reflection in adulthood* (pp. 177–193). San Francisco, CA: Jossey-Bass.

Conner, M., & Norman, P. (2005). *Predicting health behaviour.* Maidenhead, UK: Open University Press.

Cramer, P. (1991). *The development of defence mechanisms: Theory, research, and assessment.* New York: Springer.

Cressey, P., Boud, D., & Docherty, P. (2006). The emergence of productive reflection. In D. Boud, P. Cressey, & P. Docherty (Eds.), *Productive reflection at work* (pp. 11–26). Abingdon, UK: Routledge.

De Geus, A. (1992). *The living company.* Boston, MA: Harvard Business School Press.

Desforges, C. (2000). Putting educational research to use through knowledge transformation. Keynote lecture to the Further Education Research Network Conference, Coventry, 12 December 2000. London: Learning and Skills Development Agency.

Dewey, J. (1933). *How we think.* New York: Heath.

Docherty, P., Boud, D., & Cressey, P. (2006). Lessons and issues for practice and development. In D. Boud, P. Cressey, & P. Docherty (Eds.), *Productive reflection at work* (pp. 193–206). Abingdon, UK: Routledge.

Dopson, S., Locock, L., Gabbay, J., Ferlie, E., & Fitzgerald, L. (2005). Evidence-based health care and the implementation gap. In S. Dopson & L. Fitzgerald (Eds.), *Knowledge to action? – evidence-based health care in context* (pp. 28–47). Oxford: Oxford University Press.

Edquist, C. (Ed.). (1997). *Systems of innovation. Technologies, institutions and organizations.* London: Pinter.
Edquist, C. (2005). Systems of innovation. Perspectives and challenges. In J. Fagerberg, D. C. Mowery, & R. R. Nelson (Eds.), *The Oxford handbook of innovation* (pp. 181–208). Oxford: Oxford University Press.
Ellström, P.-E. (2001). Integrating learning and work: Conceptual issues and critical conditions. *Human Resource Development Quarterly, 12*(4), 421–435.
Ellström, P.-E. (2006a). The meaning and role of reflection in informal learning at work. In D. Boud, P. Cressey, & P. Docherty (Eds.), *Productive reflection. An anthology on reflection and learning at work* (pp. 43–53). London: Routledge.
Ellström, P.-E. (2006b). Two logics of learning. In E. Antonacopoulou, P. Jarvis, V. Andersen, B. Elkjær, & S. Høyrup (Eds.), *Learning, working and living. Mapping the terrain of working life learning* (pp. 33–49). London: Palgrave Macmillan.
Ellström, P.-E. (2010). Practice-based innovation: A learning perspective. *Journal of Workplace Learning, 22*(1/2), 27–40.
Ellström, P.-E. (2011). Informal learning at work: Conditions, processes and logics. In M. Malloch, L. Cairns, K. Evans, & B. N. O'Connor (Eds.), *The SAGE handbook of workplace learning*. London: SAGE Publications.
Ellström, P.-E., & Kock, H. (2009). Competence development in the workplace, concepts, strategies, and effects. In K. Illeris (Ed.), *International perspectives on competence development* (pp. 34–54). London: Routledge.
Elmholdt, C., & Brinkmann, S. (2006). Discursive practices at work – constituting the reflective learner. In D. Boud, P. Cressey, & P. Docherty (Eds.), *Productive reflection. An anthology on reflection and learning at work* (pp. 170–180). London: Routledge.
Engeström, Y. (1988). *Learning by expanding*. Helsinki: University of Helsinki.
Eraut, M. (2004). Transfer of knowledge between education and workplace settings. In H. Rainbird, A. Fuller, & A. Munro (Eds.), *Workplace learning in context* (pp. 201–221). London: Routledge.
Feist, G. J. (1999). The influence of personality on artistic and scientific creativity. In R. J. Sternberg (Ed.), *Handbook of creativity* (pp. 273–296). Cambridge: Cambridge University Press.
Fenwick, T. J. (2003). Professional growth plans: Possibilities and limitations of an organization-wide employee development strategy. *Human Resource Development Quarterly, 14*(1), 59–77.
Festinger, L. (1957). *A theory of cognitive dissonance*. Evanston, IL: Row and Peterson.
Fitzgerald, L., & Dopson, S. (2005). Knowledge, credible evidence, and utilization. In S. Dopson & L. Fitzgerald (Eds.), *Knowledge to Action? – evidence-based health care in context* (pp. 132–154). Oxford: Oxford University Press.
Flavell, J. H. (1979). Metacognition and cognitive monitoring: A new area of cognitive-developmental inquiry. *American Psychologist, 34*(10), 906–911.
Frese, M., & Zapf, D. (1994). Action as the core of work psychology. A German approach. In H. C. Triandis, M. D. Dunnette, & L. M. Hough (Eds.), *Handbook of industrial and organizational psychology* (pp. 271–340). Palo Alto, CA: Consulting Psychologists Press.
Garvin, D. A. (2000). *Learning in action*. Boston, MA: Harvard Business School Press.
Gersick, C. J. G., & Hackman, J. R. (1990). Habitual routines in task-performing groups. *Organizational Behavior and Human Decision Processes, 47*, 65–97.
Giddens, A. (1984). *The constitution of society*. Berkeley, CA: University of California Press.
Gigerenzer, G., Todd, P. M., & ABC Research Group. (1999). *Simple heuristics that make us smart*. New York: Oxford University Press.
Gladwell, M. (2005). *Blink: The power of thinking without thinking*. New York: Little Brown.
Hannan, A., Enright, H., & Ballard, P. (2000). Using research: the results of a pilot study comparing teachers, general practitioners and surgeons. http://www.leeds.ac.uk/educol/documents/000000851.htm. Accessed March 2010.

Hansen, M., Nohria, N., & Tierney, T. (1999). What's your strategy for managing knowledge? *Harvard Business Review, 77*(2), 106.

Hemsley-Brown, J., & Sharp, C. (2003). The use of research to improve professional practice: A systematic review of the literature. *Oxford Review of Education, 29*, 449–470.

Hislop, D. (2005). *Knowledge management in organizations*. Oxford: Oxford University Press.

Hommen, L. (2000). *Innovation and qualification: Institutional challenges*. Working paper, Tema T. Linköping: Linköping University.

Hunter, L., Beaumont, P., & Lee, M. (2002). Knowledge management practice in Scottish law firms. *Human Resource Management, 12*(2), 4–21.

Hörberg, C. (1996). *Att våga bryta sina ramar. Om kunskap i arbetet. Lärares kunskaps- och forskningsanvändning i skolan*. Linköping: Linköping University. (In Swedish.)

Höyrup, S., & Elkjaer, B. (2006). Reflection: Taking it beyond the individual. In D. Boud, P. Cressey, & P. Docherty (Eds.), *Productive reflection at work* (pp. 29–42). Abingdon, UK: Routledge.

Lam, A. (2005). Organizational innovation. In J. Fagerberg, D. C. Mowery, & R. R. Nelson (Eds.), *The Oxford handbook of innovation* (pp. 115–147). Oxford: Oxford University Press.

Langer, E. J. (1989). *Mindfulness*. Reading, MA: Addison-Wesley.

Lehrer, J. (2009). *How we decide*. Boston, MA: Mariner Books.

Lorenz, E., & Lundvall, B.-Å. (Eds.). (2006). *How Europe's economies learn. Coordinating competing models*. Oxford: Oxford University Press.

Loughran, J. (1996). *Developing reflective practice: Learning about teaching and learning through modelling*. London: Palmer Press.

Lundvall, B.-Å. (Ed.). (1992). *National systems of innovation: Towards a theory of innovation and interactive learning*. London: Pinter Publishers.

Lundvall, B.-Å., & Nielsen, P. (1999). Competition and transformation in the learning economy illustrated by the Danish case. *Revue d'Économie Industrielle, 88*(2), 67–89.

March, J. G. (1991). Exploration and exploitation in organizational learning. *Organization Science, 2*, 71–87.

McGrath, R. G. (2001). Exploratory learning, innovative capacity, and managerial oversight. *Academy of Management Journal, 44*(1), 118–131.

Mezirow, J. (1991). *Transformative dimensions of adult learning*. San Francisco, CA: Jossey-Bass.

Miner, A. S., Bassoff, P., & Moorman, C. (2001). Organizational improvisation and learning: A field study. *Administrative Science Quarterly, 46*(2), 304–337.

Moon, J. A. (1999). *Reflection in learning and professional development: Theory and practice*. London: Kogan Page.

Nevin, J. S. (2006). *Psychology – concepts and applications*. Boston, MA: Houghton Mifflin.

Nilsen, P. (2010). Praktiknära innovationer genom integration av praktik- och forskningsbaserad kunskap. In *Implementation – teori och tillämpning i hälso- och sjukvård* (pp. 125–144). Lund: Studentlitteratur. (In Swedish.)

Nonaka, I., Toyama, R., & Byosière, P. (2001). A theory of organizational knowledge creation: Understanding the dynamic process of creating knowledge. In M. Dierkes, A. Berthoin Antal, J. Child, & I. Nonaka (Eds.), *Handbook of organizational learning and knowledge* (pp. 491–517). Oxford: Oxford University Press.

Nutley, S. M., Walter, I., & Davies, H. T. O. (2007). *Using evidence: How research can inform public services*. Bristol, UK: The Policy Press.

Ohlsson, J. (2002). Forskaren som kritisk reflektör. In *Interaktiv forskning – för utveckling av teori och praktik* (pp. 89–104). Stockholm: Arbetslivsinstitutet. (In Swedish.)

Olsen, S. E., & Rasmussen, J. (1989). The reflective expert and the prenovice. In L. Bainbridge & S. A. Ruiz Quintanilla (Eds.), *Developing skills with information technology* (pp. 9–33). Chichester, UK: John Wiley and Sons.

Prochaska, J. O., & DiClemente, C. C. (1983). Stages and processes of self-change of smoking: Toward an integrative model of change. *Journal of Consulting and Clinical Psychology, 51*, 390–395.

Redmond, B. (2006). *Reflection in action.* Aldershot, UK: Ashgate Publishing.

Rosell, M. (2002). Ett möte mellan forskning och praktik i socialtjänstens barnavårdsarbete. In *Interaktiv forskning – för utveckling av teori och praktik* (pp. 105–118). Stockholm: Arbetslivsinstitutet. (In Swedish.)

Rosén, A. (1994). Knowledge use in direct practice. *The Social Service Review, 68*, 561–577.

Rushmer, R., & Davies, H. T. O. (2004). Unlearning in health care. *Quality & Safety in Health Care, 13*(Suppl II), ii10–ii15.

Schein, E. H. (1985). *Organizational culture and leadership.* San Francisco, CA: Jossey-Bass.

Schenkel, A. (2006). Disciplined reflection or communities of practice? In D. Boud, P. Cressey, & P. Docherty (Eds.), *Productive reflection at work* (pp. 69–79). Abingdon, UK: Routledge.

Schön, D. A. (1983). *The reflective practitioner – how professionals think in action.* New York: Basic Books.

Seibert, K. W., & Daudelin, M. W. (1999). *The role of reflection in managerial learning.* Westport, CT: Quorum.

Shani, A. B. R., & Docherty, P. (2003). *Learning by design: Building sustainable organizations.* Oxford: Blackwell Publishers.

Strampel, K., & Oliver, R. (2007). Using technology to foster reflection in higher education. In *ICT: providing choices for learners and learning. Proceedings Ascilite Singapore 2007.* http://www.ascilite.org.au/conferences/singapore07/procs/strampel.pdf. Accessed 18 August 2010.

Svensson, L. (2002). En analys och blick framåt. In *Interaktiv forskning – för utveckling av teori och praktik* (pp. 173–204). Stockholm: Arbetslivsinstitutet. (In Swedish.)

Svensson, L., Ellström, P.-E., & Åberg, C. (2004). Integrating formal and informal learning at work. *Journal of Workplace Learning, 16*(8), 479–491.

Trinder, L. (2006). A critical appraisal of evidence-based practice. In L. Trinder & S. Reynolds (Eds.), *Evidence-based practice – a critical appraisal* (pp. 212–241). Oxford: Blackwell Publishing.

Verplanken, B., & Wood, W. (2006). Interventions to break and create consumer habits. *Journal of Public Policy and Marketing, 25*, 90–103.

Weick, K. E., & Quinn, R. E. (1999). Organizational development. *Annual Review of Psychology, 50*, 361–386.

Wertsch, J. V. (1998). *Mind as action.* Oxford: Oxford University Press.

Chapter 10
The Role of Reflection, Reflection on Roles: Practice-Based Innovation Through Theatre-Based Learning

Anne Pässilä, Tuija Oikarinen, and Russ Vince

Abstract A key issue for practice-based innovation is: how can organisations generate innovation in the midst of action? In order to answer this question, this chapter discusses the relationship between learning, reflection, and practice-based innovation. Reflection is seen as an important organisational process that can create spaces for generative learning. The authors demonstrate how theatre-based learning can offer an effective strategy for the creation of reflective spaces that reveal the dynamics of innovation, both in terms of what promotes and what prevents innovative behaviour and practice. Through research and intervention in three organisations, the authors show that viewing roles and relations 'acted out' in theatre helps to reduce the unconscious acting out of entrenched emotional and political dynamics in practice. The struggle to create innovation in the midst of action can be seen in the reflexive tension between the radical possibility of such interventions and the political purpose they may serve for established power relations. There will always be a tension in organisations between dynamics that support innovation and dynamics that undermine it.

10.1 Introduction

A key issue for practice-based innovation is represented in the question: how can organisations generate innovation in the midst of action? Our aim in this chapter is to address this question from the perspective of learning – both individuals' learning in the organisation and organisational learning. In order to do this we utilise the literature on reflection and reflexivity, with a particular emphasis on

---

A. Pässilä (✉) • T. Oikarinen
Lahti School of Innovation, Lappeenranta University of Technology, Lahti, Finland
e-mail: anne.passila@lut.fi

R. Vince
School of Management, University of Bath, Bath, UK

H. Melkas and V. Harmaakorpi (eds.),
Practice-Based Innovation: Insights, Applications and Policy Implications,
DOI 10.1007/978-3-642-21723-4_10,

moving beyond the individual reflective practitioner to consider reflection as an organising process; one that supports learning and innovation at the group and organisational level. We are making a contribution to an understanding of the dynamics of practice-based innovation in organisations, and we assert the importance of critical reflection as an organisational practice that can create spaces for generative learning. We argue that such spaces are necessary to encourage innovative knowledge and innovative work as a continuous aspect of everyday practice in organisations.

Understanding the dynamics that support innovation involves an understanding of the dynamics that get in the way of innovation. Innovation is unlikely to occur without committed, mutual endeavour; and without a strong desire to transform existing knowledge. However, existing knowledge may be seen as essential to the effective functioning of an organisation, as a key value, or as an integral part of 'the way we do things here'. Attempts to transform existing knowledge are not easy because they get caught up in (for example): the messy everyday politics of organisational life; the complex feelings and emotions that are generated through interaction with others; established habits and attachments to routine ways of working that are resistant to change; individual and collective blame, apathy, cynicism and the silos that such behaviour creates; the tensions within hierarchies; and the contradictory role that managers and leaders have, which asks us to be both the champions of change and the guardians of the status quo. (Vince and Reynolds 2009)

Theatre-based learning offers an effective strategy for the creation of reflective spaces that can reveal the dynamics of innovation, both in terms of what promotes and what prevents innovative behaviour and practice. This approach encourages reflection on the difference between knowledge that is experienced from within a role (as a worker, manager, leader) and knowledge that is generated from collective, critical reflection on roles. In other words, seeing roles and relations 'acted out' in theatre helps to reduce the unconscious acting out of entrenched emotional and political dynamics in practice. This approach helps to create a space in the mind that underpins engagement with collective spaces of learning and innovation. Of course, theatre-based intervention is also conducted within the political realities of organisational life, and as a consequence may well serve to reinforce resistance to the transformation of knowledge. The struggle to create innovation in the midst of action can be seen in the reflexive tension between the radical possibility of such interventions and the political purpose they may serve for established power relations (Coopey 2007).

10.2 Reflection, the Organisation of Reflection and Reflexivity

Imagine that the following dialogue happens in an organisation where a manager, an employee, a researcher and an artist are at the beginning of a theatre-based learning process of reflection and learning.

Manager Smith: We are interested in organising reflection, of course. We need to improve our practices.
Artist Brown: Great! Understanding is such an important element of personal and professional growth.
Manager Smith: Yes but we don't want to spend too much time or employee resources on it.
Employee Davies: Yes, yes I hope that reflection generates change, and we'll get ready and usable solutions from it.
Researcher Jones: The nature of reflection is participatory interaction and you have to create time and space for conversations.
Manager Smith: Yes, yes but we have the problem of how to deal with reclaims. Could you make employees understand how important it is to co-operate together?
Artist Brown: Make them understand?
Employee Davies: As you can see managers assume that we leave our brains at home when we come to work.
Manager Smith: Sorry I didn't mean that, I was just pointing out that we have new challenges in our business which can't be solved in old ways. So the Executive Group of our company has pointed out a new direction, and I have to implement it.
Researcher Jones: What if you reflect on this new situation together? The idea of reflection is that reflectors formulate their own questions and through questioning share problems and emotions related to issues.
Manager Smith: Emotions?
Artist Brown: Emotions!
Employee Davies: Emotions... I don't know about it. We are doing things in a rational way.
Manager Smith: Exactly, I am sure that emotions are most important but I think that our employees have possibilities to share their emotions if they want to. Some employees want to express their emotions somewhere other than in workplace.
Employee Davies: Like in the pub. (laughs)
Artist Brown: But organisations as a community are full of emotion.
Manager Smith: We do business here. I have a financial responsibility!
Employee Davies: And the rest of us have a practical and operational responsibility.
Researcher Jones: Yes, but organising is a potential learning process consisting of various elements. Emotions are one element in reflection and everyday work practices. Organising reflection involves shared interpretation. It involves making sense of things together, solving problems together.
Manager Smith: But do we need a shared interpretation? Isn't there a big risk if you can't facilitate the right solution?
Employee Davies: Yes indeed, we want concrete action rather than speculations.
Artist Brown: But the uncertainty and incompleteness is the driving force in finding the multiple solutions and possibilities.
Researcher Jones: There is always a chance that things will go wrong.
Artist Brown: Reflection is rather an empowering process to make sense of a situation in different ways, not a tool for a problem solving.
Manager Smith: Why on earth should we take that risk then?
Researcher Jones: Because you need a different kind of approach to discover your ways of working that are not productive; and practices that are taken-for-granted as well as the assumptions behind them.
Employee Davies: It must be confusing!
Artist Brown: Sharing experiences and feelings might be that too. But it is also a good way for you to learn about what is really happening in your organisation.

In this dialogue Manager Smith takes a rational and managerial approach to reflection. Manager Smith wants to benefit quickly and effectively from reflection.

From Manager Smith's point of view, gaining knowledge through reflection is like any other production factor, a commodity that can be possessed, stored, accumulated, converted, or exchanged (Van de Ven and Johnson 2006). Employee Davies points out the need to change actions (Ellström 2010) and he sees the practical impacts of reflection. However, he thinks that confusion is the result of bad practice. Artist Brown has an interpretative and artful approach (Mienczakowski 1995; Meisiek and Hatch 2008). Artist Brown is inspired by the notion that reflection as part of learning is not only about possessing knowledge, but also about knowing as action. Researcher Jones's approach is informed by practice-based innovation (Harmaakorpi and Melkas 2005) in situated learning (Brown and Duguid 2001; Lave and Wenger 1991). Researcher Jones considers the organisation of reflection (Vince 2002) an integral part of knowledge, learning, and innovation, and that these develop simultaneously within a course of action (Gherardi 2006).

One problem that organisations face is that 'quick fix', ready-made 'solutions' tend to deny tensions inherent in the complexity of organisations and in turn, deny access to the emotions and politics that both promote and limit learning in organisations (Dehler 1998). These characters: Manager Smith, Employee Davies, Artist Brown and Researcher Jones made a commitment to discover and work on the dynamics that support and block innovation in the organisation. This raises an important question – what is it that prevents members of this organisation sharing their ideas, communicating and cooperating in the midst of action?

In order to make sense of dynamics that get in the way of innovation, organisational members may have to articulate their past experiences and their desired future action. Within an organisation, various work groups can reflect on their way of doing things, and try to make hidden assumptions visible. It may be important to reflect on what seems to work and what doesn't; being aware of associated feelings; of judgments made and on what basis; and of the ideas, values and assumptions which influence the interpretations made of the experience (Vince 1998). However, the messy everyday politics of organisational life complicates knowledge sharing. The possibilities of learning in and from experience require reflection-based experiential learning process linked to everyday practices. In this study, we explain how a reflection-based learning process was constructed through applied theatre with employees, managers, customers, theatre artists, and researchers within three organisations. In this context theatre is understood as a way to articulate experiences, and as a learning process both of drawing on and developing new ideas.

10.2.1 Practice-Based Innovation

From an organisation's point of view, practice-based innovation refers to employees' and managers' reflection on and renewal of their own operations. This may involve the development of new working methods, routines, products or services (Harmaakorpi and Mutanen 2008; Harmaakorpi and Melkas 2005).

Renewal is based on learning in and through work processes within the operations concerned (Ellström 2010). It focuses on workplace learning as a fundamental mechanism behind practice-based innovation and highlights work processes as well as wide participation in learning. The driving force behind these practices is the interplay between the explicit and implicit dimensions of work (Ellström 2010). It is this tension that creates the potential for practice-based innovation.

A key challenge for practice-based innovation is how to capture knowledge embedded in ongoing practices (Lundvall 1992; Jensen et al. 2007). Three critical social dynamics of creating and sharing practice-based knowledge for innovation have been proposed: boundaries, embeddedness of practice and collective sense-making (Dougherty 2004). To change taken-for-granted knowledge in organisations, interaction between diverse views and experiences is needed. Knowledge embedded in practice during an innovation process does not exist without social action. Innovations emerge through the interactions between the practices of heterogeneous groups in the social contexts in which they are located (Newell et al. 2002). Practice-based innovation in the midst of action is dependent on two fundamental social, inter- and intrapersonal processes – analysis and interpretation. Theatre-based learning is especially suited to social dimension of interpretative innovation process (where learning takes place among and through other people according to Gherardi et al. 1998) and to foster organisational and process innovations. Because of its reflective nature it allows learners to negotiate the meaning of actions and situations in a midst of innovation process. Through it, people gain common knowing what promotes and what prevents innovative behaviour and practice.

An important issue for organisations (from the perspective of practice-based innovation) is to find out what an interactive and interpretative approach to innovation actually means and involves in a specific context. This raises the question: what prevents and enhances interaction and interpretation in organisations' own socio-cultural systems? Knowing in the midst of innovation can be a blind spot for managers as well as employees, and organisations may need to "look at the world simultaneously through both analytical and interpretative lenses and flip back and forth between them as conditions required" (Lester and Piore 2004: 74).

10.2.2 Reflection on Practice-Based Innovation

Now we could imagine again; Researcher Jones argues that learning through attempts at innovation is more a process of reflection than a project related to technical problem solving in a company's everyday working life. Researcher Jones leads us into an ongoing and open-ended reflective process (Schön 1983; Vince and Reynolds 2009), the thrust of which is to discover new meanings (Lester and Piore 2004: 97). Manager Smith agrees with this, but seeks clarity and also struggles with letting go of control, as well as with accepting ambiguity and confusion. In effect, learning challenges, it undermines and it destabilises, it creates anxiety as well as the potential for change. It creates resistance to change and the desire for change at the same time. Manager Smith reminds us that reflection is a symptom of something deeper. Manager Smith, Employee Davies and the rest of the members of the organisation have to collectively reflect (Reynolds and Vince 2004b) on what is happening in their own organisational life. They have to ask – what is the nature of their own organisation and their own innovation process, how are relationships formed, what are the familiar or taken-for-granted assumptions and what other assumptions are present?

Within an organisation, experience is embedded in practice, and members share it implicitly, and that kind of implicit or tacit knowing cannot be transferred without reflection on practice (Vince 2002; Dougherty 2004). The point here is that what people already 'know' can get in the way of innovation unless they also seek to call into question the value of existing knowledge to changing circumstances. Reflection on underlying assumptions invites analysis of our own organisational world and therefore our own actions, practice, and experience. Reflection is one of the key building blocks of learning (Vince and Reynolds 2009) and it allows organisations to surface the assumptions that inform day to day action (and inaction). The concept of reflection has been elaborated as an activity undertaken by the individual (reflective practitioner) in the organisation. Recently, there has been a shift in perspective which emphasises the critical and organisational dimensions of reflection in both educational and organisational contexts. Four perspectives have helped to improve our understanding of reflection as an organisational process, beyond the notion of the 'reflective practitioner'. These are: critical reflection; public reflection; productive reflection; and organising reflection (see Table 10.1 below).

In the context of practice-based innovation, a 'critical' perspective is the starting point for understanding the shift from individual to collective reflection. Critical reflection involves a dynamic and dialectical movement between 'doing' and 'reflecting on doing' (Freire 1998: 43). In the next section of the chapter we describe how theatre-based learning can help us to interrogate this movement between action and reflection, as well as offering a way to think about current practices within organisations in order to transform them.

10.2.3 Organising Reflection via Theatre-Based Learning

Theatre instructor Augusto Boal applied Freire's (1974) educational ideas to communities with the help of theatre. His approach is centred upon making experiences, attitudes, feelings, and relationships visible by observing institutionalised habits and practices in a new way (Boal 1992, 1995, 1996). Theatre-based learning offers one way of organising reflection: "to raise questions about purpose and intent and about the assumptions and taken-for-grantedness on which organisational policies and practices are based" (Reynolds and Vince 2004b: 4). Learners can explore everyday matters and, through performance as well as narratives, learn from themselves (Heikkinen 2002: 129–133). We understand Boalian theatre as a way to articulate experience. In this context, narratives or performances are a way of

Table 10.1 Perspectives of reflection (developed from Vince and Reynolds 2009)

	Critical reflection	Public reflection	Productive reflection	Organising reflection
Characteristics	To identify and question taken-for-granted beliefs and values	To become collectively aware of and transform own behaviour	To develop work and learning activities that change work practices and personal engagement	To take account of emotional and political processes in the workplace
Focus	To question and challenge existing structures and practices	To improve policy and practice	To improve productivity and quality of working life	To generate collective knowledge
Aim	To unsettle established ways of working	To create alternative interpretations	To improve competence linked to productivity and work satisfaction	To organise structure which allows reflection
Key-element	Emancipatory	Transformational	Agency building	Organising process
	Reynolds 1998; Alvesson 1995; Alvesson and Willmott 1992; Parker 1995, 2002; Cunliffe 2009	Raelin 2001; Vince 2004; Cunliffe 2009; Reynolds and Vince 2004a	Boud et al. 2006	Vince 2002; Reynolds and Vince 2004b; Nicolini et al. 2004

framing everyday experience and digging into it. Entering the imaginative world while playing a role, as well as reflecting on the real from the point of view of drama, offers the possibility of learning (Heikkinen 2002: 139–141). Narrative and theatrical approaches revolve around the artful reframing of intransigent problems. It allows the learners to observe, analyse, define, and find out from their own experiences. In artful framing, the members of an organisation as learners express and address their own questions, perceptions, and interpretations by doing and by acting (Meisiek and Hatch 2008: 415), which leads them to problem shifting (Barry 2008: 36).

Freire (1970) has designed a specific generative approach which includes generative themes and generative questioning (open ended questions) to underpin learning. Generative questioning in particular, when used in Boalian (1992; 1995; 1996) applied theatre practices, provides a learning space for reflection. It allows learners to investigate their assumptions and for people to gradually delve deeper into their own assumptions with the help of aesthetic distance. Theatricality allows people to locate themselves within and to see the world around them as a reality in a process of ongoing transformation. (Freire 1970: 12)

Now you could imagine an interactive theatre performance where one theatrical scene is centred on one theme and one codified situation illustrates how power relations are woven into an organisation. Employee Davies and Manager Smith are in the middle of a process of dialogue with Artist Brown and her theatre company. Researcher Jones explains that the development of generative themes takes place in a codified situation on stage. Members of the organisation are encouraged into critical thinking by acting, as well as reflecting by acting. Employee Davies and Manager Smith can stop the performance and suggest changes to the events on stage. They ask 'why?', construct various scenarios together; and reveal multiple meanings. The guiding principle is that they all generate the theme through praxis and interaction with the help of a codified situation.

One vivid aspect of reflection in this theatre-based approach is 'assumption breaking': when organisational events are presented with a help of codified existential situation, in which members of organisations are liable to share and interpret it critically with others. Boal's (1996: 49) metaphor 'theatre is telescopic' emphasises the possibility of assumption breaking through clarity of focus and attention. The 'telescope' helps observation, it brings close that which can seem to be hidden far away. It helps organisational members to contextualise an event and specific places in order to apply them to the learning process (Boal 1995). Describing the situation helps people to make sense of the parts in relation to the whole, as well as any important relationships and collective dynamics (Freire 1970). It is important to remember that what is seen on the stage and through the stage is always an interpretation that is embedded in the performance and realised through it (Clark 2008: 403). Theatre itself does not engender social change but it can allow members of organisations to confront hidden conflicts, behavioural patterns or critical routines (Schreyögg, quoted in Clark 2008: 405) in order to support attempts at change.

Assumption-breaking is one key aspect of practice-based innovation. We suggest that a 'Reflection Zone' (R-zone) is created in a context of artful framing, and assumption breaking occurs between imitation (as is) and imagination (as if) of everyday action. (See Fig. 10.1.)

Reflection is understood as an investigation and evaluation process in which learners can search for new perspectives from their own experiences through their

Fig. 10.1 The 'reflection zone' of imagination and reality

attempts at challenging existing assumptions. Reflection broadens one's ability to see and understand one's own actions, as well as reactions. It is about learning in, with and through theatrical and imagined being (Heikkinen 2002: 104; Errington 1992: 50). Critical reflection as an element of theatre-based learning encourages organisational members to be able to sense hidden ideas and ideals and to generate knowledge together. In the context of innovation, different work units and professional groups within an organisation can create mutual awareness, show how they feel, what they are afraid of and what their beliefs are. Assumption breaking as a part of a reflective practice includes a commitment to ask questions that in turn may be neither comfortable nor welcome (Reynolds and Vince 2004b: 4). In this way, members of organisations are able to create collective knowledge as a base for innovations in practice. We illustrate this process with an example from research.

10.3 Reflection and Actionable Knowledge: Case Studies

The aim of the study we are presenting is to elaborate how organisations can generate innovation in the midst of action in a context of practiced-based innovation; to create a reflective process in which the people seek to understand how they are in relation to and interact with others in a specific organisational context. Our research orientation emphasises social interaction between people, as well as changing practices. We create forums in which people are able to work as co-participants and develop practices related to their everyday working life during research. Methodologically, the study follows a specific artistic orientation of action research, namely, research-based theatre (Pässilä and Oikarinen forthcoming; Mienczakowski 1995; Mienczakowski et al. 1996; Mienczakowski and Morgan 2001). In our case, theatre is applied to participatory action research and we modified it for the micro-level practice-based innovation activities of organisations (Carr and Kemmis 1986; Kemmis and McTaggart 1988; Kemmis and Wilkinson 1998; Kemmis and McTaggart 2000; Kemmis 2001; Matsson and Kemmis 2007). The method of research-based theatre in organising forums for reflection proceeds through four phases as illustrated in Fig. 10.2.

Firstly each community's or work group's own views and practices are articulated and reflected. The articulation is facilitated by storytelling methods and theatrical pictures. Then the stories are analysed and devised as performative theatre scenes. The scenes are played back to the narrators as a theatrical performance in an organisational theatre session. It is a participatory session where all participants work in mixed groups. The focus is to construct common understanding by using a practical problem as a boundary object. Later, there will be one or more action planning sessions focusing on agreement on development needs. Several techniques are used: image theatre, forum theatre and the interpretation of codified event, interpretation with the help of the distancing character,

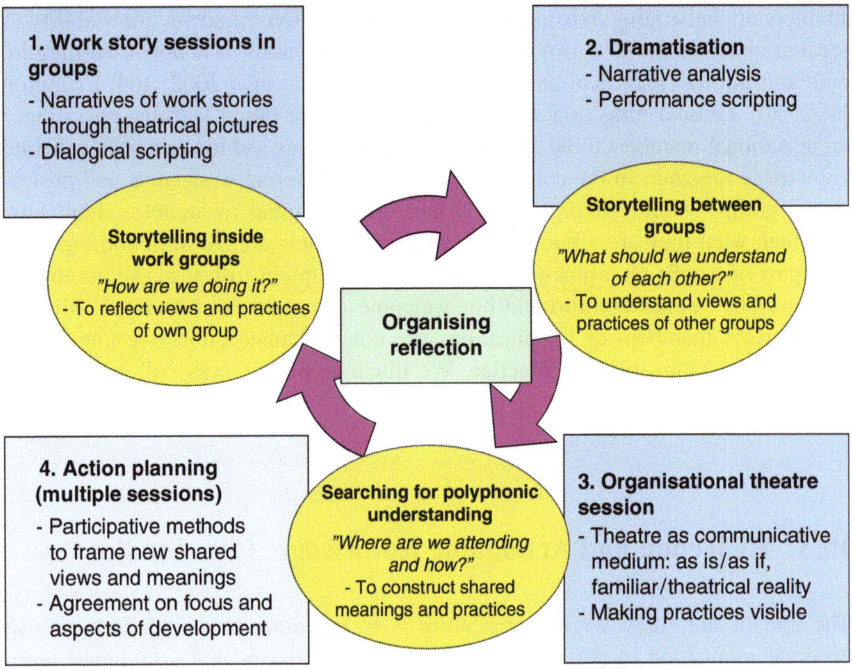

Fig. 10.2 Organising reflection through theatre-based learning

interpretations of stories, collective scripting, visual arts and creating a joint collage, and flagging the turning points in a codified event.

10.3.1 The Research Design and Research Process

The empirical data in this chapter comes from action research projects in three different organisations. All of the organisations are located in the Päijät-Häme region in Finland and they all operate in fields which have faced major structural changes during this decade. All the projects began with a development need identified in the organisation. However, in the course of planning together with the managers and employees of the organisations and researchers the aim of development was widened to issues of innovation and learning. The organisations involved in the research were one unit of a multinational company from the field of industrial engineering ('the Factory'), a public sector health care unit ('the Care Unit'), and a public health centre from a public sector enterprise ('the Public Corporation'). In each organisation there were multiple participating work groups, and in the case of the Care Unit, customers were also involved. Table 10.2 (below) illustrates the research process in each organisation.

Table 10.2 Case organisations

Cases	Case Factory	Case Care Unit	Case Public Corp.
Field	Multinational forest industry	Public sector health care unit	Public health centre
Boundary object	Customer reclamations	Teenagers' no-show to dental care	Emergency duty re-organisation
Participating work units	Operators Sales managers Sales assistants Designers	Dentists Nurses Assistants Customers	Nurses Doctors Collaborators
Participants	70 persons	36 persons	25 persons
Sessions	Storytelling (6 sessions) Theatre session Action planning session	Storytelling (7 sessions) Theatre session Action planning session	Storytelling (2 sessions) Theatre session Action planning session
Steering group	Managers	All occupations represented	Managers
Empirical data	4 hr videotape + researchers' notes	5 hr videotape + researchers' notes	3 hr videotape + researchers' notes
Time frame	2008–2010	2009–2010	2009–2010

10.3.2 Data Collection and Interactions

The data collection was focused on reflective practices in theatre-based learning interactions in the three case organisations. In every organisation, we proceeded through four phases (see Fig. 10.2): (1) work story sessions in various work units, as well as customer story sessions or interviews; (2) dramatisation and analysis of the stories; (3) an organisational theatre session between various work units; and (4) action planning session between various work units. All interactions were, in keeping with action research practices, recorded or videotaped. The researchers took part in every phase and made research notes of them. The artist took part in some work story sessions and in the dramatisation and organisational theatre session.

The storytelling sessions were organised into occupational groups so that each group could first reflect their own views and practices with like-minded others. The theme of the work stories (i.e. boundary object) in each organisation was such that it was of interest to the whole organisation and to each work unit involved, but one on which they had differing views. For example, in the factory, sales assistants, sales managers, operators, and designers had different problems so they also needed different solutions but they needed to find ways to articulate their view points to each other. They would then be able to discuss and generate ideas from different perspectives and different needs, and synchronise their solutions.

After the storytelling phase, the researcher and the theatre director analysed the experiences, tensions, and relationships between various work units and professional groups. Common themes were surfaced using drama analysis, and in the later stages these themes were transformed into codified events. We categorised the

Table 10.3 Levels of assumption

Level of systems	Level of practices	Level of interactions	Level of emotions
Assumption about organisational structure	Assumption about everyday working life and operations	Assumption about relationships between professional groups, customers and stakeholders	Assumption about feelings between different professional groups and about own individual feelings
"Managers should organise..."	"We do not have enough information/ There is so much information"	"We do not know anything about our customers "	"There is always such a hurry, I feel guilty when I . . ."
Power relations			
Power is hidden; nobody seemed to have the feeling that I or we have the power to change circumstances.			

events of stories into four levels. These levels are like different layers around assumptions. We discovered events which reflected the organisation through established organisational practices, as well as common interactions and emotions. These levels were related to power relations in the organisation. From these levels, actors and the researcher created codified events that could be performed on stage. Table 10.3 illustrates the different levels of assumption.

Through their analysis of the themes, the researcher and the artist started to script theatrical scenes (see Fig. 10.2, phase 2 – Dramatisation). The manuscript was initially a rough draft of the themes, to which the actors began creating scenes (codified events) with the help of theatre techniques. At this stage the researcher, actors and the theatre director moved functionally within the stories. This could be called a performative analysis of the material. In this context, the narrative form (text: original stories and themes) was turned into acting (characters, gestures, and images). Generative themes were codified into performative action. As the dramatisation emerged, the researcher generated reflective questions from the performative action. The rehearsals of the theatre performance began when a codified event had been formed from all the themes and when the language of the performance (reflecting the language of the organisation) had been discovered. In the 'Case Factory' two actors and one researcher worked together for six months. They had 20 rehearsals and meetings. In the 'Care Unit', two researchers and an actor worked with four students, and it took four months. They had 15 rehearsals. In the 'Public Corporation' five actors and a researcher worked for two months and they had 10 rehearsals and meetings.

In the organisational theatre session (see Fig. 10.2, phase 3) employees and managers worked in mixed teams and interpreted theatrical scenes. The discussion was guided by the researchers and the artist working as facilitators. A theatrical scene illustrated the situations that various work units had told in their storytelling sessions. In the theatre session, members of mixed work groups explained to each other why the characters in the scene were acting in such a way. "That man does not understand the instructions; the other man is explaining things from his own perspective, which is not relevant for the other man." "Maybe he does not even understand what the other man is talking about?" This was how participants interpreted their own actions. Simultaneously, through interpretation they started to suggest how to do things differently. "This man could say to that other man that he doesn't understand his language or his point of view, and maybe then they might find out how they should communicate?" (The data is from the Case Factory.) They were able to see their own part in the organisations they had created ("That is how we act and react") and through reflection they were able to express their feelings, joys, and sorrows. Some issues were more pleasant than others. Hidden assumptions emerged gradually in the artfully framed events when the participants articulated what happened on stage and why. It was a sensitive and physical event in which the participants organised the beliefs behind their own actions through the images and events on stage. The results gained from the theatre session were made into a plan of action for distribution and construction of information between different work units in the action planning sessions (see Fig. 10.2, phase 4). The aim of these sessions was not reflective and they are out of the scope of this chapter.

10.4 An Example from the Case 'Care Unit'

In this example we describe how the assumption breaking process emerged and how it enabled members of the organisation to generate ideas. Because of the rich nature of our data from three organisations we selected only one analysed interaction from the Care Unit organisation for deeper observation. This example illustrates how people negotiate the meaning of actions and situation.

During an organisational theatre session in 'Care Unit', members of the organisation performed, examined and deconstructed a codified event which could have happened or had happened in their organisation. The event was about a teenage-customer at the dentist. The problem of the organisation was that many teenagers did not show up for their appointments. With the help of the main facilitator guiding the discussion, the participants started to recognise and define problems behind the event, especially when they jumped onto the stage and took on a role. Making changes in this context means that you have to act. The stage becomes a simulation of working life where it is possible to try different types of action and to make sense of them. For example Tiina, who is a nurse, took a role of a doctor and started to make changes in the events of the scene. Tiina's performance allowed her colleagues to examine and interpret their own assumptions and to form new ideas through generative questioning.

> In a role, Tiina explained to her teenage patient Netta what will happen during an operation. During their conversation Tiina bent towards Netta and comforted her by touching Netta's shoulder. Tiina said to Netta:
> "Dear Netta, there is no problem, this is going to be an easy operation, you don't have to worry at all."

The rest of the employees and managers cheered Tiina:
"Good Tiina, very good."
They examined what happen in the codified event:
"Fear, I sense fear, Netta dreads the dentist."
They helped Tiina by advising:
"Comfort Netta because she is afraid of you."
Someone from audience asks:
"Why is she so afraid, it is such an easy operation?"
The audience tried to make sense of what was behind her fear:
"Say to her that she has taken good care of her teeth. You could also explain the operation you will do."
Then someone pointed out that
"It is a unique situation to Netta even if it is routine to us" and "we should not deny her experience, if she feels that it is a scary operation. Then it is to her. We have to respect her feeling."
It was at this stage that the employees and managers were able to transform their understanding of the problem:
"Sometimes doctors use power, they oppress their patient."
And then someone added:
"And they oppressed nurses too."
Now they were dangerously near a difficult issue about themselves as oppressors. So in this stage it was not surprising that they changed the conversation to a more comfortable level:
"Tiina, maybe you could say to Netta that she can listen to her favourite band through her own iPod. Maybe it would help her to feel more comfortable and relaxed."
At this phase Tiina just acts and reacts, and the rest of the members of the organisation observe. Then another employee, Henrik, comes to stage, and events on stage continue further.

Employees and managers posed questions to each other with the help of the codified event: What actually happened in that scene? What was that story about? What other changes could characters have made? And what would they have done in a similar situation in a real life? During the reflective conversation, employees and managers pointed out that it was a unique situation to Netta but routine to themselves and behind this event is expertise (nurses or doctors) that creates power over their patient. They critically reflected their own behaviour: "We quite often think that we know better and what is best for our patient." They also realised that acting in their professional role may mean that they ignore the fact that their patient is a person. Such realisations take them to the heart of the 'R-zone', identifying and questioning their own taken-for-granted beliefs and values. Critical reflection helped them to challenge both the structure and current practice of health care, and through this to challenge assumptions about their professional identity and power relations. They reflected that they do not encounter Netta as a person, just as a set of teeth to take care of. They noticed that seeing the patient more personally would be beneficial for the operation. Because of the sensitive nature of this issue they started to invent solutions for the problem and to suggest new ideas:

"There could be posters and magazines for young patients."; "And positive feedback. We could give some positive feedback for all patients."; "We could have a feedback system for customers."; "Employees also need a feedback system."; "We could send her a friendly text message."

Of course in this type of a reflective conversation there is also resistance:

> "We never get that kind of a technology. It is such an investment for our organisation."; "We have such busy schedules already now, I don't need more tasks to do."; "Netta doesn't need our comfort, she should take care her own teeth, but you know teenagers, they just ignore us and drink Coca Cola."

In this example, members of the organisation surfaced problems and possibilities, as well as generated new ideas through a codified event. Between 'as is' and 'as if' in the R-zone, the participants distanced themselves from an event in order to makes sense of it. It provided a conscious and thoughtful connection between ideas and experiences, past experience and future action. This type of reflection is an incremental way to surface one's own preliminary or intuitive assumptions, and the consequent emphasis on problem shifting can become an established way to reflect. This approach encourages reflection on the difference between knowledge that is experienced from within an individual role (as a worker, manager or leader) and knowledge that is generated from collective, critical reflection on roles. We see this as a significant shift of perception in relation to the successful implementation of collective and critical reflection within organisations.

10.5 Summary and Conclusion

Theatre-based learning offers an effective strategy for the creation of reflective spaces that can reveal the dynamics of practice-based innovation, both in terms of what promotes and what prevents innovative behaviour and practice. Reflection of one's own and the others' assumptions takes place between the imitation and the imagination (Heikkinen 2002) of real life situations.

In this study, the codified events created critical reflection (the R-zone) where learners study their own action, practice, feeling, and experience. Employees and managers interpreted situated practices by sharing their ideas about how different work units in the company interact with one another. By using theatre and narratives, it was possible to surface delicate and difficult questions within the organisation. Distancing the participants from actual events helped them to move beyond feeling threatened or accused. In organisational theatre, the actors performed codified events back to the original storytellers, who then interpreted the situations and reflected on actions through their performance. Reflection emerged through seeing, hearing and feeling actual behaviour: "Oh, do we really behave like that?" or "We do not act like that!" Such reflection assisted members of a work unit to challenge themselves as to whether their perceived behaviour and practice really was how they did things. Interpretation of performance, with the aid of aesthetic distance, helped those involved to handle issues related to emotions and power relations. Interpretations showed that different managerial skills and different ways of working together were necessary (which resonates with the research findings of Lester and Piore 2004).

The results point towards the need for managers to look at the potential of bottom-up perspectives on innovation. The most important insights arose when people learned to pose new questions to themselves and to each other across work units. The following phrases from the 'Case Factory' illustrate how operators and salespersons tried to find out common learning practices in everyday working life: "You could come to me, because I am stuck with the machine, you could come and ask me what the relevant issues are for me. Come and see how I do it." From a perspective on power this was an invitation to reorganise relationships between work units. Reflection on roles helped to create a space in the mind that could underpin engagement with collective spaces of learning and innovation.

Of course, theatre-based intervention is also conducted within the political realities of organisational life, and as a consequence may well serve to reinforce resistance to the transformation of knowledge. When new practices were implemented into working life practice only one female salesperson changed immediately her pattern, and spent half a day with the operator. Clearly, it is easier to discuss difficult issues, thoughts, intentions, and feelings around the codified events than it is to carry this forward and to implement changes in practice. Common knowledge was built gradually through participants' conscious and unconscious acts, and new ideas were generated through assumption breaking. Previous knowledge (existing cultural assumptions) blocked participants' ability to generate radical suggestions. The potential for innovation here may be in the silences produced; in that which participants did not dare to observe or refused to see. The importance of cultivating a tolerance for ambiguity was also pointed out in each of our cases. After their reflections and theatre-based learning actions, all the organisations involved started to convert some ideas into new services, practices or products.

We have argued that reflection underpins the potential of practice-based innovation because, in the midst of action, people are unable to separate knowledge and knowing from practice. In practice-based innovation, knowledge and knowing arise from (internal and external) interaction in the situated context. Through the role of reflection and through reflection on roles it is possible to share and to address a wide range of tensions, relationships, and misunderstandings. However, it is very difficult to create reflection in the midst of action. Theatre-based reflection helps in two ways. Firstly, it allows participants to model behaviour that can help them to comprehend and implement reflection-in-action. Secondly, it generates spaces for innovation.

Resistance is part of critical consciousness, because reflective learning is difficult even in an atmosphere of openness and trust. Quite often, social norms, roles and hierarchy create blocks to new ways of seeing what is happening in the organisation as well as new ideas that may help to transform it. Organising reflection offers one way to balance learning between analytical and interpretative approaches in a midst of practice-based innovation. Both these approaches are needed to support innovation in organisations, but the interpretative and interactive dimensions are particularly important because they shed light on the emotional and political dynamics that both support and suppress innovation in organisations.

For all the potential effectiveness of this approach it is also important to acknowledge its limitations. Members of an organisation reflected on and reconstructed assumptions within aesthetic settings, not within the workplace itself. Taking everyday dramas and placing them on a stage affords opportunities to examine practice, but it does not necessarily change practice. On a stage it is possible to talk, observe or challenge; in the everyday conflicts of organisational life it is not. The core of critical reflection in the context of practice-based innovation involves the questions: 'why do we do things the way we do?' and 'what is important to us, and why is it important?' Such questions are asked in relation to the actions that occur within a scene and to the diverse points of view that emerge. Such questions encourage the examination of underlying emotional and political issues in the organisation – if there is a political will for such an examination to be made. Organisational members may start with a desire to make change happen but they may also come to feel that change is going to undermine the ways in which 'things have always been done'. This is especially true for organisational members in positions of power who are often (understandably) reluctant to have their assumptions and ways of working challenged and their authority 'undermined'. The struggle to create innovation in the midst of action can be seen in the reflexive tension between the radical possibility of such interventions and the political purpose they may serve for established power relations (Coopey 2007). There will always be a tension in organisations between dynamics that support innovation and dynamics that undermine it.

References

Alvesson, M. (1995). The meaning and meaninglessness of postmodernism: some ironic remarks. *Organization Studies, 16*, 1047–1075.

Alvesson, M., & Willmott, H. (1992). On the idea of emancipation in management and organization studies. *Academy of Management Review, 17*, 432–464.

Barry, D. (2008). The art of... In D. Barry & H. Hansen (Eds.), *The SAGE Handbook of new approaches in management and organization* (pp. 31–41). London, Thousand Oaks, New Delhi, Singapore: Sage.

Boal, A. (1992). *Games for actors and non-actors* (A. Jackson, Trans.). London: Routledge.

Boal, A. (1995). *The rainbow of desire* (A. Jackson, Trans.). London: Routledge.

Boal, A. (1996). Politics, education and change. In J. O'Toole & K. Donelan (Eds.), *Drama, culture and empowerment* (pp. 47–52). Brisbane: IDEA Publications.

Boud, D., Cressey, P., & Docherty, P. (Eds.). (2006). *Productive reflection at work*. London: Routledge.

Brown, J. S., & Duguid, P. (2001). Knowledge and organization: A social-practice perspective. *Organization Science, 12*, 198–213.

Carr, W., & Kemmis, S. (1986). *Becoming critical: Education, knowledge and action research*. Basingstoke: The Falmer Press.

Clark, T. (2008). Performing the organization. In D. Barry & H. Hansen (Eds.), *The SAGE handbook of new approaches in management and organization* (pp. 401–411). London, Thousand Oaks, New Delhi, Singapore: Sage.

Coopey, J. (2007). Theatre in management and organization development: a critique of current trends. In M. Reynolds & R. Vince (Eds.), *The handbook of experiential learning and management education* (pp. 169–186). Oxford: Oxford University Press.

Cunliffe, A. (2009). Reflexivity, learning and reflexive practice. In S. J. Armstrong & C. V. Fukami (Eds.), *The SAGE handbook of management learning, education and development*. London: Sage.

Dehler, G. (1998). 'Relevance' in management research: a critical reappraisal. *Management Learning, 29*, 69–89.

Dougherty, D. (2004). Organizing practices in service: capturing practice-based knowledge for innovation. *Strategic Organization, 2*(1), 35–64.

Ellström, P.-E. (2010). Practice-based innovation: A learning perspective. *Journal of Workplace Learning, 22*(1/2), 27–40.

Errington, E. (1992). *Towards a socially critical drama education*. Victoria: Deakin University Press.

Freire, P. (1974). *Pedagogy of the oppressed*. New York: Seabury Press.

Freire, P. (1998). *Pedagogy of freedom ethics, democracy, and civic courage* (P. Clarke, Trans., Foreword by D. Macedo). Lanham and Oxford: Rowan & Littlefield Publishers.

Gherardi, S. (2006). *Organizational knowledge: The texture of workplace learning*. Malden, MA: Blackwell Publishing.

Gherardi, S., Niccolini, D., & Odella, F. (1998). Towards a social understanding of how people learn in organizations. *Management Learning, 29*(3), 273–297.

Harmaakorpi, V., & Melkas, H. (2005). Knowledge management in regional innovation networks: The case of Lahti, Finland. *European Planning Studies, 13*(5), 641–659.

Harmaakorpi, V., & Mutanen, A. (2008). Knowledge production in networked practice-based innovation processes – interrogative model as a methodological approach. *Interdisciplinary Journal of Information, Knowledge, and Management, 3*, 87–101.

Heikkinen, H. (2002). *Draaman maailmat oppimisalueina: draamakasvatuksen vakava leikillisyys* [Drama Worlds as Learning Areas – the Serious Playfulness of Drama Education]. University of Jyväskylä Studies in Education, Psychology and Social Research. Jyväskylä: University of Jyväskylä. (In Finnish.)

Jensen, M., Johnson, B., Lorenz, E., & Lundvall, B.-Å. (2007). Forms of knowledge and modes of innovation. *Research Policy, 36*, 680–693.

Kemmis, S. (2001). Exploring the relevance of critical theory for action research: Emansipatory action research in the footsteps of Jürgen Habermas. In P. Reason & H. Bradbury (Eds.), *Handbook of action research* (pp. 94–105). London: Routledge.

Kemmis, S., & McTaggart, R. (1988). *The action research planner* (3rd ed.). Geelong, Victoria, Australia: Deakin University Press.

Kemmis, S., & McTaggart, R. (2000). Participatory action research. In N. K. Denzin & Y. S. Lincoln (Eds.), *Handbook of qualitative research* (pp. 567–606). London: Sage Publications.

Kemmis, S., & Wilkinson, M. (1998). Participatory action research and the study of practice. In B. Atweh, S. Kemmis, & P. Weeks (Eds.), *Action research in practice. Partnership for social justice in education* (pp. 21–36). London: Routledge.

Lave, J., & Wenger, E. (1991). *Situated learning – legitimate peripheral participation*. New York: Cambridge University Press.

Lester, R. K., & Piore, M. J. (2004). *Innovation – the missing dimension*. Cambridge, MA: Harvard University Press.

Lundvall, B.-Å. (1992). *National systems of innovation: Towards a theory of innovation and interactive learning*. London: Pinter.

Mattson, M., & Kemmis, S. (2007). Praxis-related research: serving two masters? *Pedagogy, Culture & Society, 15*(2), 185–214.

Meisiek, S., & Hatch, M. J. (2008). This is work, this is play: Artful interventions and identity dynamics. In D. Barry & H. Hansen (Eds.), *The SAGE handbook of new approaches in management and organization* (pp. 412–422). London, Thousand Oaks, New Delhi, Singapore: Sage.

Mienczakowski, J. (1995). The theater of ethnography: The reconstructing of ethnography into theater with emancipatory potential. *Qualitative Inquiry, 1*(3), 360–375.

Mienczakowski, J., & Morgan, S. (2001). Ethnodrama: constructing participatory, experiental and compelling action research through performance. In P. Reason & H. Bradbury (Eds.), *Handbook of action research* (pp. 176–184). London: Sage Publications.
Mienczakowski, J., Smith, R., & Sinclair, M. (1996). On the road to catharsis: A theoretical framework for change. *Qualitative Inquiry, 2*(4), 439–462.
Newell, S., Robertson, M., & Swan, J. (2002). Interactive innovation processes and the problems of managing knowledge. In B. Renzl, K. Matzler, & H. Hinterhuber (Eds.), *The future of knowledge management* (pp. 115–136). Basingstoke: Palgrave Macmillan.
Nicolini, D., Sher, M., Childerstone, S., & Gorli, M. (2004). In search of the 'structure that reflects': promoting organizational reflection practices in a UK health authority. In M. Reynolds & R. Vince (Eds.), *Organizing reflection* (pp. 81–104). Aldershot: Ashgate.
Parker, M. (1995). Critique in the name of what – Postmodernism and critical approaches to organization. *Organization Studies, 16*, 553–564.
Parker, M. (2002). *Against management*. Cambridge, UK: Polity.
Pässilä, A., & Oikarinen, T. (forthcoming). Research-based theatre as a facilitator for organizational learning. In P. Meusburger, & A. Berthoin-Antal (Eds.), *Knowledge in organizations – learning organizations*. Vol. 6 of the series Knowledge and Space. Heidelberg: Springer.
Raelin, J. A. (2001). Public reflection as the basis of learning. *Management Learning, 32*(1), 11–30.
Reynolds, M. (1998). Reflection and critical reflection in management learning. *Management Learning, 29*(2), 183–200.
Reynolds, M., & Vince, R. (2004a). Critical management education and action-based learning: synergies and contradictions. *Academy of Management Learning and Education, 3*(4), 442–456.
Reynolds, M., & Vince, R. (Eds.). (2004b). *Organizing reflection*. Aldershot: Ashgate.
Schön, D. (1983). *The reflective practitioner: How professionals think in action*. London: Maurice Temple Smith.
Van de Ven, A., & Johnson, P. (2006). Knowledge for theory and practice. *Academy of Management Review, 31*(4), 802–821.
Vince, R. (1998). Behind and beyond Kolb's learning cycle. *Journal of Management Education, 22*(3), 304–319.
Vince, R. (2002). Organizing reflection. *Management Learning, 33*(1), 63–78.
Vince, R. (2004). *Rethinking strategic learning*. London: Routledge.
Vince, R., & Reynolds, M. (2009). Reflection, reflective practice and organizing reflection. In S. J. Armstrong & C. V. Fukami (Eds.), *The SAGE handbook of management learning, education and development*. London: SAGE.

Chapter 11
From the Artists to the Managers: Responsible Collective Innovation Practices, Inspiration Flowing Through Hosting and Harvesting Profound Change

Isabelle Mahy

Abstract The aim of this chapter is to demonstrate how the artists can inspire a collective process supporting and leading to practice-based innovation. The underlying hypothesis is that enabling innovation through art is a powerful means to foster practice-based innovation. After having presented innovation as new situated knowledge, relevant at a micro-level work process activity, the concept of *ba* is brought forward to describe organisational contexts that invite, sustain, and foster innovation. The Art of Hosting and Harvesting are then presented as appropriate collective facilitation and information gathering processes to nurture innovation. Considering innovation as knowledge emerging from collective intelligence, two cases illustrate how collective intelligence can be nurtured by artistic practices. The results are focused on the specifics of the principles and practices at work, which are creative, artistic, playful, sensible, involving concerns for ethics and aesthetics, and helpful in creating meaningful experiences.

11.1 Introduction

As it is stated in the OECD 2009 report on innovation, it is the very nature of innovation that is changing and this transformation will require 'completely new, multidisciplinary skills and competencies, and the demand for these new human resources will be immense' (OECD 2009: 8). The report also states that the current models of research and development do not appear to be strong providers of

I. Mahy
Department of Social and Public Communication, Université du Québec à Montréal, Montréal, Canada
e-mail: mahy.isabelle@uqam.ca

H. Melkas and V. Harmaakorpi (eds.),
Practice-Based Innovation: Insights, Applications and Policy Implications,
DOI 10.1007/978-3-642-21723-4_11,

innovation anymore.[1] Furthermore, new business models and organisational structures are expected to increase companies' innovation, so that 'the entire company becomes more innovative'.[2] The need for a renewal of the business models appears to impose itself, at least for models, which would bring more creativity and provide organisational contexts designed to invite and sustain the emergence of innovation from various sources, coming both from inside and outside the organisation.

Considering the limits of the current capitalistic model and the global turmoil the world is facing, these concerns are not surprising. What has actually changed, according to the OECD,[3] is that innovation will depend on a greater social responsibility assumed by companies whose survival will depend in return on their openness and ability to collaborate with all stakeholders.

Along with this concern for a more responsible approach to innovation in society, the knowledge and skills required to tap into users' knowledge, as well as to start co-creating new knowledge with them, will soon come from multiple sources and disciplines. The narrow disciplinary view on innovation providers, expected to come from hard sciences, business, and engineering now give place to different expectations. Meeting these needs require a new mindset resulting from an interdisciplinary competency portfolio based on human and social sciences experts, artists, as well as natural scientists and engineers, all working together in an interdisciplinary mode.[4]

Such patchwork of competencies is presented by the OECD as a need and a means, a relevant and appropriate way of addressing the innovation challenge society and market are facing. Central to this challenge are the co-creation[5] of new knowledge and advanced methods for tapping into user knowledge in the pre-innovation phase where new business opportunities are explored (OECD 2009: 23).

Creativity is a well-known state of mind and an ability, and it contributes to fostering innovation but it does not constitute the container, process or context required to enable it. There is a need to explore alternative practices to R&D that could support the knowledge co-creation process, i.e., in the perspective of a greater collaboration, openness, and synergy with different stakeholders involved at different stages of the process, and in various dimensions of a more organic conception of innovation.

---

[1] More accurately, the report underlines that 'according to Philips, only 20% of the outcome from Philips' R&D department has recently formed the basis for innovation in the company. [...] Hence, it is not surprising that the future role of R&D departments is under evaluation.' (p. 60).

[2] *Ibid.*

[3] *Idem.* The authors' injunctions are clear: 'In the future, companies will need to become more open; i.e. to learn from their customers, collaborate with others and assume greater social responsibility. If they fail to do so, they will not survive. While companies will still optimise their business, globalisation and the digital technology have changed the rules of the game.' (p. 9).

[4] *Idem,* p. 13.

[5] Co-creation is referred to as 'co-production' of knowledge, in Boyle et al. (2010).

11.2 Defining the New Nature of Innovation

To explore the dynamics of innovation in the making, we mobilise a definition drawn from a learning perspective, useful to detect the collective processes, which are susceptible to leverage innovation. This chapter focuses on a micro view of innovation, namely a rather radical implicit organisational work process innovation, as defined by Ellström (2010). He qualifies this type of innovation as subjective, contextual, local, individually bound, made of improvisations, deviating from the formal structure, involving considerable creativity and ability to improvise, of unofficial status and happening largely behind the scenes. We retain this definition for its pertinence to describe the very nature of the artists' work, by revealing typical dimensions of the artistic creative work processes.

According to Ellström (2010), the counterpart of this backstage informal creative work is the explicit work process, subsumed to the formal structure, principles, goals, rules, and standards, which act as borders by limiting the initiatives and efforts of redefining the workplace reality. By doing so, it counters spontaneous initiatives of doing something differently, while preserving from potential loss of identity and energy outside the boundaries of what is defined as the goals to be achieved.

Because the 'new nature of innovation is taking place at the micro level'[6] and because 'people should be empowered to innovate' (OECD 2010: 1), the practice-based innovation explored in this chapter is not so much considered for its contribution to the economy, which would be a macro external business view on the activity, but rather from an internal and organic process view, a more micro perspective embodied by workers and facilitators involved in the practices at stake.

In this view, the innovation process is understood as comprising three collective *spaces* to hold, to host and to harvest from: co-sensing, co-inspiring or co-presencing, and co-enacting or co-creating (Arthur et al. 1999; Scharmer 2007). These spaces act as generative contexts or *ba* (Nonaka and Konno 1998), designed to invite and sustain organisational or practice-based innovation.

According to these authors, for such a *ba* to be effective, the following five characteristics[7] must be in place. First, self-organisation must be granted, so the collective body involved, which we will name 'organisation' – be it a group, a team, a community of practice, or the like – has its own intention, direction, and mission, with active participants. For the authors, a good *ba* needs creative chaos, care, and love, as well as intention and direction. Secondly, the 'organisation' has an open boundary to allow for developing one's own context and openness to other contexts. Thirdly, the *ba* lets participants share time and space and transcend their own limited perspectives or boundaries; it acts as an open environment. The fourth characteristic of a good *ba* is the existence of multi-disciplinary and multi-viewpoint dialogues

---

[6] *Ibid*, p. 9.

[7] The five characteristics are described in detail here: http://www.dialogonleadership.org/WhitePaper.html

where participants can learn from one another. The authors underline the quality of the conversations as one of the most important measures of the quality of place and the health of an organisation. The fifth characteristic consists of all participants having equal access to the centre and maximum capacity with minimum conflict. Thus, every participant is at the same distance from the centre, and anyone has the potential to be a centre.

Nonaka and Konno (1998) actually suggest that such an 'organisation' is self-organising, autonomous, and autopoïetic. Fully democratic, based on a collective leadership shared by all, it grants autonomy and equal rights to everyone; collaboration is based on free will, it is network-based rather than hierarchical; meaningful relationships and creativity are encouraged, and there is no external forceful power controlling the process. In other words, a good *ba* would be a democratic, nurturing, physical, relational, creative, spiritual, and poetic place (Mahy 2008). These qualities actually correspond to the conditions that artists look for when they want to shift their state of mind, in order to act as a medium between themselves and the universe. Their creative work consists of sensing, feeling, translating, expressing, and evoking poetically their experience of the world. Artists actually transform feeling into artwork. Reframing this statement in the scope of this chapter, we suggest that practice-based innovation consists of transforming experience-based knowledge into knowledge from the future as it emerges (Scharmer 2007). In other words, innovation is seen as a collective process very similar to the artistic creative and transformative process. Needless to say, this audacious model of management is quite innovative in its very nature, not to say provocative, compared with current mainstream management practices, still far apart from the renewal principles proposed in the last decade by management research.

This chapter focuses on the actual collective dynamics from which innovation emerges. Posed as a tension (Ellström 2010) between tacit and explicit dimensions of work, and between chaos and order (van Eijnatten and Putnik 2004; Putnik 2009), these practices foster learning as the key process enacted to explore this complex territory between the known and the unknown. Distinct from the explicit processes unfolding from the formal structure, like training, the practices explored in this chapter are collective learning practices, often referred to as 'collective intelligence' practices (Garrido 2009; Penalva 2006). Framed by such a definition, the collective practices considered here differ from hierarchical decision-making processes. Multiple and eventually dissonant voices from participants inside and across organisations can be considered, and to achieve such a purpose, the underlying facilitation and information gathering processes are also framed differently.

11.3 Defining Practice

In this chapter, the concept of practice is understood from a phenomenological perspective, as a *bricolage* made in between habit and action (Gherardi 2009), or as the union between thought and action, where mind/body, knowing/doing, micro/macro, and nature/culture are not opposed anymore. Here, the duality between rational and sensible – or aesthetic – knowledge becomes poles suddenly related to one another by playfulness, described as the ability humans have to jump from feeling their world through experience to thinking and reflecting on their experience, in an ongoing movement of sensemaking, swinging back and forth (Guillet de Monthoux 2004; von Schiller 1992) in their knowledge building process.

Besides being personal, the internal creative process, which contributes to forming a practice, is also a social activity, a collective process of learning, enacted in a specific context, and from which situated knowledge emerges (Mahy 2005). This *bricolage* is also an embodiment of the organisation's culture, and in this sense, considering it as a *ba* with its five main characteristics reveals a certain ideological standpoint, as well as an ethics of work, with regard to management, power, leadership, personal and collective responsibility. This echoes quite naturally with the way many artists work when they develop projects in teams. This being said, we must refrain from believing that all artists are dedicated collectivists, and that they would neglect their personal benefits to consider others' welfare first. Divas, stars and V.I.P. do contribute to creating social turbulences in a *ba* (Mahy 2005, 2008).

Whether or not specific artists live up to the ideal of a good *ba* is not the point in this chapter, but because the goal of artists' work is to innovate, their practices are geared to sustaining innovation. This is why such practices can prove to be inspiring for other organisations outside the creative industries. In an in-depth ethnographic study showing practices from the artists of the Cirque du Soleil (Mahy 2005), it appears that the key practices that foster innovation are more social than technical. They are mainly focused on the art of encounter, storytelling, creative presence, organic leadership, interdisciplinarity, and poetic memory. Where managers would develop business speeches, artists will invent stories. Where managers would strategise, plan, and forecast, they will be present to the moment, trying to catch its spirit. Where managers would order, control, coach, and put pressure on a team, artists will have an organic view of the project, with roles and responsibilities blurred and somewhat confused. Where managers would make efforts to assimilate the artists' culture and try to behave like them, artists will either invite or refuse access to their own community but would not try to enter the managers' world. Lastly, where managers would try to manage knowledge and build corporate memory, artists will expand and reframe reality by creating memory processes serving their purpose. A good example of this is the idea of going through the looking glass, like in 'The Imaginarium of Doctor Parnassus' directed by the British film director Terry Gilliam. On the other side of what you believe to be reality, nothing feels the same.

Cirque du Soleil at a glance

- From a group of 20 street performers at its beginning in 1984, the Canadian *Cirque du Soleil* is a major Québec-based circus organisation providing high-quality artistic entertainment.
- Since 1984, '*Cirque*' has reached more than 100 million spectators in more than 300 cities on five continents. It presents simultaneously nineteen different shows around the world.

- In 1984, 73 people worked for *Cirque*. Today, the business has 5,000 employees worldwide, including more than 1,300 artists. The company's employees and artists represent close to 50 nationalities and speak 25 different languages. More than 100 types of occupations can be found at *Cirque*.
- It hasn't received any grants from the public or private sectors since 1992.
- Its citizenship principles are founded on the conviction that the arts, business, and social initiatives can, together, contribute to making a better world. It is involved in communities, with youth at risk and is focusing on global issues such as the fight against poverty and access to water, through its 'One Drop' foundation. In cooperation with its partners, *Cirque* is involved in nearly 80 communities worldwide, in some 20 countries spanning five continents.
- In the past few years, *Cirque* has been developing business initiatives based on its shows: products for retail sale, content for movie theatre, television and DVD, organising private gatherings as well as major public events like Shanghai World Expo 2010 and Expo Zaragoza 2008. With business partners, *Cirque du Soleil* develops innovative projects, particularly in the field of hospitality (restaurants, bars, spas, etc.), for instance The Revolution Lounge at The Mirage Hotel in Las Vegas.[8]

Backstage, nothing is that simple either. Radical innovation requires audacity, and the culture of an organisation can be more or less open to innovating. In the case of Cirque du Soleil, with a culture open to worldwide influences and capturing social and cultural trends from everywhere, the ability it has of not losing sight of its own roots enables the organisation to keep strong bonds locally and historically. Tapping into the world's memory and adopting strong aesthetic influences – like surrealism – fuels and propels the organisation's imagination toward defining audacious visions, accomplishing complex projects and creating inspiring artworks. Not only does the Cirque worship its own creation, it also promotes itself as an artwork (Guillet de Monthoux 2004). It considers its own culture as one of its jewels, and it masters the art of managing the public's desire to come closer to what is packaged and sold like a V.I.P.'s privilege. This tribal culture is the object of myths and gossips, and the tribe members living in their *ba* do experience the conflicts of trying to gain power by moving closer to the centre of their universe – where the founding president 'sits and shines' like the sun – to get benefits and privileges from the relationship.

These backstage maneuvers are typical of group behaviour, and without surprise, we are inclined to conclude that for the *ba*'s characteristics to be embodied and enacted according to Nonaka and Konno's (1998) expectations, a group needs to develop reflexive abilities, maturity, and coherence between discourse and practice. There is actually more to it than creativity or artistic inspiration.

Scharmer (2007) proposes a useful path toward such an ethics of work in his *social grammar of emergence*, where he describes the social system – including the

[8] A subset of public information displayed by Cirque du Soleil on their website as of January 2011.

organisation – in seven levels of depth and ten categories or dimensions from which the system operates. In a summarised table, he describes what he refers to as 'the footprints of collective human creativity' using the following dimensions: collective consciousness, the places from which the system operates, the different corresponding worldviews, the nature of knowledge created and shared, the social time and space in which the social system operates, the perception the system has of its own complexity, the primary causal mechanisms or sensemaking process, the underlying episteme, and the system's perception of the Self.

This chapter being limited in its scope and space, each aspect of this grammar is not developed further[9] here. Let's just mention that Scharmer's social grammar of emergence could very well act like a path finding process, an enabler to trigger and leverage a *ba*. Furthermore, according to it, not all practices can actually qualify to its seven levels of depth or the ten dimensions. The different practices he identifies have specific qualities, episteme, operating systems, and potential, like the 'Art of Hosting', which acts from systems thinking perspective and is ethically grounded to sustain profound change and innovation. The next section describes this practice and considers how it could foster practice-based innovation.

11.4 The Hosting and Harvesting Process

According to the facilitators who practice this art, the Art of Hosting and Harvesting is 'a pattern and a practice that allows us to meet our humanity in ourselves and in each other – as opposed to trying to be machines when meeting'.[10] It is considered an art because the process does not imply working from – and with – pre-determined methods. Practitioners rather approach each situation from a design perspective, sketching out the intervention that is the most appropriate for each context.

This situated and personalised approach to facilitation and information gathering obviously differs from traditional ways of working, which are primarily based on rational planning and summarising mechanisms, which often aim at establishing control in order to manage outcomes.

The Art of Hosting principles of self-organisation, participation, and non-linearity are presented by the practitioners as key to both individual and collective learning and discovery. Becoming a host means to deepen competency and confidence in hosting group processes[11] like the Circle practice (Baldwin and Linnea 2010), World Café (Brown and Isaacs 2005), Open Space Technology (Owen 1997), Appreciative Inquiry (Cooperrider and Srivastva 2000), and other forms of collective reflective processes.

[9] For further details, see Scharmer (2007: 366–367).

[10] http://www.artofhosting.org/theart/

[11] http://www.artofhosting.org/thepractice/coremethods/

These facilitation skills differ from train-the-trainer approaches that often unfold from a more rational behaviourist perspective. The specific competencies of becoming a practitioner in the Art of Hosting reveal their phenomenological roots when at first, the facilitator learns to be present, to sense from the whole field, to sit in the chaos and keep the space open, while not-knowing. These basic socio-psychological communication skills lead to developing the ability of being willing to listen fully, respectfully, and without judgment. Then only, being courageous, the facilitator will invite others to participate and will be willing to initiate conversations that matter, i.e., find and host powerful questions with the stakeholders, and make sure he or she harvests the answers, the patterns, insights and learnings into wise actions. Once acquired from the community, this series of competencies also open the doors of an inner circle to the facilitator. He or she can now take part in the larger community of practitioners who co-create and co-host with others, blending knowing, experience and practices, and work in partnership. A second level of learning consists of applying hosting and harvesting to more complex situations and contexts of larger scales.

The practitioners say that each of these processes generates connection and releases wisdom within groups of people. They foster synergy and provide ways for people to participate in intention, design, and outcomes/decisions/actions. Based on what has been presented above, this art appears to be an interesting framework for sustaining the emergence of innovation.

Focused on sustainability, humanism, democracy, generative spiral dynamics, and community, the Art of Hosting culture invites facilitators to act as *process* experts, not content experts. It promotes and provides the expertise of facilitators who act as social midwives of spontaneous *gemeinschafts*, where participants gather around common interests, critical situations that can only be understood and eventually solved in an innovative way by reflecting collectively on them. To create such communities, members of the Art of Hosting network send a call, and the persons who feel concerned become participants. As only the ones who feel called actually participate, motivation and personal interest do not stand in the way of the commons.

This network-facilitating innovation inquiry process (Harmaakorpi and Mutanen 2008) is structured in five phases called 'breaths,' framed to support and anchor the collective reflection thinking and feeling process into a very intuitive field. The quality of such a field resembles what the artists describe as their ability to resonate with the universe (Mahy 2008).

The five breaths[12] are described by the practitioners as follows.

1. Name the issue – calling the core question
   Wise Action: Focus the Chaos of holding the collective uncertainty and fear – step into the centre... of the disturbance.
   What not to do: Move too fast.

[12] http://www.artofhosting.org/thepractice/5breaths/

Question: What is real[ly] at stake here? What if some of us worked together to surface the real question and need that matter to the community?

2. Creating collective clarity of purpose and principles
   Wise Action: Engagement.
   What not to do: Make assumptions.
   Question: How to get from need to purpose? What is our purpose? How to see and feed the group value?

3. Design and invitation process
   Wise Action: Check in again and again to see if your design and invitation serve the purpose.
   What not to do: Try to do too much in your design (match it to the purpose).
   Question: How do we create an inspired invitation that moves people to come? How do we let go of our expectations that certain people need to be there?

4. Meeting – hosting the group, purpose, and questions
   Wise Action: —[13]
   What not to do: Go alone.
   Question: How can I best serve as the instrument/container to allow the collective wisdom?

5. Follow-up – continued learning and leading from the field
   Wise Action: Always come back to purpose.
   What not to do: Lose sight of the purpose or it won't be embodied.
   Question: How do we sustain the self-organisation?

Consistent with the phenomenological episteme and the concern for sustainability, the five breaths of the Art of Hosting offer a rich translation of Scharmer's Theory U (2007) turned into a process, and as such, the Art of Hosting is a useful embodiment of this theory, whose purpose is to innovate.

The practitioners describe the harvesting part of the art as the second side of the same effort to amplify the collective intelligence emerging through conversations. It is presented as a gateway from complexity to simplicity, capturing the wisdom, remembering, seeing patterns, and making meaning 'visible' through graphic recording, drawings, images, dance, collages, social theatre, photographs, videos, and poetry. Harvesting also tries to embed the insights and learning, i.e., to make them as relevant and useful for a specific context as possible (Nissen and Corrigan 2009), mainly through art-based inquiry approaches (Knowles and Cole 2008; Broussine 2008). The harvesting process unfolds in eight phases, presented with a metaphorical pastoral discourse (Nissen and Corrigan 2009), where new knowledge or innovation is considered to be what emerges from a meaningful conversation.

[13] In this case, the Art of Hosting authors provide no information to inform of any wise action. I suggest deep presence (Mahy 2006) and sensing from the whole field as mandatory states of being to support the practice.

1. Sensing the need
2. Preparing the field
3. Planning the harvest
4. Planting the seeds
5. Tending the crop
6. Picking the fruits: recording – or creating a collective memory
7. Preparing and processing the fruits: creating collective meaning
8. Planning the next harvest: feeding forward

As the Art of Hosting and Harvesting framework appears to be relevant and helpful to sustain and foster innovation, we will now describe two different cases where the practice has been applied in different organisational settings, both focused on creating collective intelligence and innovation. This description will be the source from which the value of the Art of Hosting will be studied as a container and collective process to foster practice-based innovation.

11.5 Methodology

The research presented here is art-based participatory action research. The research comprises different fieldworks. Two illustrations of the Art of Hosting chosen to foster collective intelligence and innovation are described as fieldworks or case studies. The key milestones of the research unfold as follows.

The researcher is the facilitator or host. She is contacted by a group and following an initial conversation, she starts gathering a hosting team. The hosting team starts to reflect with the group on how to state the needs, goals, or problems, not from a functionalist perspective but from a phenomenological one. The activity is understood as being a shared quest.

In an iterative movement, the core team sketches out the facilitation and harvesting process, by co-creating it with the group-to-be, i.e., the persons who responded to the call and feel involved. In the course of designing the event, the hosting team has identified specific practices (circle, open space, world café, etc.) useful to draw on, in relation with the demand. The researcher and her team prepare the event or series of gatherings. The team facilitates the unfolding and hosts the conversations. The team members also harvest the conversations. When the group is active online, the hosting team can open up a social network website or the equivalent; it becomes present online before, during and after the event, and by doing so, the conversation actually starts before the physical event. Thus, it can become a virtual process unfolding during weeks or months, with physical events happening in parallel. Depending on the context, one or many physical gatherings can be held. The harvest is edited and offered back to the group, eventually made available live online and afterwards, as a memory of what emerged during the process.

11.5.1 Fieldwork 1: The Urge of Becoming, Geneva, June 2010

A group of 15 professionals who have not met before, at least not all together, gathered in one place for two half days and an evening. They have the project of becoming an association, which aims at supporting social responsibility on its three pillars – economy, ecology, society – to provide and promote sustainable services in various domains, from design to education, gardening, arts, and engineering. In this inception process, this gathering will be the very first moment these people spend together to reflect on what they could become and could do as a group. Due to distance and time constraints, the participants have not taken part in designing the process, but the person in charge of making the call to the community and mobilise the group has seen the researcher at work in a similar context in the months preceding the gathering. Based on this experience, she has asked for the researcher's hosting and harvesting skills and discussed the need.

The researcher has adapted and enriched a media-based collective poetic memory process from an artistic ethnographic practice she found at the Cirque du Soleil and nurtured since (Mahy 2005, 2010). Named 'the 3rd Eye', this harvesting practice becomes an active witness of the collective learning process, and its outputs consist of performances and visual-based artworks that the group creates together. The researcher has proposed it to the group that integrates it into the gathering process.

The actual agenda for the gathering unfolded like presented here. The first half day started with an Opening Circle followed by a World Café, an Open Space, time for editing input from the 3rd Eye, and a Closing Circle. During the evening, an informal large group conversation took place. The second half day started with an Opening Circle followed by a presencing exercise; then a SWOT (Strengths, weaknesses, opportunities, threats) analysis took place, followed by a reflection on the group's mission, vision, identity, and goals. Project identification was then started, followed by project planning, time for editing the 3rd Eye input, and a final Closing Circle.

The 3rd Eye memory artwork has not been edited, but subsets of photographs have been made available to the group on the web after the event.

11.5.2 Fieldwork 2: The Emerging Future Workshop, Montreal, October 2009

A community of more than 100 people was gathered around a call on 'what collective intelligence could be, and how we could see the future emerge in the making'. From being together and reflecting on this very question, an online community emerged at the end of the summer, in 2009. A hosting team was formed and it started to design the hosting and harvesting process, which took the shape of a one-day event. 33 people out of the 100 members from the online community attended the gathering, while the others could follow the event online, through blog posts. After a few weeks of invitations, the one day event started with a lunch and an opening Circle, followed by an Open Space Forum, which lasted nearly half a day. A plenary Circle enabled everyone to share on what the different groups had learned. A 3rd Eye activity was active in parallel, and the participants were invited to contribute to it in different collective art performances facilitated by artists of the hosting team. After the summary Circle where all learning and discoveries were presented and captured, dinner was served, followed by more collective art performances done again by the participants and facilitated by the artists. The closing Circle took the format of a dance. In the days and weeks that followed, more people joined the community, and images, videos, and comments were posted online in the social network website of the community. During the same weeks, a member of the hosting team held two focus groups with participants to get an understanding of the meaning that the participants had given to the experience.

Another view on this project shows that:

- More than 1000 people had been invited to the gathering through various social networking websites;
- Hundred and thirty-eight members had joined the community social network site by October 29;
- Forty-five persons had confirmed their presence at the gathering;
- Thirty-three persons actually came to the gathering;
- Among them, ten contributed to its organisation, as part of the design (hosting and harvesting) team;
- After the gathering, six participants took part in a first focus group, and among them, one was a member of the design team;
- Seven participants took part in a second focus group, among them five were members of the design team;
- One participant chose to contribute to the data gathering process in an individual interview.

11.5.3 Research Design

This art-based participatory action research adopts and applies the principles of Theory U (Scharmer 2007) and the Art of Hosting in its different field interventions. The researcher and her team design, facilitate and harvest the outputs of the gatherings. The team members may change according to each fieldwork's characteristics, as the expected competencies may change, depending on the group processes that are mobilised and the specific media activated in the 3rd Eye process (video-based, performance-based, etc.).

11.5.3.1 Results

The aim of this chapter being to demonstrate how the artists inspired the process, we focus the results on the specifics of the principles and practices at work, which are creative, artistic, playful, sensible, involving concerns for ethics and aesthetics, and helpful in creating meaningful experiences. In other words, what is presented here are the aspects of the Art of Hosting and Harvesting practice that turn a collective intelligence experience into a unique moment of going through the looking glass and expanding reality, creating a different future along the way. The underlying hypothesis is that enabling innovation through art is considered a powerful means to foster practice-based innovation.

The Sense of Place or *Ba*

In each of the two cases, the physical location has proven to be important and is considered a key enabler. In its relational definition, the *ba* created was designed to invite participants to get acquainted, share and build relationships. In both cases, including meals in the design contributed to it. On the creative level, both gatherings offered a large place for artistic creativity process, through the 3rd Eye and the collective performance facilitated by the artists. On a spiritual level, participants from both gatherings expressed their experience as unique, rich and moving, meaningful, somewhat transformative, and positive. 'I never like the reality I see' stated one participant in a very emotional moment, 'but in these two days, I liked what I saw'. The quality of the experience gave him the feeling that he could trust some people and there could be some hope for the future. At a poetic level, the artworks from the 3rd Eye created collectively after the gatherings show the atmosphere, the delicate texture of relationships, the playfulness, the fragile quality of the present, and the great source of meaningful emotions they reveal.

Playfulness, Ethics and Aesthetics

In such contexts, ethics and aesthetics are central concerns, because they are instrumental to building trust, showing patterns emerging from the group, and to letting the quality of emotions flow in from – and to – the participants during the whole event (Csikszenmihalyi 1990). These invisible and intangible abilities are core competencies for the hosting and harvesting team.

Playfulness is present through the activity that the artists facilitate, and it augments the reality by bringing in and revealing powerful metaphors into the room. In the Montreal event, an artist decided to show the network between people in the making, during the opening Circle, by standing in the centre of the circle and weaving a blue thread from one person to the other. Each time a person would share the reason of his or her presence and explain what he or she was offering to the group, the artist would slowly move from one person to another and very softly bind the thread to an arm, a leg, or around the chest. From an ethical and aesthetical standpoint, by demonstrating the potential linkages between people in such a simple metaphorical way, the artist offered a fragile ephemeral artwork to the whole group. The softness of her gesture, respectful of every emotion, granted her the necessary trust to go along and succeed in her endeavour. She was allowed to complete the performance because everybody felt at ease to be part of such a social sculpture, as this creation could be qualified, referring to Joseph Beuys.

To create this quality of experience, the hosting team must sense and integrate the personal and inter-relational tensions in order to offer a nurturing space, which welcomes the implicit, silent and behind the scene experiences, and turn this complex meshwork into a positive experience.

For the hosting team to be pertinent, accurate and timely, everyone had to act in synchronicity with the whole system, and thus every member had to be fully present (Senge et al. 2004) to all aspects of the event and, more specifically, to all participants' experience in the making.

Artists can also create a breach in reality by inviting people to behave in ways they would never accept in other circumstances. In the Montreal event, two artists from the core team were the first persons participants met after they had retrieved their nametag at the desk. These artists had created a beauty salon where they invited everyone to sit down and put bright red lipstick on. Sitting on a small chair, with a pocket mirror in hand, every man and woman went along, while images were shot of them trying to put make up on. After this first potentially intimidating moment, they were invited to go to a large white paper *murale* and leave their trace by kissing the paper. When all participants had arrived, the *murale* had become an artwork, filled with red kisses making waves on the wall. Later on at the dinner, the pictures taken at the beauty salon were displayed on a large screen and all participants felt happily surprised to discover that they actually shared a very intimate moment with people they had never met before. This playful complicity acted as a pattern that connected people and revealed that intimacy could also be shared when trust was present.

The Art of Innovating Together

Based on the Montreal event data,[14] the collective process experienced by the participants of the gathering can be described by a spiral (Fig. 11.1). The spiral represents time and the complexity of social meshwork in an ongoing state of becoming. While the event is unfolding, states of mind shift from curiosity to

[14] The author wishes to thank Caroline Durand for her very active contribution to this study, as a member of the hosting team and as a graduate student completing her master's degree in communication at Université du Québec à Montréal on participatory practices. The Montreal event was her fieldwork.

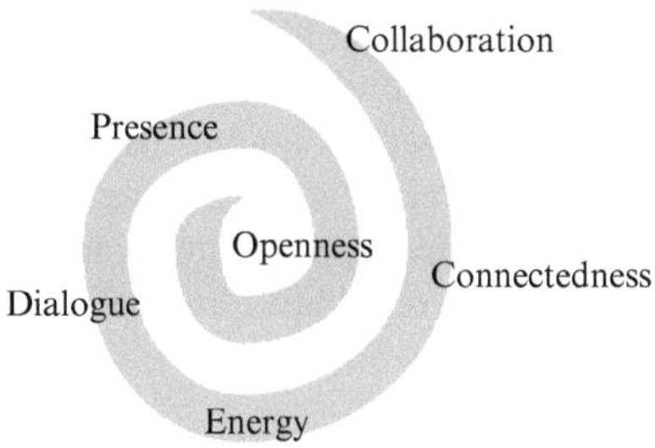

Fig. 11.1 The collective process of innovating together

openness, and from there to a more peaceful inner state of presence, followed by authentic dialogue, which releases energy, shared purpose, connectedness, and a spontaneous will to collaborate in various projects.

Each of the states is illustrated by a few examples chosen from the two focus groups held after the gathering.

Openness

One participant mentioned that she felt an open mind, an implicit complicity with the others at the gathering. Someone also mentioned that she felt authorised to say what she wanted in a simple way – things we believe are true or important. Another one mentioned that when the first artistic performance started, he felt everyone was open and ready; it was lively, colourful, and authentic. One participant mentioned that her desire to let go came alive when she heard the reasons why the other people had come to the gathering.

Presence

A participant mentioned that he felt the state of presence mainly through the silent moments proposed in the artistic rituals. He felt that the group had succeeded to create peace among itself and that he felt connected to everyone. Another participant said he felt that silence invited different emotions, like gratitude. He felt the spiritual dimension silence gave access to, as well as deep listening. Before coming to the event, a participant said that she felt her life was getting everywhere and nowhere. The gathering very quickly lifted the weight on her shoulders and she felt lighter. She was suddenly able to focus on being present here and now.

Dialogue

A participant mentioned that ideas were popping from everywhere, everyone was listening to the others, and that the listening was beautiful. She felt it was inspiring and that participants had developed a consciousness of the problem they were addressing. A bubble or privileged space had been created where everyone shared their experience and dived into the problem. A participant said that it was not a debate but an authentic sharing, where no-one tried to win against the other. Everyone could express himself freely, without being judged. Another participant said that this space was serene and appropriate to foster trust. Another person mentioned that true listening was contributive and collaborative.

Energy

Two-thirds of the participants have mentioned that something very special had emerged from the gathering, something they described as magical, energising, and inspiring. Someone mentioned the crossroads of energy as being the source of this powerful emotion. Another participant mentioned that individualities had finally met together. Two other persons felt nurtured, irrigated by the energy they had experienced.

Connectedness

A participant mentioned that the intimacy of the dialogue had created trust. Another was surprised by the very short time it took to address serious and difficult questions during the dialogues. Two participants expressed their view by saying they felt they had taken mental shortcuts, taking less time for superficial chat and getting more rapidly closer to people they didn't know before. They said they had discovered proximity very quickly. Some said it was because no hierarchical mentions had been made; no status had been used to identify and qualify the participants. A participant mentioned that he felt the linkages between people would persist over time.

Collaboration

A participant mentioned that collaboration depended on explicit sharing of values of openness and generosity, with passion. Horizontal relationships also invited collaboration, as well as deep listening. One person said that when we stop focusing on what can disconnect us from the others and stop our sharing, when we try to enter directly into a relationship, collaboration has a great potential to develop quickly.

Qualifying this collective intelligence process, participants mentioned that it also becomes possible when mutual respect and admiration among the participants is present. Furthermore, it has been mentioned that sustainable solutions to a collective problem can only be generated collectively. It is worth noticing that some participants have added that it is only through collective intelligence that organisational practices have a chance to evolve and transform. Some participants went further by saying that collective intelligence contributes to innovative and creative solutions, and that time has come to cross the borders of rationality and dare explore the unknown.

The Art of Hosting process being a generative container, what emerges from it cannot be planned or controlled. This characteristic of complex social systems also applies to innovation. It is therefore interesting to mention other positive impacts resulting from this type of gathering. Among them, participants have identified that the connections created with people during the event and the projects they decided to do together for their future have been persistent and fruitful. Friendships and initiatives have emerged from a *ba* perceived as offering a creative and nurturing ground, irrigated by positive energy, trust, and hope.

11.6 Summary and Conclusion

The results presented here have been chosen from a larger data set to illustrate the influence and inspiration an artist can have on a practice-based innovation process when we choose to draw upon different facilitating and information gathering practices to foster innovation. The Art of Hosting and Harvesting practice sheds light on the rich potential of involving art in different ways in practice. The results presented here are obviously limited and are not meant to constitute a picture of the truth, which could be generalised. The knowledge presented here is context-based, local, and meaningful for the communities who have created it. Nevertheless, the research shows the powerful lever the Art of Hosting and Harvesting practice can become when called upon to foster innovation.

In a larger perspective, this practice may be of help in different contexts, to engage the collective body in meaningful conversations held beside the limitations that hierarchy, status or power impose on group processes, maintaining them in today's thinking rather than inviting them to imagine the future, not from today's evaluation but rather from the future itself.

By doing so, innovation does not only constitute a resulting outcome expected and hoped for by everyone, but the collective reflective process itself becomes innovative and is a source of potential new knowledge. When perspectives and frames shift from the present to the future, today's limitations dissolve into new possibilities never perceived before.

The growing interest for generative community practices like the Art of Hosting and Harvesting show that they already represent a social technology of innovation for an increasing number of communities. This emerging social movement appears to hold a good potential for the renewal of democracy and a more sustainable approach to social justice, as design thinking practices tend to show (Brown and Wyatt 2010).

These last considerations lead to further questions on the very identity and interest of these practices. How do these practices migrate to different organisational settings where the *gemeinschaft* is threatened or controlled externally? Can it do so without becoming subsumed to hierarchical power, and can it survive without becoming a fashionable superficial tool? Can its ethics and aesthetics anchoring contribute to the becoming of a more sustainable world, and if so, under what conditions?

As the Art of Hosting practice is bringing back the issue of power on the front stage in the organisation and society, and while it is trying to maintain its identity, fighting not to succumb under the power of the business and market, it has already proven to be of great interest. Not only does it invite art, aesthetics and creativity to play a central role in the process but it also offers a sustainable perspective to explore the XXIst century innovation challenge. The artists face the same issues of integrity (Mahy 2008). For the Art of Hosting and Harvesting practice, the choice between being perceived as radical but meaningful or becoming the next empty buzzword in the business world is not fair. Rather than spoiling this practice by emptying it from its political contents, managers could envisage learning new paradigms and perspectives, more collaborative, collective and democratic (von Hippel 2005), with shared leadership and knowledge. For the last 10 years, many authors have invited managers to act like artists and to manage in a creative and innovative way (Guillet de Monthoux 2004). To innovate by letting the workers redefine how the work could be done would be a good start.

References

Arthur, W. B., et al. (2000). *Illuminating the blind spot: Leadership in the context of emerging worlds, 20 propositions based on conversations among the authors and dialogue-interviews with thought leaders on knowledge and leadership*. Summary paper on an ongoing research project. McKinsey—Society for Organizational Learning (SoL) leadership project. http://www.dialogonleadership.org/WhitePaper.html. Accessed 31 January 2011.

Baldwin, C., & Linnea, A. (2010). *The circle way: A leader in every chair*. San Francisco: Berrett-Koehler Publishers.

Boyle, D., Coote, A., Sherwood, C., & Slay, J. (2010). *Right here, right now*. NESTA: Taking co-production into the mainstream. Discussion paper.

Broussine, M. (2008). *Creative methods in organizational research*. London: Sage.

Brown, J., & Isaacs, D. (2005). *The world café: Shaping our futures through conversations that matter*. San Francisco: Berrett-Koehler Publishers.

Brown, T., & Wyatt, J. (2010). Design thinking for social innovation. *Stanford Social Innovation Review*, (Winter). Stanford, CA: Stanford Graduate School of Business.

Cirque du Soleil website. http://www.cirquedusoleil.com. Accessed 31 January 2011.

Cooperrider, D. L., & Srivastva, S. (2000). Appreciative inquiry in organizational life. In D. L. Cooperrider, P. F. Sorensen Jr., D. Whitney, & T. F. Jaeger (Eds.), *Appreciative inquiry: Rethinking human organization toward a positive theory of change* (pp. 55–97). Champaign: Stipes.

Csikszenmihalyi, M. (1990). *Flow, the psychology of optimal experience*. New York: Harper & Row.

Ellström, P.-E. (2010). Practice-based innovation: A learning perspective. *Journal of Workplace Learning, 22*(1/2), 27–40.

Garrido, P. (2009). Business sustainability and collective intelligence. *The Learning Organization, 16*(3), 208–222.

Gherardi, S. (2009). Knowing and learning in practice-based studies: An introduction. *The Learning Organization, Special issue, 16*(5), 352–359.

Guillet de Monthoux, P. (2004). *The art firm. Aesthetics management and metaphysical marketing from Wagner to Wilson.* Stanford, CA: Stanford Business Books.

Harmaakorpi, V., & Mutanen, A. (2008). Knowledge production in networked practice-based innovation processes – interrogative model as a methodological approach. *Interdisciplinary Journal of Information, Knowledge, and Management, 3*, 87–101.

Knowles, J. G., & Cole, A. L. (2008). *Handbook of the arts in qualitative research, perspectives, methodologies, examples and issues.* London: Sage.

Mahy, I. (2010). *Practicing the 3rd eye: Reflexion on a collective visual harvesting process.* Paper presented at the 28th Standing Conference on Organization Symbolism, Lille, France.

Mahy, I. (2008). *Les coulisses de l'innovation.* Québec: Presses de l'Université Laval.

Mahy, I. (2006). *Le rôle de la présence créatrice dans le processus d'innovation.* 74e Congrès de l'ACFAS (Association francophone pour le savoir), 15–19 mai, Montréal, Québec. http://www.archipel.uqam.ca/378/. Accessed 30 January 2011.

Mahy, I. (2005). *Innovation, artistes et managers, ethnographie du Cirque du Soleil.* Thèse de doctorat. Montréal: Faculté des Études Supérieures, Université de Montréal.

Nissen, M., & Corrigan, C. (2009). *The Art of Harvesting* (2nd ed.). http://www.artofhosting.org/thepractice/artofharvesting. Accessed 31 January 2011.

Nonaka, I., & Konno, N. (1998). The concept of 'Ba': Building a foundation for knowledge creation. *California Management Review, 40*(3), 40–54.

OECD. (2010). *The OECD innovation strategy: getting a head start on tomorrow – OECD executive summary report on innovation.* Paris: OECD.

OECD (2009). *New Nature of Innovation.* Copenhagen: OECD. www.newnatureofinnovation.org. Accessed 31 January 2011.

Owen, H. (1997). *Open space technology: A user's guide.* San Francisco: Berrett-Koehler Publishers.

Penalva, J.-M. (2006). *Intelligence collective: Rencontres 2006.* École des mines d'Alès du 22 au 24 mai 2006. Paris: Presses de l'École des mines.

Putnik, G. D. (2009). Complexity framework for sustainability: An analysis of five papers. *The Learning Organization, 16*(3), 261–270.

Scharmer, C. O. (2007). *Theory U – leading from the future as it emerges. The social technology of presencing.* Cambridge, MA: Society for Organizational Learning (SoL).

Senge, P. M., et al. (2004). *Presence: Human purpose and the field of the future.* Cambridge, MA: Society for Organizational Learning.

The Art of Hosting practice website. http://www.artofhosting.org/home/. Accessed 31 January 2011.

The InterChange website on hosting and harvesting practices. http://www.interchange.dk/practices/. Accessed 31 January 2011.

van Eijnatten, F. M., & Putnik, G. D. (2004). Chaordic systems thinking for learning organizations. *The Learning Organization, 11*(6), 415–417.

von Hippel, E. (2005). *Democratizing innovation.* Cambridge, MA: MIT Press.

von Schiller, F. (1992). *Lettres sur l'éducation esthétique de l'homme, Briefe über die ästhetische Erziehung des Menschen.* Paris: Aubier.

Weisbord, M., & Janoff, S. (2000). *Future search – an action guide to finding common ground in organizations & communities* (2nd ed.). San Francisco: Berrett-Koehler Publishers.

Guillet de Monthoux, P. (2004). *The art firm: Aesthetic management and metaphysical marketing from Wagner to Wilson*. Stanford, CA: Stanford Business Books.
Harmaakorpi, V., & Mutanen, A. (2008). Knowledge production in networked practice-based innovation processes – interrogative model as a methodological approach. *Interdisciplinary Journal of Information, Knowledge, and Management, 3*, 87–101.
Knowles, J. G., & Cole, A. L. (2008). *Handbook of the arts in qualitative research: perspectives, methodologies, examples, and issues*. London: Sage.
[illegible]

Chapter 12
Collective Intelligence and Practice-Based Innovation: An Idea Evaluation Method Based on Collective Intelligence

Juho Salminen and Vesa Harmaakorpi

Abstract Users and customers are becoming increasingly important sources of knowledge due to changes in innovation policies and paradigms. Simultaneously innovation is becoming more of a networking activity. New methods are needed for processing information and ideas coming from multiple sources more effectively. For example, the whole personnel of an organisation are seen as a great potential for innovation. The recent development of communication technologies such as the Internet has increased interest towards the multidisciplinary field of collective intelligence. To investigate the possibilities of collective intelligence, the nest-site selection process of honeybees was used as model for an idea evaluation tool, a prototype of which was then tested in a case organisation. The results were promising; the prototype was able to evaluate ideas effectively, and it was highly accepted in the organisation.

12.1 Introduction

This study focuses on the front end of innovation, and more specifically on utilising the principles of collective intelligence in idea evaluation. The topic is very relevant in rapidly changing environments, where new sources of innovation are needed. For instance, the emphasis of Finnish innovation policy is changing from a science-based approach to a more practice-oriented one (Government's Communication... 2009). The change can be described as a shift from the Science, Technology and Innovation (STI) mode towards an approach including also the Doing, Using and Interacting (DUI) mode. While the STI mode relies strongly on the use of codified knowledge and science-based learning, in the DUI mode the focus is on tacit knowledge, organisational learning, and user needs (Jensen et al. 2007). The shift

J. Salminen (✉) • V. Harmaakorpi
Lahti School of Innovation, Lappeenranta University of Technology, Lahti, Finland
e-mail: juho.salminen@lut.fi

H. Melkas and V. Harmaakorpi (eds.),
Practice-Based Innovation: Insights, Applications and Policy Implications,
DOI 10.1007/978-3-642-21723-4_12,

of focus is very welcome indeed, as it is suggested that only 4% of enterprises consider universities important for producing innovations (Office for... 2004). The knowledge environment has become more fragmented, and as a result, the traditional approach to innovation with centralised R&D is replaced by other forms of creating innovation.

These changes in innovation policies and paradigms have major effects on the front end of innovation, which is defined as the activities taking place before the formal, well-structured development process begins. Activities in the front end of innovation process are often unpredictable and unstructured, and therefore hard to manage. Nevertheless, these activities have a major role in determining which projects to execute, and they affect strongly the definitions of quality, costs and time frame of the project. (Herstatt et al. 2004) Effective documentation and evaluation of ideas at the front end are necessary for the innovation process to be successful (Harmaakorpi and Melkas 2005; Tidd et al. 2005; Forssen 2001; Desouza et al. 2009). Changing paradigms are increasing the importance of customers, users, and shop floor employees as sources of knowledge and ideas, which makes these tasks challenging. In the STI mode of innovation, the transition from research to development is relatively easy and straightforward, but under the DUI mode the matters are more complex. The large amount of information increases the strain on idea processing mechanisms (Bjelland and Wood 2008; Bothos et al. 2009). These issues could be relieved by decentralising the tasks; in principle, anyone could evaluate ideas. Furthermore, the emerging fifth generation innovation processes demand flatter, more flexible decision making structures (Rothwell 1994).

The present study acknowledges the great and often hidden potential of the entire personnel of organisations for innovation. This potential remains often unexploited due to lacking methods of fostering different phases of innovation. There have been serious efforts to assess different kinds of, for example, High-Involvement innovation approaches (see e.g. Bessant 2003; Hallgren 2008), but the subject is still to a great extent untouched.

The aim of this study is to develop an idea evaluation tool capable of utilising collective intelligence. Employees on the shop floor form the main target group for the tool. Relevant literature on collective intelligence is first reviewed and theoretical insights are used to develop a prototype of an idea evaluation tool, which is based on the nest-site selection process of honey bees. The prototype is then tested in a case organisation. The study contributes to knowledge on the applicability of collective intelligence on innovation processes.

12.1.1 Collective Intelligence

Collective intelligence is an age-old phenomenon, but it's the recent changes in technology that make it highly relevant now. Constantly decreasing costs of communication enable new forms of decentralisation and collaboration in organisations (Malone 1997). The vast possibilities of collective intelligence are demonstrated by the indisputable successes of such systems as the Google search engine and Wikipedia (MIT Center... 2009). As the aim of collective intelligence is to aggregate the knowledge of large groups, it seems like a promising approach to dealing with current issues of front end of innovation.

Collective intelligence can be defined broadly as "the general ability of the group to perform a wide variety of tasks" (Woolley et al. 2010) or as "groups of individuals doing things collectively that seem intelligent" (Malone et al. 2010). The term is closely related to swarm intelligence, which means collective behavior emerging from self-organising groups of insects (Bonabeau and Meyer 2001). Even if one individual is not capable of much, collectively a swarm of insects can solve difficult problems of nest-site selection and nest building, foraging, task division, and route optimisation (e.g. Bonabeau and Meyer 2001; Camazine et al. 2001; Conradt and Roper 2005; Visscher 2007). Many artificial systems have been designed on the basis of swarm intelligence, including Internet traffic routing algorithms and logistics and production line management systems (Bonabeau and Meyer 2001). Wikis, open source software development communities and prediction markets are further examples of the potential of collective intelligence (e.g. Bonabeau 2009; Tapscott and Williams 2006; Wolfers and Zitzewitz 2004).

Even though various human networks and communities are often very effective at their tasks, it should not be assumed that the performance of a group is automatically intelligent. Information processing and decision making of human beings are susceptible to many biases, which can take place both on individual and group levels (Sunstein 2006; Hinsz et al. 2007; Iandoli et al. 2008; Asch 1955). Although complete theoretical framework does not yet exist, certain features of decision-making systems such as the diversity of viewpoints, independence of opinions, decentralised decision making and self-organisation are considered to facilitate collective intelligence.

The diversity of opinions is considered to be a critical factor in collective intelligence. Adding new perspectives to a subject matter is valuable as it brings in new ideas and viewpoints, which would otherwise probably remain absent in a group. (Surowiecki 2004; Bonabeau 2009) Homogenous groups tend to be good at what they do, but they often lack the capability to explore for new solutions (Surowiecki 2004). Simulation models have shown that diverse groups of problem solvers can outperform homogenous groups of highly skilled problem solvers (Hong and Page 2004).

Diversity can also help to reduce the negative effects of individual and group-level biases trough the addition of perspectives and by making it easier for people to voice their opinions (Surowiecki 2004). Nevertheless, the right balance of diversity and expertise is required when tapping into the collective intelligence of crowds. While some problem-solving situations benefit from adding more perspectives, this beneficial effect is prevented if the problem solvers do not have any knowledge about the topic whatsoever. (Bonabeau 2009; Page 2007)

A certain level of independence is another major factor enhancing collective intelligence. Independence produces random errors in individual estimates, which can be filtered out through aggregation. Individual assessments always contain some errors, but unless the mistakes the people make do not become correlated and are not systematically pointing in the same direction, the errors do not harm the collective decision making. Independent individuals are also more likely to have new information, which increases the diversity of the group. (Surowiecki 2004)

To avoid the group decision-making biases, the collective intelligence system should support the independence of the estimations produced by the group members. For a group to be collectively intelligent, the availability of diverse information and the ability to make individual aggregations are required. (Surowiecki 2004)

The decentralisation of decision making has potential to yield many benefits through increased efficiency, motivation and creativity of individuals, and higher flexibility of organisations. Good decisions require accurate information, and information technology makes the structures requiring high levels of communication feasible; it is now economically feasible to transfer large amounts of data to decentralised decision makers, for example through the internet. As the transferability of knowledge varies, it would make sense to transfer the easily communicated explicit knowledge to the decision makers possessing sticky and tacit knowledge (Malone 1997). Decentralisation also allows many minds to work on the same problem simultaneously (Malone 2004). The energy and creativity of people often depend on who makes the decisions about their work. Even though physically hard or routine work may benefit from supervision from above, creative work flourishes when the external control is reduced (Amabile 1998; Malone 1997). This way the decentralised decision making can increase motivation and creativity.

Self-organisation can be defined as "a process in which [a] pattern at the global level of a system emerges solely from numerous interactions among the lower-level components of the system. Moreover, the rules specifying interactions among the system's components are executed using only local information, without reference to the global pattern" (Camazine et al. 2001). Self-organisation is considered to be one of the facilitating factors of collectively intelligent phenomena because it increases the flexibility and robustness of a system (Bonabeau and Meyer 2001; Gloor 2006). A flexible system is able to adapt to changes in the environment, and robustness means that even if a part of the system fails, it is still able to perform its tasks. Social insects use self-organising systems extensively to solve complex problems, such as foraging route optimisation, nest-site selection and division of work (Camazine et al. 2001). Self-organisation results from a set of mechanisms relying on easy-to-use components at hand. The main components include simple rules of thumb, multiple interactions, positive and negative feedback loops, and the constraints of the environment (Camazine et al. 2001).

12.1.2 Assessment of Existing Systems

Various approaches have been used to capture the collective intelligence of groups of different sizes. Some well-known examples of such systems are listed in

Table 12.1 Examples of systems utilising collective intelligence (Handbook of Collective Intelligence 2009; Bjelland and Wood 2008; Bonabeau 2009; Tapscott and Williams 2006)

Example	Description
Amazon Mechanical Turk	A marketplace for small tasks developed by Amazon.com
Collective Intellect	Summarises information from blogs and web pages to produce knowledge for marketing
Digg	A news discovery service integrating submissions and evaluations of many people
Google	Search engine based on aggregated implicit human evaluations
Gwap	Games with the purpose of training computers to solve problems
IBM Innovation Jam	A massive-scale collaborative innovation session
InnoCentive	A marketplace for innovations connecting solution seekers to problem solvers
Lego Mindstorms	A robot toy. Users can participate in development through a forum
My Starbucks Idea	A web page for collecting ideas from customers of Starbucks
Intrade	A prediction market in which people buy and sell predictions about the future
reCHAPTCHA	CHAPTCHA service that helps to digitise books
Salesforce IdeaExchange	An open and direct channel of communication for salesforce customers
Sermo	A closed community of health care professionals
SourceForge	One of the leading open source software community environments
Threadless	Clothing retailer selling t-shirts designed and chosen by users
Wikipedia	Free encyclopedia that anyone can edit
YourEncore	Connects companies with retired scientists and engineers

Table 12.1. After that, a few selected approaches for idea generation and evaluation are discussed in more detail.

Various group decision support systems have been used to improve the quality of teamwork. Group decision support systems are electronic systems designed for supporting meetings and group work (Dennis et al. 1988). They have been shown to be capable of facilitating opportunity identification, idea generation and assessment, customer need assessment, and concept evaluation in small groups. Especially various voting features have been described as very useful, allowing the users to see which ideas are viewed as the most important. (Elfvengren 2006)

On the other end of the scale is the IBM Innovation Jam. It is a massive-scale (up to 150,000 participants) collaborative innovation session held over the Internet. Advantages of the IBM Innovation Jam are that it takes every comment seriously and is capable of aggregating many mediocre ideas together. The Innovation Jam has been successful in generating significant new businesses for IBM, but it can only be used periodically due to the ineffective evaluation of ideas. The evaluation process is time consuming and requires lots of resources; at IBM, the generated content was first evaluated by volunteers, and then top management spent a week reviewing the results. (Bjelland and Wood 2008)

Relatively simple information aggregation methods can be employed to improve both the accuracy and capacity of the evaluation process. Prediction markets have been found capable of producing accurate predictions about future events (Berg et al. 2008; Cowgill 2005; Wolfers and Zitzewitz 2004). Applying them for collecting and evaluating ideas has also yielded encouraging results, but information overload has been found to be an issue. Deciding which suggestions by users to introduce on the market is difficult. Democratising the decision process thoroughly is suggested as a possible solution for this issue. (Bothos et al. 2009)

A fully democratised approach has already been adopted by websites such as Digg and Salesforce IdeaExchange, which ask visitors to vote on the content submitted to the pages continuously. The items getting the most votes are displayed at the top of the page and visitors are encouraged to comment on them. The main advantage of these systems is that when a good idea is submitted, the discussion can evolve fast as to how the idea could be put into practice. These systems can also be scaled up easily and can handle large amounts of information. The moderate accuracy of the evaluations is an issue though, as these systems are known to produce plenty of false positives, and on the other hand ignore some of the valuable content. (Bjelland and Wood 2008) A balanced compromise between the speed and accuracy of evaluations is needed to increase the usability of such idea management systems.

12.1.3 Nest-Site Selection by Honeybees

Human beings are not the only species on our planet who face challenging decision-making tasks. Nest-site selection is a crucially important decision for social insect colonies. While in most cases this decision is made individually by the founding female, in some species of ants, bees, polistine wasps, and stingless bees, the colony moves to a new nest as a whole. In these cases, the decision about the nest-site is made collectively. Studies in the field of biology have found striking similarities between the nest-site selection processes of these species, although they have all evolved the mechanisms of social behaviour independently (Visscher 2007). Separate species are converging towards the same solution from differing initial conditions. The decision-making process is also scalable and fits well for different sizes of insect colonies (Franks et al. 2006).

On honeybees, the nest-site selection process has been studied thoroughly and it is also one of the most complex known examples of self-organising group decision making among social insects (e.g. Seeley and Buhrman 1999; Seeley et al. 2006; Passino et al. 2007; Visscher 2007). Like other species, honeybees use cognitively demanding weighted additive strategy in their decision making (Seeley and Buhrman 1999; Visscher 2007). All the relevant attributes of each alternative are evaluated and given different weights, depending on their relative importance. Then the weighted evaluations are combined and the best overall option is chosen.

The phases of the nest-site selection process of honeybees are the following (Seeley and Buhrman 1999):

1. Swarming. A colony of bees forms a cluster on a tree branch.
2. Search. Scouts start searching for suitable nest sites in the environment.
3. Evaluation. When a scout bee finds a possible nest site, it evaluates the site and returns to the swarm to announce the finding by means of a waggle dance.
4. Competition. Scouts following the dance fly to the described site and repeat the evaluation process.
5. Decision. A final decision is made at a nest site when the number of scouts reaches a threshold value.

During the nest-site selection process, individual evaluations are aggregated to the collective opinion on the dance floor. The total number of dance rounds a scout bee performs depends on the quality of the site on a near-exponential scale: 15, 45, 90, 150, 225, and 315 dance rounds on average (Seeley et al. 2006). The higher the number of dance rounds, the higher the probability that another scout will follow the dance. As a result traffic to high-quality nest sites increases faster than traffic to lower-quality nest sites. When the number of scout bees evaluating a particular nest site simultaneously reaches a threshold value of about 15, the scouts interpret that the decision has been made. After that the scouts return to the swarm and stimulate it to take off and fly to the new nest site. (Seeley and Buhrman 1999)

Simulations have shown that natural selection has tuned the different parameters of this process near the optimum compromise between the speed and accuracy of the decision (Passino and Seeley 2005). The attention of the swarm turns quickly to better-quality sites, while the poor-quality sites are dropped from consideration relatively fast (Passino et al. 2007). That is, the resources of the swarm are directed to the evaluation of the higher-quality sites. At the same time, the likelihood of a bad decision remains low (Passino and Seeley 2005).

The decision-making process has many interesting features from the point of view of collective intelligence. During the process, individual scouts use only local information and no direct comparison of nest sites is necessary. All the available information is taken into account in the process, but none of the bees has to hold all the information. The exponential scale used in the evaluation amplifies the perceived differences in nest site qualities (Seeley et al. 2006). Even while each individual follows only simple rules of thumb and uses only local knowledge, the self-organising system manages to integrate available information in a meaningful way. (Conradt and Roper 2005; Visscher 2007)

12.2 Methods

This study aims to develop an effective idea evaluation tool for the front end of innovation processes capable of utilising collective intelligence. The study is carried out as a case study research using a constructive research approach (see Kasanen et al. 1993), where the focus is on designing new constructs and testing them in real-life applications. After a tool for evaluating ideas emerging from the course of everyday work is developed, it is tested in a case organisation. Typically for a case study, multiple sources of evidence are used to increase the reliability of the results (Yin 1994).

12.2.1 Building the Construct

A clear need exists for effective tools for collecting and evaluating ideas emerging from multiple sources. Such a tool should enable easy documentation and evaluation of large amounts of ideas in order to avoid the common issue of information overload. Evaluations should be accurate enough to discriminate promising ideas from mediocre ones. A more complete list of the requirements derived from the relevant literature is presented in Table 12.2.

A tool utilising the collective intelligence of large groups could be able to fill these requirements. In this study, the nest site selection process of honeybees is used as a model for the idea evaluation tool. The issues in nest site selection and idea evaluation are very similar. In both cases the number of alternative options is high,

Table 12.2 Requirements for an idea evaluation tool

Requirement	References
Users as a source of knowledge	Government's Communication... 2009; Jensen et al. 2007; Kallio and Bergenholtz (forthcoming); Von Hippel 2005; Bessant 2003
Effective documentation, evaluation and processing of ideas	Harmaakorpi 2004; Forssen 2001; Tidd et al. 2005; Bessant 2003
Support for bringing up problems and incomplete ideas	Bessant 2003; Paalanen and Konsti-Laakso 2008
Capacity to process a large amount of ideas	Bothos et al. 2009; Paalanen and Konsti-Laakso 2008
Feedback from co-workers	Dominic et al. 1997
Integration of the organisation and systems	Rothwell 1994; Gloor 2006
Flat and flexible organisational structures	Rothwell 1994; Gloor 2006; Malone 1997; Surowiecki 2004
High connectivity and interactivity	Rothwell 1994; Gloor 2006
Decentralised decision making	Rothwell 1994; Gloor 2006; Malone 1997; Surowiecki 2004
Motivation to participate	Harmaakorpi and Melkas 2005; Gloor 2006; Malone et al. 2010; Bonabeau 2009; Gloor and Cooper 2007; Tapscott and Williams 2006; Bessant 2003
Diversity	Surowiecki 2004; Bonabeau 2009; Hong and Page 2004; Page 2007
Independence	Surowiecki 2004; Asch 1955
Modularity	Malone 2004
Self-organisation	Bonabeau and Meyer 2001; Camazine et al. 2001; Gloor 2006

and a good decision requires accurate assessment and integration of multiple attributes of each option. In nest site selection, important attributes are at least the sizes of the nest cavity and the entrance, while idea evaluation should take into account economic, technical, and organisational viewpoints. It is likely that abilities and knowledge needed to evaluate different attributes are dispersed among various individuals.

The nest site selection process of honeybees is an effective solution to attention-perception-choice type of problems, as demonstrated by simulation models and the convergent evolution of various social insect species. The nest site selection process also utilises all the facilitating factors of collective intelligence: diversity of opinions, independence of evaluations and decentralisation of decision making in a modular, self-organising process.

12.2.2 Features of an Idea Evaluation Tool

A constructed evaluation tool uses a strictly standardised procedure with a random allocation of evaluation tasks to ensure the equal treatment of all the ideas, to optimise the use of resources and to minimise the negative effects of systematic biases.

The phases of the idea collection and the evaluation process are the following:

1. A user generates an idea and feeds it into the system.
2. The system selects evaluators for the idea randomly.
3. Selected users evaluate the idea on a linear scale from 1 to 5 (1, 2, 3, 4 and 5).
4. The system transforms the evaluations to a non-linear scale (0, 0.5, 1.5, 3 and 5). The ratio of the scale is the same as in the nest site selection process of honeybees
5. The scores on the non-linear scale are averaged and ideas listed in the order of superiority.

In theory, the advantages of such an evaluation tool are numerous. Emphasis on the ease of use and on a light workload enables documentation and evaluation of all the ideas emerging during everyday work. The diversity of users increases both the likelihood of new ideas emerging and the accuracy of evaluations by aggregating information from multiple sources. The randomised selection of evaluators makes evaluations independent of each other, which should reduce the typical issues of group decision making, such as conformity to opinions of others in group decision-making situations. At the same time, the accuracy is improved by introducing a random error in evaluations, which can then be eliminated through aggregation. Furthermore, the independent evaluations and users being unaware of the origin of the ideas or opinions of other users should reduce the influence of politics, hierarchies and harmful cascade effects.

By decentralising the decision making, *tacit knowledge otherwise difficult to transfer can be utilised.* Dividing the evaluation tasks helps to avoid overloading the system and allows faster feedback to users about their ideas. Decentralisation also spreads ideas further in the organisation, which could stimulate new ideas from other users. Combined with fast feedback and appreciation of employees' opinions demonstrated by empowerment, the tool could help to activate employees to participate more enthusiastically in the innovation process.

The use of a non-linear evaluation scale borrowed from bees improves the results of evaluations in two ways: the differences at the top of the scale are emphasised and the variance of the evaluations is automatically accommodated. Because of the non-linearity of the scale, ideas with high variance receive higher scores than the ones with low variance. This should be beneficial for the accuracy of evaluations, because logically the ideas dividing the opinions should be more likely to be valuable than the ideas unanimously assessed as average. In other words, it is assumed that evaluators have knowledge about the value of ideas, and low variance in evaluations implies that all available knowledge points to the same value. High variance implies that in some viewpoints the idea has high value, and further investigation might be needed to assess the actual potential of the idea.

12.2.3 Case Study

The developed construct was tested in a multidisciplinary research organisation. The organisational structure of the unit is typical for scientific units: researchers work more or less independently with centralised management. Formal tools for idea collection or evaluation have not been used in the organisation before, but instead ideas have been put forward more or less randomly. As Fig. 12.1 shows, in recent years the unit has grown. As a result, a need for more formal methods for promoting ideas has been recognised by the management of the unit. This finding is

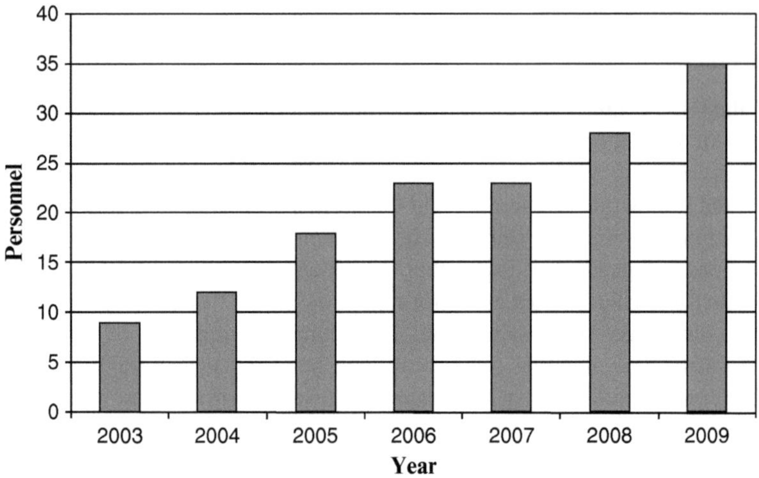

Fig. 12.1 Development of staff resources at the case organisation

in line with the notion that the marginal limit for the size of collaborations in team form is typically around 25 (Lipnack and Stamps 2000).

In order to test the developed construct in a research organisation, a manually operated prototype of the idea evaluation tool was set up. Email was used to facilitate the idea collection and evaluation processes. The types of accepted ideas were not limited in any way and evaluation criteria were left vague in order to ensure diversity. The instructions given to participants can be found in Appendix 1. Items submitted by the users were divided in four categories: ideas, problems, observations and development ideas. The ideas category consisted of ideas and general suggestions for improvements. The problems category was used for announcing detected issues demanding further attention and solutions. Interesting findings from newspapers, television, Internet or other sources were posted in the observations category. The development ideas category was used for suggestions concerning improvements for the idea evaluation tool. The items were collected via email and from informal conversations. Each idea, problem or observation was sent to ten randomly chosen users for evaluation. The evaluations of the users on a linear scale (1, 2, 3, 4 and 5) were transformed to a non-linear scale (0, 0.5, 1.5, 3 and 5) according to the specifications of the tool and the average scores for each idea were calculated. Finally the ideas, problems and observations were listed in order of superiority in the corresponding categories. The updated list was published weekly.

Even though further development or implementation of ideas was not in the scope of this study, it was found necessary to take actions on surfacing issues during the testing period. A couple of the best new ideas were discussed at weekly meetings with all employees, and decisions about further actions were made.

The effectiveness and efficiency of the prototype was evaluated using three methods. The performance of the prototype was observed, users were asked to fill in a questionnaire after the test period and two representatives of the unit's management were interviewed.

12.3 Results and Discussion

The prototype of the idea evaluation tool was tested from March 17 until June 15, 2009. 29 employees of the research organisation participated in the test by evaluating ideas at least once. Ideas, problems and observations originated from 24 different employees. 31 ideas were submitted via email, while the rest came up during informal face-to-face conversations and were submitted for evaluation by the facilitator of the prototype. Each item was sent for ten randomly chosen employees for evaluation. On average, seven of them responded. The quantitative results from the idea evaluation prototype testing are summarised in Table 12.3.

The idea evaluation tool managed to collect and evaluate 48 ideas, 14 problems, seven observations and five development ideas, totalling 74 items altogether. The idea and problem categories yielded the most of useful content. Observations did not lead to any further actions and due to the limited number of development ideas,

Table 12.3 Summary of quantitative results from the idea evaluation tool prototype testing

	Ideas	Problems	Observations	Development ideas
All ideas	48	14	7	5
Average score	2.95	2.82	2.52	2.83
Standard deviation	0.91	0.95	0.60	0.93
Ideas in development	15	6	0	3
Percent of all ideas	31.25	42.86	0.00	60.00
Average score	3.51	3.17	0.00	3.43
Standard deviation	0.64	1.21	0.00	0.44
Ideas with no action	33	8	7	2
Percent of all ideas	68.75	57.14	100.00	40.00
Average score	2.70	2.56	2.52	1.92
Standard deviation	0.90	0.66	0.60	0.59
One-tailed *t*-test	0.0005	0.1474	0.0000	0.0527

Table 12.4 Summary of qualitative results from the idea evaluation tool prototype testing

Type of innovation	Ideas	Development ideas	Total
Product/Service	2	5	7
Incremental	2	2	4
New to organisation	0	3	3
Radical	0	0	0
Process	44	0	44
Incremental	30	0	30
New to organisation	13	0	13
Radical	1	0	1
Position	2	0	2
Incremental	0	0	0
New to organisation	1	0	1
Radical	1	0	1
Paradigm	0	0	0
Incremental	0	0	0
New to organisation	0	0	0
Radical	0	0	0

a separate category for them was quite unnecessary. The average score of all evaluated items on the non-linear scale was 2.88. At the end of the test period, 24 of the items collected by the tool had proceeded to the implementation phase or were seriously considered for further development. 19 of these items had scored at least three. The average score of the items in the implementation phase was 3.42, as opposed to 2.62 for the items that had not caused any actions. The difference between the scores is statistically significant (one-tailed t-test, $\alpha = 0.0001$). An item was considered to have proceeded to the implementation phase when a decision about implementation had been made, the proposed actions had been taken or the search for a solution to an announced problem had begun.

Table 12.4 summarises the qualitative results from the prototype testing. In the case organisation, the vast majority of ideas were related to processes. 64% of all items suggested incremental improvements, 32% dealt with matters new to the organisation and only two ideas could be considered to be quite radical. It should be noted that these categorisations of ideas are only suggestive, providing a view on what sort of ideas were generated during the test.

Based on the results of the prototype testing, the idea evaluation tool seems to be able to bring up process innovations in particular. It is difficult to tell, however, whether this is a feature of the tool or of the case organisation. The main focus of the work done in the organisation is on research and it is possible that the results would have been different in an organisation aimed strictly at product or service development. The degree of novelty involved in the ideas was generally low. The majority of them were incremental improvements or solutions that are broadly used outside the organisation. Even though radical innovations should be expected to be much rarer than incremental ones, the lack of wild and outstanding creativity among submitted ideas suggests that the idea evaluation tool might be better suited for facilitating gradual improvements. This is in line with the literature stating that different methods are needed to support incremental and radical innovations (e.g. McDermott and O'Connor 2002).

The efficiency of evaluations was estimated qualitatively by using three factors: the accuracy of evaluations, the speed of evaluations and the processing capacity of the system. The results of testing the idea evaluation tool show that ideas receiving high scores are implemented much more frequently than ideas with low scores. Considering the number of participants, the idea evaluation tool managed to collect a reasonably high number of ideas. Many of them were unfinished, indicating a low threshold for submitting ideas. The idea evaluation tool seems to be capable of recognising promising ideas and important issues with sufficient accuracy.

The results from the evaluations were usually gathered in a few days after submitting an idea. An updated list of evaluated items was published every Monday; users got feedback about their ideas within a week. The number of ideas did not load the idea evaluation tool notably during the testing period. The sufficient accuracy and speed of evaluations with adequate processing capacity indicate that the idea evaluation tool is efficient in its task.

It should be noted that technical functionality alone is not enough for a tool to be useful. Users' willingness to participate in idea generation and evaluation is one of the critical factors determining the success of the idea evaluation tool. In the case organisation, the lack of motivation to participate was not found to be an issue.

The documentation and evaluation of ideas was found to be of major importance for the successful management of the front end of the innovation process. This was demonstrated when several old issues surfaced during the testing of the idea evaluation tool prototype. Many of the issues had been known for years and were discussed every now and then in the organisation, but without effective documentation and evaluation they were forgotten before any actions were taken. Additionally, without prioritisation it can be difficult to decide on which issues to concentrate on. The introduction of the idea evaluation tool managed to provide some formalisation to the chaotic front end of the innovation process.

The idea evaluation tool was found to be able to stimulate idea generation. Many times the participants responded to evaluation requests by submitting new ideas for evaluation. Unfinished ideas were improved and solutions were suggested for problems. On one occasion, submitting an idea for evaluation generated a chain of new ideas about varying topics. Some of these ideas were then developed further by suggesting improvements for them. The end product of this emergent process was a fully developed concept ready for implementation. The people with the necessary skills and willingness to take care of the execution were also found during the successive idea evaluations. In this anecdotal case, the whole process seemed to emerge very naturally.

The acceptance of the idea evaluation tool in the case organisation was evaluated with a short questionnaire for employees and with more detailed interviews with two managers: one in charge of the unit and another responsible for the operational management of the unit. The questionnaire and interviews took place after the prototype testing was finished. 19 participants (including the two managers) responded to the questionnaire, giving the response rate of 66%. The results of the questionnaire are summarised in Fig. 12.2.

Generally the results of both the questionnaire and the interviews were very positive and encouraging and indicate acceptance of the idea evaluation tool in the organisation. In consequence of the implementation of the idea evaluation tool, the whole issue of idea generation has surfaced, leading to many beneficial changes in idea management. The process was considered much more visible, giving everyone a possibility to participate in idea generation. The idea evaluation tool allowed the development of ideas to be spread further in the organisation; tasks that previously

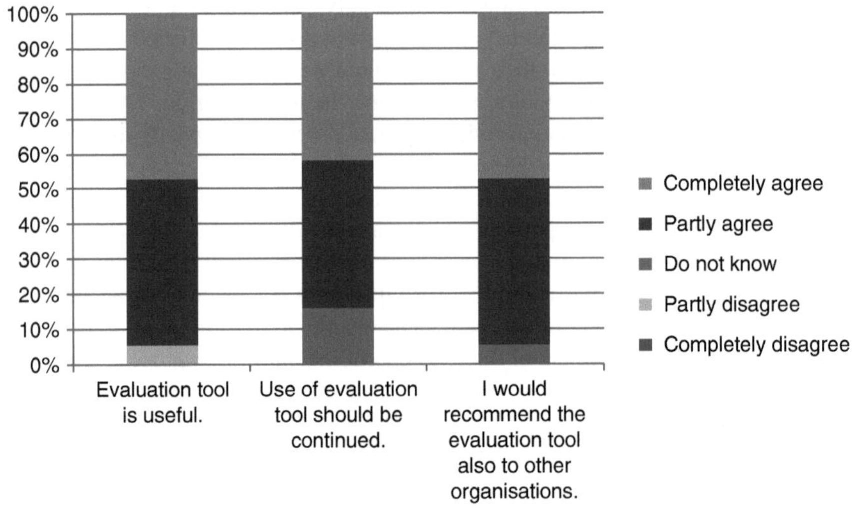

Fig. 12.2 Results of the questionnaire

were in the responsibility of a single individual could now be divided. The documentation of the ideas made it possible to observe the types and the amount of ideas the organisation was able to produce.

The implementation of the idea evaluation tool was not considered to have serious disadvantages. A few participants viewed aggregating personal knowledge to a single numeric evaluation as difficult and possibly too simplified to capture the often complex nature of problems. Evaluating ideas in simplified form was also suspected to increase the risk of misunderstandings, especially if the user is not familiar with the context of a particular idea. Some of the evaluations were viewed as unnecessary; duplicates, minor issues or ideas that can be put into practice straight away would not require a formal evaluation process. This raised suspicion about the organisation losing some of its spontaneity and creativity if things are not just done anymore. All these previous issues were each mentioned only a few times in the interviews and the questionnaire. On the other hand, the lack of effective implementation after the evaluation process was brought up in both interviews and it was the most often mentioned issue in the questionnaire. Ineffective implementation makes even the best ideas worthless and can reduce the motivation to participate in idea generation. The important question here is what happens after the front end of the innovation process. Answering the question is not in the scope of this study, but it would nevertheless be interesting to examine this topic more thoroughly.

The continuation of the use of the idea evaluation tool in the case organisation is supported by the personnel and the management of the unit. The prototype is viewed as very promising but it naturally requires further development. Idea generation and evaluation should be made even more visible and implementation issues should be resolved, allowing a smoother transition from the front end to the later phases of the innovation process.

The findings of this study are in line with the theoretical background of collective intelligence. The diversity of users contributed to the emergence of new ideas and viewpoints. The modularity of the system allowed the decentralisation of the evaluation tasks and the combining of results in a meaningful way. Self-organisation was demonstrated when only two simple rules (1. submit an idea if you get one, 2. evaluate an idea if you are requested to do so) were needed to collect and organise ideas emerging from the organisation. In this case, the intrinsic motivation of users was enough to ensure sufficient participation, as no forms of external rewards were offered. Still, as the feedback from the users suggests, the lack of effective implementation could greatly reduce the motivation to submit ideas. This issue should be contemplated more thoroughly. In theory, the independence and decentralisation of evaluation tasks should reduce the biases of decision making and utilise tacit knowledge, but without comparison to a more centralised and dependent system these effects could not be measured.

At the time of writing (late 2010), the first software version of the idea evaluation tool has just been installed at the case organisation. Simultaneously, some new features could be added. Users now have the possibility to comment on ideas, as the feedback from users indicated that this feature would be essential. A search function and alternative listings of submitted ideas based on category, time of posting and ratings are now also available, allowing easier browsing of the collected content.

There are also some features that have been considered but have not been implemented yet. A mechanism for grouping or combining ideas could be useful. Many ideas do not work as stand-alone solutions, but they could be feasible when combined with other similar ideas. Tags or categorising could help facilitate this function, as well as improving the usability of the tool in larger organisations. In addition to collecting only emerging bottom-up ideas, it might be useful to be able to request ideas on pre-specified topics and to use identified problems to direct idea generation. In large organisations, it could make sense to create user profiles and divide the employees in different groups. Instead of purely random allocation, users with specific profiles could be chosen to carry out evaluations; an idea connected closely to a production line could be evaluated solely by the production line employees, or a certain amount of evaluators from each group could be chosen to ensure sufficient representation of all the relevant viewpoints. In this case, the idea evaluation tool was used in a rather small organisation, but it could presumably be scaled to fit organisations of different sizes by changing the number of users evaluating each idea and the number of evaluation rounds. Similarly the trade-off between the speed and accuracy of evaluations could be adjusted.

So far the idea evaluation tool has been shown to be usable in collecting inputs inside an organisation. An interesting question is whether it could also be used as a link to external knowledge from customers, users and other organisations. Due to its low need for cognitive or ICT skills, the idea evaluation tool could prove to be valuable in capturing the insights of a large user group and in facilitating their more active participation in practice-based innovation activities.

12.4 Summary and Conclusion

The focus of this study was on the front end of the innovation process. The recent shift from the STI mode towards the DUI mode and the changing paradigms demand new approaches to the management of the front end of innovation. The main objective of this study was to construct an effective tool for collecting and evaluating ideas at the front end of the innovation process. The problem of idea evaluation was approached from the point of view of collective intelligence. Interest in this relatively new multidisciplinary field is rising and it is not surprising that many alternative approaches are being explored at the moment. Only time will tell which solutions are the best, but one thing seems obvious already; attention to details is crucial when designing a system aiming at facilitating collective intelligence. Most systems include at least some self-organising properties and as a result, apparently trivial changes in details may generate huge differences on the system level. Millions of years of evolution have optimised the decision making process of bees, which was therefore chosen as a model on which the development of the idea evaluation tool was based.

The prototype of the idea evaluation tool was tested in a case organisation and the results were encouraging. The prototype managed to distinguish promising ideas from mediocre ones and to point out important issues. Additionally, spreading ideas around the organisation was found to stimulate the generation of new ideas. The common wisdom about the importance of proper implementation was strengthened up once again; even the best ideas are worthless if they are not properly put into practice. The usability of the developed construct was demonstrated by the acceptance of the idea evaluation tool in the case organisation.

Appendix: Instructions for the Users of the Idea Evaluation Tool Prototype

Translation from the original Finnish instructions used in the test.

The Idea Evaluation Tool

Instructions

The purpose of the tool is to collect all the occurring ideas, as they are most of the time forgotten instantly. Forget self-criticism, because all thoughts are accepted. The ideas are divided in four categories:

1. Ideas: all general development suggestions and ideas
2. Problems: observed problems and things requiring improvements
3. Observations: observations about the surrounding world that others might find interesting, for example newspaper articles
4. Development ideas: suggestions and observations concerning this evaluation tool

Submitting an Idea

Follow these instructions when you are submitting an idea for evaluation:

1. Write either IDEA, PROBLEM, OBSERVATION, or DEVELOPMENT IDEA in/as the topic of a message.
2. Describe the idea, problem, observation or development idea briefly in the message.
3. Send the message to (email address of facilitator).

Evaluating an Idea

When you receive a message with IDEA, PROBLEM, OBSERVATION or DEVELOPMENT IDEA in the topic, follow these instructions:

1. Evaluate the importance of the idea, problem, observation or development idea according to your best knowledge on a scale of 1–5. 1 means poor/unimportant and 5 means good/important.
2. Send your reply to (email address of facilitator).

Results

The evaluated ideas can be viewed in a common network folder.

References

Amabile, T. (1998). How to kill creativity. *Harvard Business Review, 76*(5), 76–87.

Asch, S. (1955). Opinions and social pressure. *Scientific American, 193*(5), 31–35.

Berg, J., Nelson, F., & Rietz, T. (2008). Prediction market accuracy in the long run. *International Journal of Forecasting, 24*(2), 283.

Bessant, J. (2003). *High-involvement innovation: Building and sustaining competitive advantage trough continuous change*. Chichester: Wiley.

Bjelland, O., & Wood, R. (2008). An inside view of IBM's 'Innovation Jam'. *MIT Sloan Management Review, 50*(1), 32–40.

Bonabeau, E. (2009). Decisions 2.0: The power of collective intelligence. *MIT Sloan Management Review, 50*(2), 45–52.

Bonabeau, E., & Meyer, C. (2001). Swarm intelligence: A whole new way to think about business. *Harvard Business Review, 79*(5), 106–114.

Bothos, E., Apostolou, D., & Mentzas, G. (2009). Collective intelligence for idea management with Internet-based information aggregation markets. *Internet Research, 19*(1), 26–41.

Camazine, S., Deneubourg, J.-L., Franks, N. R., Sneyd, J., Theraulaz, G., & Bonabeau, E. (2001). *Self-organization in biological systems*. New Jersey: Princeton University Press.

Conradt, L., & Roper, T. (2005). Consensus decision making in animals. *Trends in Ecology & Evolution, 20*(8), 449–456.

Cowgill, B. (2005). Putting crowd wisdom to work. E-document. The Official Google Blog. http://googleblog.blogspot.com/2005/09/putting-crowd-wisdom-to-work.html. Accessed 7 May 2009.

Dennis, A., George, J., Jessup, L., Nunamaker, J., & Vogel, D. (1988). Information technology to support electronic meetings. *MIS Quarterly, 12*(4), 591–618.

Desouza, K., Dombrowski, C., Awazu, Y., Baloh, P., Papagari, S., Jha, S., & Kim, J. (2009). Crafting organizational innovation processes. *Innovation: Management, Policy & Practice, 11*(1), 6–33.

Dominic, P., Reilly, R., & McGourty, J. (1997). The effects of peer feedback on team member behavior. *Group & Organization Studies, 22*(4), 508–520.

Elfvengren, K. (2006). *Group support system for managing the front end of innovation: Case applications in business-to-business enterprises*. Lappeenranta: Lappeenranta University of Technology.

Forssen, M. (2001). Life cycles of organizational 'bottom-up' development ideas. *Knowledge and Process Management, 8*(4), 249–261.

Franks, N., Dornhaus, A., Best, C., & Jones, E. (2006). Decision making by small and large house-hunting ant colonies: one size fits all. *Animal Behaviour, 72*, 611–616.

Gloor, P. (2006). *Swarm creativity*. New York: Oxford University Press.

Gloor, P., & Cooper, S. (2007). The new principles of a swarm business. *MIT Sloan Management Review, 48*(3), 81–84.

Government's Communication on Finland's National Innovation Strategy to the Parliament (2009). E-document. Ministry of Employment and the Economy. http://www.tem.fi/files/21010/National_Innovation_Strategy_March_2009.pdf. Accessed 15 April 2009.

Hallgren, E. (2008). *Employee driven innovation: A case of implementing high-involvement innovation*. Lyngby: Technical University of Denmark, Department of Management Engineering.

Handbook of Collective Intelligence (2009). E-document. Handbook of collective intelligence. http://scripts.mit.edu/~cci/HCI/index.php?title=Main_Page. Accessed 3 March 2009.

Harmaakorpi, V. (2004). *Building a competitive regional innovation environment – The regional development platform method as a tool for regional innovation policy*. Espoo: Helsinki University of Technology, Department of Industrial Engineering and Management.

Harmaakorpi, V., & Melkas, H. (2005). Knowledge management in regional innovation networks: The case of Lahti, Finland. *European Planning Studies, 13*(5), 641–659.

Herstatt, C., Verworn, B., & Nagahira, A. (2004). Reducing project related uncertainty in the "fuzzy front end" of innovation – A comparison of German and Japanese product innovation projects. *International Journal of Product Development, 1*(1), 43–65.

Hinsz, V., Tindale, R., & Nagao, D. (2007). Accentuation of information processes and biases in group judgments integrating base-rate and case-specific information. *Journal of Experimental Social Psychology, 44*(1), 116–126.

Hong, L., & Page, S. (2004). Groups of diverse problem solvers can outperform groups of high-ability problem solvers. *Proceedings of the National Academy of Sciences of the USA, 101*(46), 16385–16389.

Iandoli, L., Klein, M., & Zollo, G. (2008). Can we exploit collective intelligence for collaborative deliberation? The case of the climate change Collaboratorium. E-document. MIT Sloan School of Management. http://cci.mit.edu/publications/workingpapers.htm. Accessed 8 March 2009.

Jensen, M., Johnson, B., Lorenz, E., & Lundvall, B. (2007). Forms of knowledge and modes of innovation. *Research Policy, 36*(5), 680.

Kallio, A., & Bergenholtz, C. (forthcoming). Generating innovation opportunities: how to explore and absorb customer knowledge. *International Journal of Technology Management*.

Kasanen, E., Lukka, K., & Siitonen, A. (1993). The constructive approach in management accounting research. *Journal of Management Accounting Research, 5*, 243–264.

Lipnack, J., & Stamps, J. (2000). *Virtual teams*. Toronto: Wiley.

Malone, T. (1997). Is empowerment just a fad? Control, decision making, and IT. *MIT Sloan Management Review, 38*(2), 23–34.

Malone, T. (2004). *The future of work*. Boston: Harvard Business School Press.

Malone, T., Laubacher, R., & Dellarocas, C. (2010). The collective intelligence genome. *MIT Sloan Management Review, 51*(3), 21–31.

McDermott, C., & O'Connor, G. (2002). Managing radical innovation: An overview of emergent strategy issues. *Journal of Product Innovation Management, 19*, 424–438.

MIT Center for Collective Intelligence (2009). MIT Center for Collective Intelligence http://cci.mit.edu/. Accessed 11 April 2009.

Office for Official Publications of the European Communities (2004). Innovation in Europe – Results for the EU, Iceland and Norway. E-document. Office for Official Publications of the European Communities. ftp://ftp.cordis.europa.eu/pub/innovationsmes/docs/results_from_cis3_for_eu_iceland_norway.pdf. Accessed 22 April 2009.

Paalanen, A., & Konsti-Laakso, S. (2008). Innovaatiohaavi organisaation innovaatiokyvykkyyden kehittäjänä [Innovation catcher as a developer of organizational innovation capability]. In V. Harmaakorpi & H. Melkas (Eds.), *Innovaatiopolitiikkaa järjestelmien välimaastossa* (pp. 186–195). Helsinki: Suomen Kuntaliitto. (In Finnish.)

Page, S. (2007). *The difference: How the power of diversity creates better groups, firms, schools and societies*. Princeton: Princeton University Press.

Passino, K., & Seeley, T. (2005). Modeling and analysis of nest-site selection by honeybee swarms: The speed and accuracy trade-off. *Behavioral Ecology and Sociobiology, 59*(3), 427–442.

Passino, K., Seeley, T., & Visscher, P. (2007). Swarm cognition in honey bees. *Behavioral Ecology and Sociobiology, 62*(3), 401–414.

Rothwell, R. (1994). Towards the fifth-generation innovation process. *International Marketing Review, 11*(1), 7–31.

Seeley, T., & Buhrman, S. (1999). Group decision making in swarms of honey bees. *Behavioral Ecology and Sociobiology, 45*(1), 19–31.

Seeley, T., Visscher, P., & Passino, K. (2006). Group decision making in honey bee swarms. *American Scientist, 94*(3), 220–229.

Sunstein, C. (2006). *Infotopia: How many minds produce knowledge*. New York: Oxford University Press.

Surowiecki, J. (2004). *Wisdom of crowds*. New York: Anchor Books.

Tapscott, D., & Williams, A. (2006). *Wikinomics: How mass collaboration changes everything*. New York: Penguin Group.

Tidd, J., Bessant, J., & Pavitt, K. (2005). *Managing innovation: Integrating technological, market and organizational change*. Chichester: Wiley.

Visscher, P. (2007). Group decision making in nest-site selection among social insects. *Annual Review of Entomology, 52*, 255–275.

Von Hippel, E. (2005). *Democratizing innovation*. Cambridge: MIT Press.

Wolfers, J., & Zitzewitz, E. (2004). Prediction markets. *Journal of Economic Perspectives, 18*(2), 107–126.

Woolley, A., Chabris, F., Pentland, A., Hashimi, N., & Malone, T. (2010). Evidence for a collective intelligence factor in the performance of human groups. *Science, 330*, 686–688.

Yin, R. (1994). *Case study research*. Thousand Oaks: Sage Publications.

Chapter 13
Users as Sources of Radical Service Innovation

A Closer Look into Opportunities for Integrating Service Lead Users in Service Development

Florian Skiba and Cornelius Herstatt

Abstract Though theory and praxis both underline the importance of radical service innovations, until now lead users in the service industry have been excluded from the scope of research. Based on explorative findings on user innovations in services and recent research on service innovations, this chapter answers the following questions: (a) can users lead trends and independently innovate in the service sector; (b) if so, do lead users in the service industry show similar characteristics to lead users from other industries, and (c) can approaches to identification and integration from the consumer and investment goods sectors be applied in the service sector as well? By illustrating the potential impact of lead users as a source of radical service innovations, we hope to raise interest in this powerful concept, not yet systematically applied in service industry.

13.1 Introduction

Most growth in modern economies is in services. For reasons of factor cost disadvantages compared to emerging market countries, mature economies are gradually shifting towards service-driven economies. In the US, the service sectors grew from 50% to 80% of the total employment over the last 50 years (Fitzsimmons and Fitzsimmons 2005); the picture is similar in OECD countries, where about two-thirds of business sector growth between 1985 and 1997 was in services (Howells and Tether 2004). In Germany, a country with a strong industrial heritage, service is now reaching levels of 70% and more of total economic value added (Destatis 2007).

F. Skiba (✉) • C. Herstatt
Institute for Technology and Innovation Management, Hamburg University of Technology, Hamburg, Germany
e-mail: Florian.Skiba@web.de

H. Melkas and V. Harmaakorpi (eds.),
Practice-Based Innovation: Insights, Applications and Policy Implications,
DOI 10.1007/978-3-642-21723-4_13,

As the service sector continues to grow, new players are gradually attracted, technologies change, and user needs shift. Continuous innovation efforts, thus, become imperative for existing service providers to reduce costs, enhance existing service quality, and expand current service offerings to increase market share in existing markets or to enter new ones (Ginzburg et al. 2007; Tether 2003).

To decrease the risk of market failure of innovations, future user needs must be anticipated and translated into innovative services early in the process of designing such market offerings.

However, radical innovations, which promise a sustainable competitive advantage in comparison to incremental innovations, seldom result from the integration of ordinary users. As sources of radical innovations, most users are unsatisfactory because they possess a limited level of imagination and abstraction – the latter being vital for the generation of truly novel ideas. Average users often misjudge fundamentally new dimensions or characteristics of performance, because they are not able to perceive the true level of benefit associated with such innovations (Herstatt 2002; von Hippel 1989). This problem exists especially for services, which are immaterial by nature, making it even harder to grasp the improved level of benefit for the market.

To overcome this problem, leading companies such as 3M, HILTI, Nortel Networks, and Kellogg are increasingly working with so-called *lead users* in the early phases of innovation projects (Herstatt 2002). The lead user approach offers a solution to the problem of user unsuitability for innovation work by targeting all integration efforts on specific users who deal with advanced usage situations, encountering needs today that the rest of the market does not yet face (von Hippel 1986).

Though the benefits of the lead user approach are unquestioned for consumer and investment goods (von Hippel et al. 1999), so far little is known about lead users in service industries. Besides studies by von Hippel and Riggs (1996) and Lüthje (2000) describing one case study each on the identification of lead users in a service setting, no service-specific insights on lead users appear to be available.

Based on multiple industry examples and recent research on service innovations, the present chapter, therefore, answers the questions: (a) can users lead trends and independently innovate in the service sector as well; (b) if so, do lead users in the service industry show similar characteristics to lead users from other industries, and (c) can approaches to identification and integration from the consumer and investment goods sectors be applied to the service sector as well?

To answer these research questions and gain insights into service lead user, case studies have been used to research the phenomenon of user-driven service innovations within real-life contexts (Flyvbjerg 2006; Yin 2002). After reviewing the existing literature to gain an understanding of radical service innovations and the characteristics of lead users, the Internet, newspapers, and company reports have been scanned for suitable cases of users being the source of a truly novel service innovation. As it is essential to understand the service user's motives for innovating to answer the research questions, only cases with adequate background information available (such as newspaper interviews with users) have been used as research material for this chapter.

Our work is organised as follows. In the next Section (13.2), we introduce the theoretical framework of service innovations in general and radical service innovations in particular. Section 13.3 highlights (a) the current understanding of the importance of a market focus for innovations and (b) the lead user approach to identifying qualified users as a source for radical innovations. Section 13.4 deals with the question of how to identify lead users and integrate them into services. In Section 13.5, we present explorative examples of service lead users to highlight their value as sources for radical innovations in services and to identify their distinct features as compared to lead users from the area of product innovation. We end this work with a conclusion outlining implications for service providers and proposals for future research.

13.2 Theoretical Framework: Innovation in Services

13.2.1 Services as Development Object

Though services have been present ever since man began to specialise in particular areas of work and thus became dependent on others and their 'share of work' for survival, it has only been a couple of years that the specific aspects and dimensions of service innovations have been systematically researched (Drejer 2004). According to Gadrey et al. (1995: 5–6), to provide a service means...

> ... to organize a solution to a problem (a treatment, an operation) which does not principally involve supplying a good. It is to place a bundle of capabilities and competences (human, technological, organizational) at the disposal of a client and to organize a solution, which may be given to varying degrees of precision.

To fully understand possible dimensions of innovations in services, one, therefore, has to first appreciate the constitutive characteristics of service production. In particular, a high level of user integration results in high visibility of all dimensions of the service creation process (Reckenfelderbäumer and Busse 2006). A strict separation of the inbound capability and process innovation and the outbound product innovation does not exist, as it does in consumer and investment goods. Any innovation in one of the following three dimensions defining services (Donabedian 1980) consequently directly impacts the user experience.

The *capability dimension* describes the potential and readiness of a company to create a service – the 'with what' of service creation. *Innovations in the capability dimension* thus comprise all improvements in the potential and readiness of a service provider to create services. All internal factors of the service provider can, for this reason, become subjects of innovation (den Hertog 2000). Such factors include human resources (e.g., better training, job enlargement, and empowerment of service staff) and information processing systems (e.g., the introduction of customer relationship management systems), but also the information available to the service provider (Kelley et al. 1990).

The *process dimension* describes the 'how' of service creation. Based on the inherent potential of internal factors, the actual process of service creation is triggered by the integration of an external factor (Mills and Moberg 1982). The external factors can be real immaterial production factors (e.g., information, work, or nominal goods), real material production factors (e.g., immobile or mobile production factors), and the active or passive collaboration of the user. *Process innovations*, therefore, involve changes in linking internally combined factors with external factors to provide more efficient solutions to existing problems (Gadrey et al. 1995). In contrast to consumer or investment goods, the creation process is already part of the overall user experience. Consequently, any improvement of the latter will directly impact user satisfaction. Process innovations can encompass the development of fundamentally new processes (e.g., online banking) or the optimisation of existing processes.

The service *output dimension* defines the 'what' of service creation. The result is immaterial in nature and manifests on the user or its object(s) only (Corsten 1989). *Innovations in the outcome dimension* are defined by a new result at the end of the service process. Such innovations are new core services for markets that have not been previously defined (Gadrey et al. 1995). The necessary prerequisites for the realisation of outcome innovations are almost always changes in technology and processes (Gadrey et al. 1995), the other two dimensions mentioned earlier. As an example of an outcome innovation, the concept of club cruises illustrates the above context. Creating a club atmosphere with entertainment aboard a cruise ship is new to the user (outcome). Realisation, however, requires a ship equipped for such purpose and a crew trained to entertain the guest on board (potential), in addition to concrete forms of entertainment and hospitality (process).

13.2.2 Radical Service Innovations

Aside from continuous incremental innovations for optimisation of existing service offerings, or closure of gaps in the existing service portfolio, companies from time to time do require a radical innovative leap to expand into new markets or cement their competitive situation in existing markets.

In general – with a comprehensive and commonly agreed on single definition of the term 'radical innovation' still not having evolved (Green et al. 1995) – breakthrough, disruptive, or radical innovation are referred to as new rather than improved products with a high, long-term sales potential. They are based on new, discontinuous technologies (Kessler and Chakrabarti 1999; Lynn et al. 1996; McDermott and O'Connor 2002; Rice et al. 1998), have a high inherent level of market uncertainty (Abernathy and Utterback 1978; Rice et al. 1998; Vreyzer 1998), and pose a significant level of risk for the innovating companies, as the usually high costs involved in developing such products might not pay off. Radical innovations, for these reasons, are usually considered to mark one extreme in the innovation continuum, with incremental innovations – which focus instead on improvements to existing products or technologies through progress along a known technology trajectory with little uncertainty regarding the potential for success or failure – marking the other.

Blinding out intra-company perceptions and implications of radical innovations (Green et al. 1995), the common denominator of defining radical innovations would be the degree of novelty as perceived by the market (Gadrey et al. 1995) – which also entails the transfer of analogies from other markets and/or market segments.

For consumer and investment goods, the construct of novelty thereby either entails the innovation's means, i.e., the product's inherent technology, its objective, i.e., the user's benefit from interacting with the product's technology, or both (Baker et al. 1967; Schumpeter 1934).

Knowing that the service outcome dimension also is a function of internal capabilities (means) and a way of integrating external resources by leveraging the internal capability potential (objective), the same scheme for characterising radical innovations also applies to services. As noted in the previous section, innovations in services can encompass three dimensions: capability, process, and outcome (Donabedian 1980). Unlike with manufactured goods, any of these is visible to the user due to a high level of integration into the service creation process (Reckenfelderbäumer and Busse 2006). The more of a service's three dimensions are perceived as novel by a particular market, the more radical the innovation is as a whole to the user.

13.2.3 Challenges in Developing (Radical) Service Innovations

With three instead of only one dimension visible to the user, service companies have more degrees of freedom to alter existing offerings and to 'out-innovate' competitors than manufacturing companies. With more dimensions to fail at, however, radical service innovations also involve a much higher degree of uncertainty with regard to market acceptance of novel services.

The intangible nature of services and the integration of the user as an external factor thereby raise additional challenges for successful development of radical service innovations for various reasons. Compared to manufactured products, the intangibility of the output dimension requires a much higher level of abstraction from the user, not only to evaluate existing services (Zeithaml 1991) but even more so for the development of novel services – and the same problems exist for innovation sources within the firm. For the same reasons, prototypes, which facilitate discussions of conceptual changes and communicate the concept to all stakeholders during the development process, can neither be provided in the same remote format as for manufactured products – i.e., without involving test users and having all capabilities in place to test-run the service concept – nor in the early phases of the product development process. Besides, in some instances, new services can also require the user to learn new protocols, for instance travellers having to use a terminal for self-check in at the airport, compared to the traditional way of having an assistant at the desk helping with the check-in procedure. These service-specific characteristics can impose an additional cognitive barrier for people to generate out-of-the-box solutions in the first place (Adamson 1952; Duncker 1945; Lunchins 1941) and evaluate the innovation's value for the market in the second place.

For those companies willing to face such challenges in the pursuit of advantages of successful, market-conforming service innovations, the key question, therefore, is which ideas to pursue to minimise the risk of market failure and how to generate such ideas.

13.3 Sources of Innovations

13.3.1 Integration of Users as a Success Factor for Innovation Management

The literature describes two drivers for innovation. Innovations based on newly developed technologies can be pushed actively into the market by companies in hopes that users will accept and buy the new product or service (technology push), or companies can wait to answer a market pull for concrete new products and services.

A third line of argumentation brings together both perspectives and argues that neither of the two extremes can be successful (Rothwell 1992). A new technology developed internally without a concrete or potential market need is useless. Successful innovation management needs rather to balance the capability to listen to and understand external demand as the primary impetus for innovations – both incremental and radical ones – with the capability to internally develop need-based technologies as the basis for a product and its production process.

Various empirical studies show that the integration of users as market spokespeople positively impacts multiple areas of the companies' innovation efforts and is an important success factor in the development of products (Baker et al. 1986; de Brentani 1995) and services (AberdeenGroup 2005; Edgett and Parkinson 1994; Martin and Horne 1995). By circumventing the existing gaps between internal beliefs regarding market needs and actual existing market needs – one of the most frequent causes of failure of innovation projects (Zollner 1995) – user-centric development of products and services saves time and money. Research, however, shows that user integration not only offers companies the prospect of probing, testing, and filtering their internally generated ideas and concepts, but also the possibility of leveraging the user as a source of innovations (Magnusson 2003; Magnusson et al. 2003).

Most companies today acknowledge these facts and integrate users into the early phase of the innovation process for both service (Tether 2003) and product development (Gurgul 2007) in multiple ways.

13.3.2 Lead Users as a Source of Radical Innovations

Average users, however, often do not qualify as a source of radically novel ideas for reasons of functional fixedness; integration of average users still results in market-conforming innovations, though incremental ones (Lettl et al. 2006).

The concept of functional fixedness stems from gestalt psychology and is based on a set of experiments conducted in the 1920s by Maier (1930). These experiments showed that individuals familiarised with the intended purpose of a set of tools are less likely to be creative in inventing new ways of either utilising or combining them to solve unforeseen problems.

Average users' insights into new product (and process and service) needs and potential solutions are constrained by their individual real-world experiences. This 'block', then, limits their ability to use the components given to them to make a specific item, as they cannot move past the original intention of the object (Duncker 1945). Users infused by the present are thus unlikely to generate novel product concepts that conflict with the familiar ones (Adamson 1952; Lunchins 1941).

On the other hand, numerous studies show that certain users are very capable of generating radical innovations. These users not only initiate innovations through request and concrete needs (Biemans 1991; Mansfield 1988; Utterback et al. 1976) – they also generate ideas and concepts (Baker et al. 1986; Voss 1985) and even develop and use prototypes (Lettl et al. 2006; Lüthje 2004; Shah 2000; von Hippel 1982, 1986, 1988). Erich von Hippel, one of the pioneers in the field of innovative users, terms this group *lead users*. Two characteristics typify this user category (von Hippel 1986: 796):

> "Lead Users face needs that will be general in a marketplace – but face them months or years before the bulk of that marketplace encounters them."
> "Lead Users are positioned to benefit significantly by obtaining a solution to those needs."

Lead users have to imagine themselves in the future application. They are able to achieve this from their usage experience and dissatisfaction with current products and services. From the perspective of innovation as an economically motivated activity (Mansfield 1968), expected economic or personal benefits associated with an innovation can become so strong that lead users become the innovators themselves.

A well-known example of a user innovation is Liquid Paper, a correction fluid for typewriters that was invented by a secretary in the US at the end of the 1950s. 3M later took over the innovation and implemented it on an industrial scale.

The sports and recreation market is another area of high lead user activity, as newly developed sports usually originate from the participants themselves. A study on innovations in skateboarding, snowboarding, and surfing showed that practically all of the basic product development was carried out by the participants and not by sporting goods manufacturers (Shah 2004). In a study on outdoor and trekking products, a high level of innovative product users was also confirmed (Lüthje 2004).

Examples from some industrial goods markets demonstrate that lead user innovations are not limited to end users. Significant advances in technology in semi-conductor and capacitor manufacturing were made by the semi-conductor manufacturers themselves, and not the developers of the respective process technologies (von Hippel 1978). Similar results of lead user activity were found for other process technologies, such as CAD and CAM systems, in the medical field (Shaw 1985), and for the fastening industry in construction (Herstatt and von Hippel 1991).

Even in the dynamic IT industry, numerous examples of user innovations exist. The Linux operating system and Apache server software are two prominent examples of the open-source movement. First initiated by individual software users, active user communities – not manufacturers – took the driver's seat in the further development of the software (Franke and von Hippel 2002).

13.4 Identification of Lead Users

13.4.1 Identification of Lead Users in the Consumer and Investment Goods Sector

Though companies already integrate users to identify ideas and needs for radical innovations, they do not necessarily integrate the *right* users.

For radical innovations, integration efforts need to focus on lead users for the reasons highlighted above. Because traditional marketing methods fall short in identifying this particular species of user, Erich von Hippel and other lead user researchers highlight two process steps as key to identifying lead users (Herstatt and von Hippel 1991; Urban and von Hippel 1988; von Hippel 1986; von Hippel and Riggs 1996).

As lead users are defined as being ahead of the market with respect to a specific key dimension, before one can identify lead users in a product category of interest, one must *identify the underlying trend* in which these users have a leading position. Though experts, trend extrapolation, and complex econometric models offer companies some formal methods of trend assessment, in the end, identifying trends cannot be fully formalised. Based on the data gathered, analysts must judge which market trends to focus on. In many cases, an analyst's industry expertise, therefore, is a crucial asset in the identification of trends.

Once a firm has identified significant trends that appear to be associated with promising new product opportunities, the next step is to begin *searching for lead users*. These are users who are (a) at the leading edge of each identified trend in terms of related new product and service needs and (b) expecting to obtain a relatively high net benefit from solutions to those needs (von Hippel 1986). To operationalise this construct, multiple empirical studies have searched for concrete user types who mirror it. Research on investment and consumer goods shows that especially extreme users are likely to unite the above two prerequisites (Lettl et al. 2006; Lüthje 2003, 2004; Lüthje et al. 2005; Shah 2000). This user type encounters extreme usage situations, thus facing needs the rest of the market does not (yet) encounter. The extremity of the usage situation can often also result in such high pressure to innovate an adequate solution not yet provided by the market that the benefit of innovation outweighs the cost (Lettl et al. 2006). Extremity of usage encompasses intensity of use with respect to both time and intention as well as duration of usage experience (Lüthje et al. 2005). These are criteria through which lead users can easily be identified via questionnaires or direct search in user communities.

An additional, very practical method for identifying lead users involves identifying those who are already actively innovating to solve problems present at the leading edge of a trend. Identifying lead users by seeking innovating users can be very economical, because the identity of users conducting R&D on a given problem area is often common knowledge to industry participants (von Hippel 1986).

When searching for lead users, it is, however, important to note that these should not necessarily be sought within the usual user base. Lead users may also be users employed by a competitor or even in other industries with analogous usage needs. 3M, for example, identified military reconnaissance as a user with analogous needs when searching for an innovation to help surgeons read body-scans. Using low-resolution satellite pictures as a primary data source, the military encountered similar problems when trying to determine the presence of specific structures, such as tanks and missiles, as surgeons did trying to identify tumours and other abnormalities in body scans (von Hippel et al. 1999).

13.4.2 Identification of Service Lead Users

Though until now only few observations on the identification of service lead users exist in scientific literature, there seems to be no reason why the existing approach validated in multiple studies in the consumer and investment goods industries should not be applied to services as well. Opening the door for further research, two studies have already successfully applied this approach in a service context.

13.4.2.1 The Work of Von Hippel and Riggs on Home Banking Services

Von Hippel and Riggs (1996) utilised the approach of identifying lead users in searching for new home-banking service offerings. Trend identification encompassed expert interviews with leading employees of banks. The underlying consensus among these was that users would be able to do progressively more and eventually all of their banking from home over their ordinary home telephone lines.

The lead users within this trend were defined as exhibiting the following two attributes: (a) they used relatively advanced electronic home banking systems now ('ahead of the trend'); (b) they had done some level of development or tinkering with respect to these (driven by expectations of high benefit). The actual process of identification relied on a bank's internal electronic mail system to send an inquiry to approximately 1,300 employees who seemed to be the most promising sources in view of the team conducting the lead user project. The message asked people to respond if they had

> "... found novel ways to take care of their personal banking service needs. For example, ... written or adapted a home software program to automate a manual procedure, found a novel way to use a service offered by the bank to achieve a purpose other than was originally intended, or devised a novel procedure for paying bills or keeping records" (von Hippel and Riggs 1996: 7).

Of 15 people who responded within 24 hours, eight passed subsequent interviews, during which the respondents were screened as to whether their home banking activities included novel solutions actually tested in practice and how well the respondents would be able to work with the project team.

The service offerings identified by the lead users were perceived by three samples of users to be significantly more attractive than a previously developed concept ($p > 0.001$). The number of respondents favouring the 'lead user system' was nearly five times greater than those favouring the alternative concept (von Hippel and Riggs 1996).

13.4.2.2 The Work of Lüthje on the Transportation Industry

Lüthje (2000) utilised the same approach in the transportation industry. The task was to develop a future tariff system for a public transport provider, including aspects of tariff structure, ticket offerings, payment systems, and provision of information to users. As in the work of von Hippel and Riggs (1996), trend exploration involved expert interviews within a group made up of experts from diverse backgrounds (company, industry, and industry-related experts). The trends identified were (a) microchip-bearing cards as a means of making payments more convenient and flexible, and (b) user-friendly information systems.

Based on the identified trends, the possible group for recruitment of lead users was narrowed down by the team using creativity techniques (e.g., brainwriting). The team ended up with four groups: students of computer science, users of Lufthansa electronic tickets (ETIX), authors of IT magazines, and teleworkers. Each group was surveyed through mailed questionnaires. The responses were once again checked for compliance with the characteristics of innovative users, i.e., dissatisfaction with existing offerings, usage experience (time and intensity), and presence of active user innovations.

The lead users identified thereby generated ideas that were in line with the identified trends. An evaluation by the initial experts used for trend identification, users, and employees of the transportation provider characterised the new service offerings as novel and very original (Lüthje 2000).

13.4.2.3 Anecdotal Findings of Integrating Innovative Service Users into the Industry

Outside academic discussion, examples of radical changes in the service industry additionally underline the fact that trend exploration and identification of users who are advanced in terms of service needs has helped and continues to help service companies identify radical new business opportunities.

For example, Frederic W. Smith, founder of FedEx, identified the trend his business model was based on during his undergraduate studies at Yale University in 1965. In the process of writing an economics paper, Smith observed a trend in society towards more automation and computerisation. It was this insight that made him identify the computer industry not only as benefactor of the trend, but also a need to ensure dependable products. Computers could not make business cheaper, faster, and more accurate unless they could be fixed rapidly. Leading the trend of automation, computer manufactures and their users were in need of a logistics network that would allow delivery of replacements and spares within the shortest time possible. After serving in Vietnam, Smith returned home in 1971 to find his predictions having come true. Computers were becoming an indispensable part of the business infrastructure, and delivery systems were not keeping up. The rest of the story is history. Smith was the first to use the hub – spoke distribution paradigm in air freight, enabling FedEx to provide an enormous number of connections much more efficient than any point-to-point network could. (FedEx 2007; Who made America 2007a)

But also companies such as SAP and Deutsche Post World Net report using trend exploration and an immanent focus on the needs of industry-leading companies as sources of service innovations.

In an interview as part of a recent study published by PricewaterhouseCoopers on service innovations (Scholich et al. 2006), Michael Lohmeier, Corporate Technology and Innovation Management of Deutsche Post World Net, outlines his company's strategy to integrate companies explicitly outside the logistics value chain (e.g., from the automotive industry) as sources of new technologies enabling service innovations in logistics.

In the same study, Prof. Dr. Lutz Heuser, Chief Development Architect at SAP, speaks about SAP's innovation strategy focusing on upcoming trends in their user base rather than innovations in today's deployed technology. SAP's innovation strategy builds on a so-called outside-in approach. This encompasses the continuous monitoring of external trends by means of a scientific network that includes industrial and public partners, leading users, independent software providers, and other service providers. In doing so, SAP seeks to develop innovations that support its users in upcoming changes to their business models.

13.5 Service Lead Users as Source of Radical Service Innovations!

But can service lead users independently innovate in service industries and create truly radical innovations, as we have seen them do in other industries? With the academic literature not yet offering a sufficient answer, outside-in cases from screening the Internet and newspapers are used to provide anecdotal evidence for the existence of innovative service lead users. As result of our explorative research, we provide here three cases of service lead users and their radical innovations.

- ***Case 1 – Weight Watchers:*** It all began when Jean Nidetch, a 5′7″ 38-year-old woman who weighed 214 pounds or – as she described herself – an "overweight housewife obsessed with eating cookies" (Who made America 2007b), became discouraged by years of fad dieting. It was then, that she attended a free diet clinic, finally managing to lose 20 pounds in 10 weeks. Yet again, she felt losing her motivation and therefore called over some friends for mutual support. The meeting became regular and showed to work. They all lost weight and realised that the success was mainly due to collaborative support and encouragement to overcome old habits. Word spread, and by the end of 1962, hundreds of people wanted to join the group. By 1968, Weight Watchers operated franchises nationwide and completed a sold-out initial public offering. Today, Weight Watchers is a global enterprise offering meal plans and cookbooks to millions of overweight people (The biography channel 2007; Weight Watchers 2007; Who made America 2007b)
- ***Case 2 – Erento.com:*** In 2001, Chris Möller needed multiple sets of garden furniture for a party. However, renting them turned out to be exhausting and time-consuming. In 2001, offers on the Internet were solely categorised by company but not by product. Identifying the desired goods and comparing prices, conditions, and availability across multiple companies was anything but easy. As Chris Möller learnt from his own experience, users are not in search for particular rental companies but for goods – and the idea to create Erento.com, a platform where users can directly search for rental goods, was born. Today, Erento.com is the biggest market place for rental goods on the Internet. It offers its service of matching demand and supply to renters and hirers. The site lists 962,000 items in 2,200 categories and encompasses a product range from trailer and cars across sporting equipment and work force to fighter planes and super yachts. (Bonow 2007)
- ***Case 3 – 2te-Zahnarztmeinung.de:*** In Germany, dental prostheses involve significant costs for the majority of people, with public health insurance reimbursing only a fraction of the price. When Holger Lehmann, by former profession a manager, needed a dental bridge, he was amazed at how much prices differed from dentist to dentist. While his dentist asked for EUR 500, his parent's dentist only charged EUR 350 for the same work – an eye-opening experience for Mr. Lehmann, as he notes (Friederichs 2005). Obviously the

market for dental services lacked transparency, as users like Mr. Lehmann did not know about dentists willing and/or or able to offer the same services for less money. The idea for an auctioning site for dental services was born. Patients upload their dentist's cost schedule and all dentists registered on the site can make a bid for it. In the end, the user can choose the dentist best fitting his needs in terms of price, the dentist's quality ratings by other users, and distance from home. The auctions are limited to implants only, as only for them can the dentist influence the cost directly through his choice of dental lab (and its cost) and the size of his fees exceeding the standardised insurance catalogues (Friederichs 2005). Savings for users are significant. For the first time, 2te-Zahnarztmeinung.de offers patients the opportunity to compare prices for dental services. By May 2007, no less than 68 health insurance companies had recognised the value to patients and now cooperate with 2te-Zahnarztmeinung.de by bearing their users' expenses for using the site's services. (Lehmann 2007)

Figure 13.1 shows that the identified service lead users do not differ from lead users as defined for the consumer and investment goods sector with regard to their most two prominent characteristics: leading a trend and expecting a benefit from innovation.

Firstly, each individual faced needs the market was not serving, either through inadequate offering or complete absence of offering. Holger Lehmann's need was neither addressed by dentists nor by health care providers. He offered the first solution to that problem ever. In case of the other two examples, Weight Watchers and Erento.com, the solutions available did not satisfy the needs of the users. Jane Nidetch's frustration with existing concepts of losing weight, and the unmet need of permanently becoming slim, made her take innovation into her own hands. Chris Möller found himself in a similar situation: unsatisfied with the existing solutions for renting goods via the Internet, he created his own superior solution, Erento.com.

The second lead user characteristic is the motivation for innovation. The expected benefits associated with an innovation can become so large that the lead user himself becomes the innovator. All of the described users benefited significantly from their innovation. Jane Nidetch eventually succeeded in permanently losing weight. The other two expected significant benefits from commercialisation of their radical ideas – and they got them in the end.

With regard to the novelty of the innovations in the respective markets, all of the examples above are clearly radical new service offerings, the first of their kind. Weight Watchers and Erento.com, aside from being global market leaders in their particular segment, also prove the innovations' very high sales potential – a criterion also often used in other lead user studies as a proxy measure for an innovations radicalness (Franke et al. 2005; Lilien et al. 2002).

Though it is difficult to clearly distinguish the different dimensions of service innovations outside theory (Evangelista and Sirilli 1998; Gadrey et al. 1995), Figure 13.2 tries to capture the most obvious innovations in the different service dimensions.

The market success of the respective innovations also underlines the magnitude of the innovative potential resting with users in the market. In many instances, this

	Industry examples	Lead user characteristics	
Theory		• Unmet needs of users who are ahead of the market	• Significant (expected) benefit from obtaining a solution to those needs
Empirical studies on lead users	**Investment goods: Example X-Y Table** (von Hippel 1978)	• IBM needed a machine to insert components into printed circuit cards which were in turn incorporated into computers	• For IBM to develop such a machine meant to significantly increase efficiency in producing a new generation of computers on industrial scale
	Consumer goods: Example windsurfing (Shah 2004)	• New needs emerged for a group of early windsurfers-called the Hawaiians - when high winds and waves, as well as their own use patterns, challenged the existing equipment	• *A member of the Hawaiians on their motivation to innovate: "The motivation was control, definitely not to get rich, but the motivation was control of the craft at speed... I was integrated in four or five other, some major, some not so major developments, and they all were pretty much for the same reason. Just trying to make the craft more efficient and more fun to sail. .."*
Own findings on service lead users	**Weight Watchers**	• Using existing dieting concepts, Jane Nidetch was frustrated at still not having succeeded in her goal to lose weight permanently	• The personal benefit of Mrs. Nidetch – to finally solve the problem of falling back to old eating habits – motivated her to call in her friends for a joint dieting concept
	Erento.com	• Chris Möller found existing web-offerings to be too fragmented and thus inconvenient and untransparent for users, when searching the web for the best rental offers	• Mr. Möller expected to benefit financially when creating a superior solution to a user community he believed to suffer from the same unmet needs he did
	2te-Zahnarztmeinung.de	• Holger Lehmann found German dentists to significantly differ in prices for dental prostheses – untransparent to patients often paying more for the same work	• Opening a platform for matching price and demand for dental prostheses, Mr. Lehmann hoped to benefit financially from earning fees from his service offerings

Fig. 13.1 Comparing service lead users' characteristics with those of lead users from other industries at the level of theory

potential is not leveraged, as firms are either not aware of or do not appreciate the importance of the emerging user needs (Herstatt 2002). As the cases illustrate, not leveraging the innovation potential of service lead users, however, can be problematic for existing service providers for two reasons. First, not identifying the advanced needs of such user groups means missing out on valuable opportunities for radical service innovations. Second, forced to innovate, service lead users are able to innovate independently of existing service providers, and are perfectly capable of outperforming the latter by developing and commercialising superior service offerings themselves.

It is, therefore, interesting to notice that, in contrast to findings from empirical work on innovations in manufactured goods (Hienerth 2006; Shah 2004), the innovations of service lead users in these case studies were not commercialised

	Innovation in service dimensions			Indicators for innovations' market novelty/ attractiveness
	Capability dimension	Process dimension	Outcome dimension	
Weight Watchers	• Bringing together group of people with same problems as self-enforcing circle to endure losing weight through mutual support and sharing of problems	• Interaction in regular group meeting format as compared to former single sessions in which dietary plans are developed	• Change of a person's intrinsic eating habits rather than pure calorie-focused diet	• World's leading provider of weight management services • USD 3 billion revenue in 2006 • Over 97% of US women recognise the Weight Watchers brand (Weight Watchers International 2007)
Erento.com	• Creating a database to match demand and supply of renter and hirer	• One-stop internet platform covering all information needs of hirers	• Direct search for rental goods by product category rather than fragmented search by rental company	• World's first, largest and fastest growing online rental market place • Present in the US, UK, Germany, Switzerland and Austria • More than 50.000 clicks per day • Used by 25% of rental companies in Germany, including major brands such as Almo and Intersport (Walsh 2007)
2te-Zahnarztmeinung.de	• Creating an internet platform for matching patient's demand with dentists' willingness and ability to realise savings on material costs and laboratory work	• One-stop internet platform covering all information needs of patients	• Auction to get best price for dental prosthesis, as compared to dentists' former untransparent pricing praxis	• Germany's first online market place for dental prosthesis • Cooperation with 68 leading German health insurance companies • Up to 50% saving for patients (Lehmann 2007)

Fig. 13.2 Evaluating the radicalness of service lead users' innovation

by an existing firm, but rather are produced and marketed by the service lead users themselves.

These findings from the consumer product sector help to explain the motivation for users to commercialise their ideas for novel products and to find possible explanations for this difference between sectors:

> "... users are most likely to commercialize their innovations when they have access to complementary assets such as distribution, and they possess informational advantages enabling them to uniquely identify opportunities that established firms would underestimate" (Shah and Tripsas 2004: 2).

With empirical findings from both the service (Kristensson et al. 2004; Magnusson 2003; Magnusson et al. 2003) and the manufactured goods sector (Lilien et al. 2002) indicating that there are no apparent differences in the ability of users and in-house resources to identify attractive ideas, the reason for the seemingly more easy commercialisation of ideas in the service sector must be related to the second aspect, namely the access to complementary resources.

Though general evidence for this aspect cannot be drawn from the present case studies, implying the need for additional research noted in the discussion section of this chapter, it seems obvious that service users are in a likely position to access the required resources more easily – a hypothesis that fits well with the picture of service companies outnumbering manufacturing companies by number and rate of establishment (Eurostat 2006). The constitutive characteristics of services allow derivation of several factors that help to explain this observation. First of all, in terms of service definition, the integration of the external factor, namely the user, makes the latter a co-producer of the service (Kelley et al. 1990). The resulting high visibility of all dimensions of the service creation process equips each service user with a much more profound knowledge of critical aspects regarding quality procedures and required capabilities than the user of a physical product who purchases and uses it after it has already left its production process. Second, with services being intangible, they are too a much lesser degree subject to intellectual property rights (Blind et al. 2003; Djellal and Gallouj 2001; Gadrey et al. 1995) than physical products, for which patents often impose significant barriers to commercialisation of user innovation (Braun and Herstatt 2007). Third, for many branches of the service industry, especially those with a high degree of human interaction – think of the emerging coffee shop industry – the technological know-how required to implement a service innovation is much lower than, for instance, tinkering with complex physical products. But even for web-based services, which have grown significantly in number over the past decade, the use of standard software and open-source program codes as well as the younger generation's general affinity for the Internet has significantly lowered technological barriers, allowing a large number of user innovations in the IT sector.[1]

Summarising the findings, then, it can be hypothesised that, though more barriers exist to generating the initial idea for a radical innovation in the service industry due to the services' inherent abstract and intangible nature, the commercialisation of a service innovation is much easier than for innovations in manufactured goods.

For this reason, the threat posed by commercialised user innovations is significantly greater for service providers than in other industries, making it an imperative to identify service lead users, understand their needs, and develop adequate offerings to meet the latter – before they do it themselves.

13.6 Summary and Conclusion

The present work illustrates examples of active user innovators in the service industry and gives insights from both theory and practice on the identification of users as a source of radical service innovations.

- **Service lead users do exist.** As shown by the cases analysed, service users not only initiate innovations on request; they also generate ideas and concepts and even develop and commercialise novel services independent of existing service providers as a response to unfulfilled needs.
- **Service lead users exhibit high degrees of benefit expectation and their needs are ahead of market needs.** With regard to advanced usage needs and

[1] The website www.programmableweb.com, for example, lists over 2,500 user-derived sites with innovative new functionalities.

expected benefits, the examples of Weight Watchers, Erento.com, and 2te-Zahnarztmeinung.de suggest that no difference exists between lead users in services and in the consumer and investment goods industries.

- **Service lead users are more likely to commercialise their innovations.** Unlike in the manufacturing industry, lead users in services are in a better position to commercialise their innovations independently of existing market players, due to easier access to the complementary resources required to do so.
- **Service lead users can be identified by the two-step approach of trend exploration and search for leading users within the trend.** Both the literature review and explorative outside-in cases exemplify the impact this approach offers to service providers in search of sources of radical new service innovations.

The cases described highlight the impact radical innovations have on the service industry and underline the role of service lead users in service innovation. Presenting insights from both the theory and the practice of lead user identification and integration, this chapter helps service providers to focus their efforts on the most promising user species early in their innovation process.

For the management of service providers, our findings suggest focusing particularly on identifying users at the leading edge of trends who are faced with inherent needs not yet addressed by the market and who possess the necessary reserve of experience to spur the cognitive process of developing innovative solutions to these needs.

The findings presented here also indicate additional areas for future research in the area of user innovation in services.

- First, in terms of answering the above questions with regard to similarities between lead users in services and other industries, indications of systematic differences between the respective parties in the commercialisation of the innovations are evident. This chapter has provided the first arguments explaining the characteristics of why service innovations face fewer barriers to commercialisation than their counterparts in other industries. Yet, the limited number of initial observation points restricts any general statements and demands further empirical work to validate whether these differences exist for services in general and whether the derived factors explaining the latter hold true on a general level as well.
- Secondly, additional research is required to better understand the impact of innovative users on the attractiveness of their innovations to the market. Both the studies of Kristensson et al. (2004); Magnusson (2003); Magnusson et al. (2003) and the findings presented here show that users are able to generate innovative ideas for new services. While this generally speaks in favour of integrating users as a valuable source of innovation into the early phases of the service development process, questions of whether the integration of average service users or service lead users provides better results, and which user characteristics actually have a significant effect on the attractiveness of the innovation, offer interesting perspectives for additional research.

- Thirdly, so far no insights on the actual practice of user integration exist. While empirical studies have shown that, as in other industries, service companies regard users as one of the most important sources for innovation (Ginzburg et al. 2007; Martin and Horne 1993; Tether 2003), it remains unclear according to which criteria these users are selected for integration. Does the service industry acknowledge the existence of lead users, or are other user characteristics of importance to service industry? If so, what are those characteristics?

With the service industry in need of innovations, we are confident that research and industry will acknowledge the inherent potential of integrating lead users as sources of service innovations.

References

AberdeenGroup. (2005). *The emergence of 'Chief Service Officer'*. Boston, MA: Aberdeen Group Inc.

Abernathy, W. J., & Utterback, J. M. (1978). Patterns of industrial innovation. *Technology Review, 80*(7), 41–47.

Adamson, R. E. (1952). Functional fixedness as related to problem solving: A repetition of three experiments. *Journal of Experimental Psychology, 44*, 288–291.

Baker, N. R., et al. (1986). Why R&D projects succeed or fail. *Research Management, 29*(6), 29–34.

Baker, N. R., et al. (1967). The effects of perceived needs and means on the generation of ideas for industrial research and development projects. *IEEE Transaction on Engineering Management, EM, 14*(4), 156–163.

Biemans, W. G. (1991). User and third-party involvement in developing medical equipment innovations. *Technovation, 11*(3), 163–182.

Blind, K., et al. (2003). *Patents in the service industries*. Final report prepared for the European Commission, EUR 20815. Karlsruhe: Fraunhofer ISI.

Bonow, T. (2007). Pressebereich erento.com – Zahlen & Fakten. http://www.erento.com/mieten/info/ger_presse_zahlen_fakten. Accessed 4 June 2007.

Braun, V., & Herstatt, C. (2007). Barriers to user innovation: Moving towards a paradigm of 'licence to innovate'? *International Journal of Technology Policy and Management, 7*(3), 292–303.

Corsten, H. (1989). Überlegungen zu einem Innovationsmanagement. Organisationale und personale Aspekte. In H. Corsten (Ed.), *Die Gestaltung von Innovationsprozessen. Hindernisse und Erfolgsfaktoren im Organisations-, Finanz- und Informationsbereich* (pp. 1–56). Berlin: Schmidt.

de Brentani, U. (1995). New industrial service development: Scenarios for success and failure. *Journal of Business Research, 32*(2), 93–103.

den Hertog, P. (2000). Knowledge-intensive business services as co-producers of innovation. *International Journal of Innovation Management, 4*(4), 491–498.

Destatis. (2007). *Volkswirtschaftliche Gesamtrechnungen – Wichtige Zusammenhänge im Überblick*. Wiesbaden: Statistisches Bundesamt.

Djellal, F., & Gallouj, F. (2001). Innovation in services, patterns of innovation organisation in service firms: Postal survey results and theoretical models. *Science and Public Policy, 28*(1), 57–67.

Donabedian, A. (1980). *The definition of quality and approaches to its assessment*. Ann Arbor, MI: Health Administration Press.

Drejer, I. (2004). Identifying innovation in surveys of services: A Schumpeterian perspective. *Research Policy, 33*(4), 551–562.
Duncker, K. (1945). On problem solving. *Psychological Monographs, 58*(5).
Edgett, S. J., & Parkinson, S. (1994). The development of new financial services: Identifying determinants of success and failure. *International Journal of Service Industry Management, 5*(4), 24–38.
Eurostat. (2006). *European business – facts and figures*. Luxembourg: European Communities.
Evangelista, R., & Sirilli, G. (1998). Innovation in the service sector: Results from the Italian statistical survey. *Technological Forecasting and Social Change, 58*(3), 251–269.
FedEx (2007). FedEx Corporate History. http://www.fedex.com/us/about/today/history/. Accessed 4 June 2007.
Fitzsimmons, J. A., & Fitzsimmons, M. J. (2005). *Service management: Operations, strategy, and information technology* (5th ed.). Columbus: McGraw-Hill.
Flyvbjerg, B. (2006). Five misunderstandings about case-study research. *Qualitative Inquiry, 12*(2), 219–245.
Franke, N., & von Hippel, E. (2002). *Satisfying heterogeneous user needs via innovation toolkits: The case of apache security software*. Working Paper No. 4341-02. Cambridge, MA: Sloan School of Management. Massachusetts Institute of Technology (MIT).
Franke, N., et al. (2005). *Finding commercially attractive user innovations: A test of lead user theory*. Working Paper 4536-05. Cambridge, MA: Sloan School of Management. Massachusetts Institute of Technology (MIT).
Friederichs, H. (2005). 3, 2, 1 . . . mein Patient. *FTD*, 09.06.2005, Hamburg.
Gadrey, J., et al. (1995). New modes of innovation: how services benefit industry. *International Journal of Service Industry Management, 6*(3), 4–16.
Ginzburg, I., et al. (2007). Looking for the locus of innovation in new service development. In Hawaii International Conference on System Science (HICSS-40 2007). Waikoloa, Big Island, Hawaii.
Green, S. G., et al. (1995). Assessing a multidimensional measure of radical technological innovation. *IEEE Transactions on Engineering Management, 42*(3), 203–214.
Gurgul, G. (2007). The customer – an untapped source of innovation. Developing a customer integration network in HP. In *Knowledge networks for business growth* (pp. 59–76). Berlin: Springer.
Herstatt, C. (2002). Search fields for radical innovations involving market research. *International Journal of Entrepreneurship and Innovation Management, 2*(6), 473–484.
Herstatt, C., & von Hippel, E. (1991). *An implementation of the lead user market research method in a "low tech" product area: Pipe hangers*. Working Paper No. 3249-91. Cambridge, MA: Sloan School of Management. Massachusetts Institute of Technology (MIT).
Hienerth, C. (2006). The commercialization of user innovations: The development of the rodeo kayak industry. *R&D Management, 36*(3), 273–294.
Howells, J., & Tether, B. (2004). *Innovation in services: Issues at stake and trends. INNO-Studies*. Brussels/Luxembourg: Commission of the European Communities.
Kelley, S. W., et al. (1990). Customer participation in service production and delivery. *Journal of Retailing, 66*(3), 315–325.
Kessler, E. H., & Chakrabarti, A. K. (1999). Speeding up the pace of new product development. *Journal of Product Innovation Management, 16*, 231–247.
Kristensson, P., et al. (2004). Harnessing the creative potential among users. *Journal of Product Innovation Management, 21*(1), 4–14.
Lehmann, H. (2007). 2te-Zahnarztmeinung – Ihr Marktplatz für Zahnarztleistungen. https://www.2te-zahnarztmeinung.de/?partner. Accessed 4 June 2007.
Lettl, C., et al. (2006). Users' contribution to radical innovation: Evidence from four cases in the field of medical equipment technology. *R&D Management, 36*(3), 251–271.
Lilien, G. L., et al. (2002). Performance assessment of the lead user idea generation process for new product development. *Management Science, 48*(8), 1042–1059.

Lunchins, A. S. (1941). Mechanization in problem solving: The effect of Einstellung. *Psychological Monographs, 54*(6), 1–95.
Lüthje, C. (2000). *Kundenorientierung im Innovationsprozess – Eine Untersuchung der Kunden-Hersteller-Interaktion in Konsumgütermärkten*. Wiesbaden: Deutscher Universitäts-Verlag.
Lüthje, C. (2003). Customers as co-inventors: An empirical analysis of the antecedents of customer-driven innovations in the field of medical equipment. In 32nd EMAC Conference, Glasgow.
Lüthje, C. (2004). Characteristics of innovating users in a consumer goods field. An empirical study of sport-related product consumers. *Technovation, 24*(9), 683–695.
Lüthje, C., et al. (2005). User-innovators and 'local' information: The case of mountain biking. *Research Policy, 34*(6), 951–965.
Lynn, G. S., et al. (1996). Marketing and discontinuous innovation: The probe and learn process. *California Management Review, 38*(3), 8–37.
Magnusson, P. R. (2003). Benefits of involving users in service innovation. *European Journal of Innovation Management, 6*(4), 228–238.
Magnusson, P. R., et al. (2003). Managing user involvement in service innovation: Experiments with innovative end-users. *Journal of Service Research, 6*(2), 111–124.
Maier, N. R. (1930). Reasoning about humans: I. On direction. *Journal of Comparative Psychology, 11*(8), 115–143.
Mansfield, E. (1968). *The economics of technological change*. New York: W.W. Norton & Company.
Mansfield, E. (1988). Industrial R&D in Japan and the United States: A comparative study. *The American Economic Review, 78*(2), 223–228.
Martin, C. R. J., & Horne, D. A. (1993). Services innovation: Successful versus unsuccessful firms. *International Journal of Service Industry Management, 4*(1), 49–65.
Martin, C. R. J., & Horne, D. A. (1995). Level of success inputs for service innovations in the same firm. *International Journal of Service Industry Management, 6*(4), 40–56.
McDermott, C. M., & O'Connor, G. C. (2002). Managing radical innovation: An overview of emergent strategy issues. *Journal of Product Innovation Management, 19*, 424–438.
Mills, P. K., & Moberg, D. J. (1982). Perspectives on the technology of service operations. *Academy of Management Review, 7*(3), 467–478.
Reckenfelderbäumer, M., & Busse, D. (2006). Kundenmitwirkung bei der Entwicklung von industriellen Dienstleistungen – eine phasenbezogene Analyse. In H.-J. Bullinger & A.-W. Scheer (Eds.), *Service Engineering – Entwicklung und Gestaltung innovativer Dienstleistungen* (pp. 141–166). Berlin: Springer.
Rice, M. P., et al. (1998). Managing discontinuous innovation. *Research-Technology Management, 41*(3), 52–58.
Rothwell, R. (1992). Successful industrial innovation: Critical factors for the 1990s. *R&D Management, 22*(3), 221–239.
Scholich, M., et al. (Eds.). (2006). *Innovation performance – Das Erfolgsgeheimnis innovativer Dienstleister*. Frankfurt: PricewaterhouseCoopers AG WPG.
Schumpeter, J. A. (1934). *The theory of economic development*. Harvard: Oxford University Press.
Shah, S. (2000). *Sources and patterns of innovation in a consumer product field: Innovations in sporting equipment*. Working Paper No. 4105. Cambridge, MA: Sloan School of Management. Massachusetts Institute of Technology (MIT).
Shah, S. (2004). *From innovation to firm formation in the windsurfing, skateboarding, and snowboarding industry*. Working paper 05-0107. Urbana-Champaign, IL: University of Illinois.
Shah, S., & Tripsas, M. (2004). *When do user-innovators start firms? Towards a theory of user entrepreneurship*. Working Paper No. 1511. Harvard, MA: Harvard Business School.
Shaw, B. (1985). The role of the interaction between the user and the manufacturer in medical equipment innovation. *R&D Management, 15*(4), 283–292.

Tether, B. S. (2003). The sources and aims of innovation in services: Variety between and within sectors. *Economics of Innovation and New Technology, 12*(6), 481–505.
The biography channel (2007). Real lives – Weight Watchers. http://www.thebiographychannel.co.uk/biography_story/1265:967/1/Weight_Watchers.htm. Accessed 4 June 2007.
Urban, G. L., & von Hippel, E. (1988). Lead user analyses for the development of new industrial products. *Management Science, 34*(5), 569–582.
Utterback, J. M., et al. (1976). The process of innovation in five industries in Europe and Japan. *IEEE Transactions on Engineering Management, 23*(1), 3–9.
von Hippel, E. (1978). The dominant role of the user in semiconductor and electronic subassembly process innovation. *IEEE Transactions on Engineering Management, 24*(2), 60–71.
von Hippel, E. (1982). Get new products from customers. *Harvard Business Review, 60*(2), 117–122.
von Hippel, E. (1986). Lead users: A source of novel product concepts. *Management Science, 32*(7), 791–805.
von Hippel, E. (1988). *The sources of innovation.* New York: Oxford University Press.
von Hippel, E. (1989). New product ideas from 'Lead Users'. *Research-Technology Management, 32*(3), 24–28.
von Hippel, E., & Riggs, W. (1996). *A lead user study of electronic home banking services: Lessons from the learning-curve.* Working Paper No. 3911-96. Cambridge, MA: Sloan School of Management. Massachusetts Institute of Technology (MIT).
von Hippel, E., et al. (1999). Creating breakthroughs at 3M. *Harvard Business Review, 77*(5), 47–57.
Voss, C. A. (1985). The role of users in the development of applications software. *Journal of Product Innovation Management, 2*(2), 113–121.
Vreyzer, R. W. J. (1998). Discontinuous innovation and the new product development process. *Journal of Product Innovation Management, 15*(4), 304–321.
Walsh, T. (2007). Are we becoming a nation of renters? The world's first online rental marketplace launches in the US as rental businesses flourish. *PR Newswire (U.S.),* 14 November 2007, Palo Alto.
Weight Watchers International (2007). Our History. http://www.weightwatchers.com/about/his/history.aspx. Accessed 4 June 2007.
Who made America (2007a). Fredrick W. Smith – Overnight delivery. http://www.pbs.org/wgbh/theymadeamerica/whomade/fsmith_hi.html. Accessed 4 June 2007.
Who made America (2007b). Jean Nidetch – Weight loss therapy. http://www.pbs.org/wgbh/theymadeamerica/whomade/nidetch_lo.html. Accessed 4 June 2007.
Yin, R. K. (2002). *Case study research. Design and methods* (3rd ed.). Thousands Oaks, CA: Sage.
Zeithaml, V. (1991). How consumer evaluation processes differ between goods and services. In J. H. Donnelly & W. R. George (Eds.), *Marketing of services* (pp. 186–190). Chicago, IL: American Marketing Association.
Zollner, G. (1995). *Kundennähe in Dienstleistungsunternehmen: Empirische Analyse von Banken.* Wiesbaden: Deutscher Universitäts-Verlag.

Chapter 14
Challenges of Bringing Citizen Knowledge into Public Sector Service Innovation

Lea Hennala, Suvi Konsti-Laakso, and Vesa Harmaakorpi

Abstract User-driven innovation is an essential part of practice-based innovation. This study, which belongs to user-driven service innovation research, attempts to find out what kinds of challenges there are pertaining to a municipal resident-involving service innovation approach in the public sector. This research question is approached from the points of view of both the municipal resident and the public authorities. With the municipal resident point of view we aim to find out how the municipal residents as service users are disposed towards having the possibility to personally participate in the service development activities, and from what kinds of positions the service users produce a voice to support service development. The public authority point of view examines how the customer-driven service development approach is perceived by the developer authorities, and from what kinds of positions the developer authorities listen to the message compiled from the users' voices. As a result we present six challenges of the service user-involving service innovation approach. The case study is based on the qualitative analysis of four research datasets.

14.1 Introduction

A remarkable amount of practice-based innovations stem from contact with clients and users (CIS 2004). Accordingly, user contributions to innovation have been acknowledged to have positive impacts in organisations' innovation activities (Alam 2002; von Hippel 2005; Jeppesen and Molin 2003). There are research results that indicate that user involvement generates valuable ideas, speeds up the innovation process and supports the diffusion of innovations, as compared to traditional innovation activities (Alam 2006; von Hippel 2005). However, the

L. Hennala (✉) • S. Konsti-Laakso • V. Harmaakorpi
Lahti School of Innovation, Lappeenranta University of Technology, Lahti, Finland
e-mail: lea.hennala@lut.fi

H. Melkas and V. Harmaakorpi (eds.),
Practice-Based Innovation: Insights, Applications and Policy Implications,
DOI 10.1007/978-3-642-21723-4_14,

user approach has not gained very much success in practice. The critics emphasise that users do not know how or are otherwise unable to foresee their future needs, and they do not know what they want. There is no clear empirical evidence about the benefits of the user approach. It is acknowledged that users can contribute to incremental innovation, but their ability to contribute to radical innovation is under debate (Janssen and Dankbaar 2008). Despite the inconsistencies in research findings, there is a growing interest to include users or customers into the innovation activities of organisations. However, organisations still struggle with how to use user information and how to benefit from the users (Kristensson et al. 2004; Janssen and Dankbaar 2008).

The research of user involvement in innovation usually takes place in an industrial context, with an emphasis on technological service innovations (Jeppesen and Molin 2003) or the development of new products. Despite of this hegemony, there is, however, some reason to apply the user involvement concept to service innovation, especially in the context of public sector services. With the private sector companies harnessing customer knowledge in order to enhance their innovation process outputs, the public sector needs innovations for its service systems in order to fulfil its main purpose, i.e., to provide value for the citizens (Hartley 2006).

Service innovation can be defined as a new implemented service or service renewal, which is perceived to be useful to its developers (Toivonen 2004). Innovation in the public sector can be better understood as social innovations and service innovations rather than technology or product innovations (Jaeger 2009; Hennala et al. 2008). Despite their differences in terms of aims, rationales, and values, the public and private sectors would seem to utilise similar innovation management processes and methods (Hasu and Saari 2009).

In this chapter we discuss user involvement in innovation in the context of public sector service innovation, as well as identify challenges that the user involvement approach brings to organisations in the public sector. The empirical part of our chapter presents a case study about incorporating customer knowledge into a service innovation process in the public sector. In the results section we present six key challenges related to incorporating user knowledge to a service innovation process.

14.2 User-Driven Innovation

14.2.1 User Roles in Innovation

The term 'user' can be understood in several ways depending on context. For example, professional users such as blue-collar workers in factories or highly skilled specialists (von Hippel 2005; Buur and Matthews 2008), hobbyists (von Hippel 2005; Raasch et al. 2008), or consumers or customers (Jeppesen and

Molin 2003; Janssen and Dankbaar 2008). In this study we focus on users who can be described as citizens, customers or 'ordinary' users. In recent business literature the consumer-users are no longer considered to be merely passive buyers to whom professionals design, manufacture or provide goods or services, but active developers who are able to provide vital information for and together with developers, hence taking part in value creation (Prahalad and Ramaswamy 2004).

Exactly why individual users or customers participate in organisational innovation activities is still unclear. The motivation has been noted to derive from experienced benefits, such as learning, social contacts, and hedonic benefits (Nambisan and Baron 2009). Another enabling factor is that the rapid development of communication technology has made the participation possible and easy (Kristensson and Magnusson 2010; Prahalad and Ramaswamy 2004).

User knowledge can be absorbed to the organisation at different points of the innovation process, for example in the idea generation, concept design or testing phase. The level of interaction varies as well. According to Cavaye (1995), user involvement can be assessed according to the six attributes presented in (Table 14.1):

The levels and several possibilities of user involvement are mentioned as one possible reason for the inconsistencies in research results concerning the benefits of user involvement (Cavaye 1995). The wide range of user involvement blurs the clear evidence of user input.

Users have been noted to have some benefits that arise from not knowing the matter at hand to a great detail. Following Granovetter's (1973) logic of weak ties, users who do not have a deep understanding of the limitations involved in the process, such as technological constraints or the service production logic, propose rather unpredictable ideas. This idea is supported by research findings that show that the 'ordinary' users, who were not technologically biased, created the most valuable ideas in a mobile service context, and the product developers and advanced users in the same context made the more realisable ideas (Kristensson and Magnusson 2010; Kristensson et al. 2004).

However, the user involvement phenomenon has not gained as much success in practice as its inherent potential would indicate. The organisations fear the risks and perceived costs of customer interaction. The ability to interact with customers, the dissemination user knowledge internally and the risks related to confidentiality have been identified as reasons why organisations hesitate to involve uses in their innovation activities (Alam 2006).

Table 14.1 Assessment attributes (adapted from Cavaye 1995)

Attribute	Description
Type	Sample or all
Degree	Responsibility: team member, leader
Content	Aspect of design: technological or social
Extent	Idea generation, concept design, testing
Formality	Official or unofficial
Influence	The influence on outcome: low, medium, high

14.2.2 Three Types of User Involvement

In general, there are two main perspectives on the methodology of user-driven innovation: the 'voice of the customer' perspective (VOC), and the 'lead-user innovation' perspective. These two schools of thought originate from the academic research and practical experiences of David Kelley, Tim Brown and Eric von Hippel, among many others (Nordic Council of Ministers 2006). The user involvement itself, according to literature, can be divided into three different types: (1) user information, (2) co-creation, and (3) user innovators (Grabher et al. 2008; Buur and Matthews 2008). The main aspects and attributes of these approaches are illustrated in Table 14.2.

The main difference between these types is the user's role and the intensity of the involvement in the product or service development processes. The user can be considered as a source or a producer of information. In the user information type the user is the target of intensive knowledge sourcing and the user is often absent. This is the field of traditional marketing research and it uncovers information like latent and stated needs, wishes, preferences, and values. It can be said that the knowledge is about the user. Some of the market research techniques enable the collection of a large quantity of information. The most important characteristic of the users is the ability to express the experiences that they have gained (Grabher et al. 2008). User information can be difficult to transfer; von Hippel (2005) referred to it as 'sticky' information. As the information is codified it can lose some valuable nuances. This can be reduced or even eliminated by having the user and the developer in direct contact between one another, something which is highlighted in the co-creation approach.

In the co-creation approach the users are knowledge co-creators: they are present and involved in some or all phases of the development process and act as knowledge sources and creators (Magnusson 2003; Grabher et al. 2008; Blazevic and Lievens 2008; Edvardsson et al. 2006). The user involvement at its best is an interactive process between the user and the developer. The user creates and evaluates ideas and develops services as an expert together with professional developers (Edvardsson et al. 2006).

Co-creation as an approach is more qualitative by nature, and the number of users involved is therefore often smaller than in the user information approach. It often requires face-to-face contact, and the knowledge thus generated is different; it

Table 14.2 Three forms of user involvement

	User information	Co-creation/participation	User innovators
Locus of information	Source	Source and creator	Creator
Role	Object	Subject	Subject
Responsibility	Producer driven	Producer driven	User driven
Perspective	Consulting	Team member	Leading
Methodology	VOC	VOC and lead user	Lead user

VOC the voice of the customer perspective

is richer and versatile but, on the other hand, the costs of the numerous relationships may be high (Gales and Mansour-Cole 1995). However, the possibilities of and opportunities provided by information technology help the coproduction of knowledge (Blazevic and Lievens 2008).

The user innovator approach is similar to von Hippel's (2005) lead user concept. The lead users are individuals who are 'ahead of their time', as it were. They are able to produce new solutions in order to fulfil their needs. A typical example of lead users could be a hobbyist who helps to develop mountain bikes (von Hippel 2005).

The role of the user can be to develop a product while the commercialisation is taken care of by other stakeholders. This can be called user's knowledge.

14.2.3 The Variety of User Involvement in the Public Sector Context

The public sector needs innovations for its service systems in order to fulfil its main purpose, i.e., to provide value for the citizens (Hartley 2006). Innovations in the public sector can be characterised as incremental rather than radical, and social rather than technological (Jaeger 2009; Hennala et al. 2008). Although public and private organisations have different premises, aims, or motivations for innovativeness, they both use and apply similar innovation methods and tools (Pekkarinen et al. 2006).

The end-user of public services, namely, the citizen, has many roles. In the Nordic countries the comprehensive public services are financed by taxpayers in the form of taxes. The legitimacy for user involvement lies in the fact that the customer of the public services is always the owner and supervisor as well (Anttiroiko et al. 2003). The public sector is also legally obliged to involve citizens in for example urban planning. The citizen perspective in the public sector is often related to terms like democracy, citizenship, and participation. The role and position of the citizen is therefore not that straightforward as a similar situation could be between a paying customer and a private sector company. Therefore involving the citizens and their ideas in the renewal of public sector organisations can be considered a type of citizen participation requiring new kinds of thinking from all stakeholders.

The involvement is executed for similar reasons as in the private sector, i.e., in order to develop better services. Public sector organisations have become more willing to listen to the voice of the users or the public than they used to be (Fuglsang 2008; Ali-Yrkkö et al. 2006). The key question is how organisation uses and adopts the user knowledge. Organisational structures, institutional practices, and professional attitudes can affect the extent to which service users can influence the change. The users will not have an effective voice until they perceive that 'the consultancy' is a regular activity and they can see observable changes (Carr 2007). The legitimacy of the citizen as a participant of a public sector innovation process is not a straightforward issue.

14.3 Research Process

14.3.1 Research Task and Research Setting

This study belongs to user-involving service innovation research. The research field is a form of user-driven innovation research. The purpose of this study is to increase understanding of a service renewal approach based on the voice of the service users in the public sector. Of the dichotomy presented in the theory section above, the methodology of this particular study is based on the approach that focuses on the voice of the customer. In addition to methodological aspects, the voice is also used as a metaphor presented and listened to by someone.

This is a case study conducted in the field of social and health care in the public sector. The voice of the customers belongs to outpatients. The party listening to this voice, i.e., the patients' message, is a developer group representing the public authorities. The research material has been collected during storytelling evenings, where the voice of the customer could be heard, as well as in a meeting between the developer group and the university actors, in which the message compiled from the customers' voices was discussed. This study is qualitative in nature.

The aim of this study is to find out what kinds of challenges there are in a public sector service innovation approach that seeks to involve the municipal residents as well. Methodologically, this task is approached from two points of view. In the so-called municipal resident point of view, issues relating to the service users' willingness of giving and ways of producing their voice will be examined. In the point of view focusing on the authorities, i.e., the public authority point of view, the developer authorities are examined as the party enabling the customers to voice their opinions and listening to them.

14.3.2 Background of the Case Study

This study was conducted in the social and health care service sector of a city in Southern Finland with approximately 100,000 inhabitants. The service strategy of the city provides that services are to be developed in a user-oriented manner. This development work has been done mainly by the public authorities themselves, consulting external experts when necessary. In addition to this, evaluations of the services have been gathered from the municipal residents and the service users mainly through feedback questionnaires. The service producers and providers have not had any established practices or methods for utilising the results of these questionnaires or gathering information or insight from the customers more extensively.

In the summer of 2008 the management of the sector decided to implement a pilot experiment together with their university collaborators. The purpose of the pilot experiment was to gather the customers' experiences of the public sector service development approach that involves the user in the social and health care sector, as well as to utilise this information in the ongoing outpatient care service process renewal. The experience gained from this pilot study would be vital when deciding whether this operational model would be used more extensively in the field. The pilot experiment forms the research context of this study.

14.3.3 Acquisition and Analysis of Study Materials

The aim of the municipal outpatient health care service process reform in question, which had begun already before the study was commenced, was to improve the functionality of outpatient care by unifying the work practices of telephone counselling, public health centres and emergency call services, as well as by implementing new innovations. In the pilot study the patients of the emergency department were included in the service development by using the storytelling method (for more information about the method, see Pässilä and Oikarinen, forthcoming). The emergency department triage nurse was responsible for recruiting patients to the customer story evenings. The aim was to recruit 15 outpatient care service users, but in the end only eight persons participated in the storytelling. The participants were of working or retirement age, and two of them were male and six female. As a reward for their participation they received an appointment with a nurse of their choosing.

A developer group with seven members was founded in order to utilise the customer experience data. All members of the group were social and health service officials. They were also in responsible positions in the outpatient care service process development. The group consisted of four women and three men. Apart from one individual, all of the members had higher academic degrees and were in managerial positions. Three of the group members had a background in medicine.

The four main phases of the development pilot were as follows:

1. Enabling the voice of the service users, i.e., the patients of the emergency unit, to be heard and known using the storytelling method on two occasions. The work in these events was directed by a methodology expert from the university so that her effect on the contents of the stories was as small as possible.
2. The recognition and conversion of the voices in the stories to a form that could be listened to by the service developers. The selected message format was a description of positive service experiences. It was structured according to the logic of the service process, and it presented the factors that came across as essential for creating a good service experience according to the story material.
3. Deepening, listening to and discussing the message compiled from the voices in a meeting between the officials' developer group and the university actors.
4. Decision-making regarding the effectiveness of the message in the development of the outpatient care process and any further measures.

The research material was acquired during the first and third phases of the development pilot. The customer stories from the first phase of the pilot study were recorded and transcribed (dataset 1). The two researchers who had participated in the sessions as observers and the methodology expert compiled observation diaries of the storytelling sessions (dataset 2). A content analysis of this material let us approach the research task from the municipal resident point of view.

In a meeting between the developer group and the university actors during the third phase of the pilot experiment, the individual developer group members defined how they understood customer-oriented public service development. A written assignment and conversation notes about the meeting form the third block of research data for this study. The fourth block of research material is comprised of the notes of the participants from the university from the time when the message compiled from the customers' voices was discussed in a meeting. A discourse analysis of this material and the results thereof aim to find answers to the research question from the public authority point of view.

In order to increase the reliability of the results, the head researcher interviewed the two other researchers involved in the collection of empirical study material. In those interviews the head researcher tested and deepened the observations and conclusions she made based on the material. This case study is qualitative in nature, and its results cannot be generalised in the same sense as those of research based on a quantitative approach.

Service innovation activities are also referred to as service reform and service development in the reports about the research results. It is often not possible to differentiate between the concepts clearly – although in the context of innovation, a clear discontinuity between the past and the future is emphasised (see, e.g., Osborne and Brown 2005).

14.4 Analysis – Citizens as Producers and Officials as Listeners of Voice

14.4.1 Municipal Residents as the Voice of the Customers

In this study, the municipal resident point of view is approached in two ways. The study examines how the municipal residents as service users are disposed towards the possibility to personally participate in the development of public services. In addition to this, the study investigates the positions from which the service users voice their views to support the service development. The position reveals the relationship of the municipal resident, i.e., the producer of the voice, to the subject of the storytelling, such as seeing oneself as a funder of a service.

14.4.1.1 Disposition Towards the Giving of a Voice

Based on the material, it seems that the participants considered it important that the municipal residents could be involved in the development of public services. The experienced importance of this issue was revealed to be one of the reasons behind the participation. When arriving at the storytelling evening, one participant stated that she "was very excited about this opportunity". This comment was an apt description of the attitude of the other participants as well. The possibility to be heard inspired and motivated many people to tell several stories about their own experiences or those of their acquaintances. All in all, the storytelling took place in a neutral and relaxed atmosphere.

Another important reason behind the participation seemed to be a desire to give feedback to the service producers regarding the functionality of public services. One storytelling session participant stated: "At last I get to give positive feedback!" In practice, giving feedback referred to giving both positive and negative feedback. The participants had prepared themselves carefully to tell their stories. The stories they told proceeded logically and were very detailed almost without any assistance from the methodology expert. What was interesting was that the participants asked questions or commented each other's tales during the storytelling even if they had not been specifically asked to do so. The participants expressed, on their own accord, a desire to get a copy of the report compiled of the sessions, in which a message was formulated to the outpatient care developer group out of the stories.

Besides their positive comments, the participants also expressed doubts regarding any real desire from the authorities' part to include the service users in the development of the health care service process in any meaningful way, as becomes evident from the following comment: "Are they (the authorities) really interested?" Besides these doubts, the participants were concerned about being included in the process in name only and that their contribution would not affect the development of the service process. "I hope this doesn't just end here", one participant commented when leaving the storytelling session.

14.4.1.2 Voice Production Positions

With the customer story method used here, the participants produced a voice for the authorities to use as support for the development work. Based on the research material, the voice production positions include: 'I as a taxpayer', 'I as a customer', 'I as an expert on illnesses and care' and 'I as an expert on the service system and process'. What is common to these positions is that reflections of societal development in which increasing individualisation, the rising of the level of education and wealth, as well as the mushrooming of information and communications technology also increase the customers' demands for quality when it comes to services (Osborne and Brown 2005; Mokka and Neuvonen 2006), can be recognised in all of them.

In the 'I as a taxpayer' position the incentive of the voice was to evaluate what the service system could offer to the customer in exchange for his or her tax money. In this position there is a clear demanding or an administrative subject mentality evident in the message. The demand mentality has many connections to the service users' role as consumers. It is not only a question of whether the municipal residents feel they have got their money's worth, but what they feel entitled to. The administrative subject mentality, on the other hand, reflects the power relationship between the taxpayer and the service provider. Even though people have paid taxes as dutifully as one possibly can, gaining access to the services does not seem to be simple. However, besides the sacrifice mentality, there is also a hint of silent defiance in the administrative subject mode of speaking, as can be seen in the following comment:

> "I have given my all and done it running, because walking would have been too slow. It would perhaps be preferable that one didn't have to think about every single penny and wonder what I could do now. Whether I could go and see a doctor and which doctor could I go to. And how much that would cost." (Female, an ill-health pensioner)

In the 'I as a customer' position the incentive of the voice is to evaluate the quality of the services that the municipal residents receive. The assessment is primarily based on the customers' own service experiences, or then the service in question is compared with one's previous service experiences in either the public or the private sector. In addition to this, evaluation refers equally to the experiences of any peer groups, hearsay and urban legends alike.

It is said that there will most likely be a shift from the customer-expert setting towards an inclusive and communal service production (Mokka and Neuvonen 2006). This trend seems to be strong in the final two voice production positions: The 'I as an expert on illnesses and care' and 'I as an expert on the service system and process' positions.

In the 'I as an expert on illnesses and care' position the incentive of the storytelling are the service user's reflections upon the contents of the service s/he has received in relation to his/her need, which in turn the service user has determined based on his or her own knowledge. This became evident particularly when discussing diagnosing illnesses. The 'expert on illnesses and care' position was also prominent when reflecting upon treatment practices, as can be seen in the following comment:

> ". . . people are working pointlessly when one must take the Marevan test once per month. And I too have had this for eight years already and you always test for it every month . . . I find this Marevan monitoring utterly pointless after the first year." (Male, pensioner)

In the 'I as an expert on the service system and service process' position, the storytelling is guided by opinions on the flow of the process and the allocation of public resources. The flow of the process is primarily examined in the light of one's own experiences, such as irritation over being directed to the wrong place or happiness over knowing the waiting period for a treatment. The allocation of resources, on the other hand, is discussed on a collective level and not contextualised into one's own service experiences. There seems to be more extensive knowledge of the society behind this dialogue, as can be seen in the following discussion.

> "... that kinds of savings, that you get things cheaper when you make the employees work under some company." (Female, in the working life)
> "That's one reason, but another quite essential one is the labour shortage. The one that has struck already and they're afraid it will strike even worse in the future. It's that the private sector pays the doctors better so it's a more attractive employer than the public sector." (Female, in the working life)

14.4.2 Authorities Enabling and Listening to the Voices

The public authority point of view is approached by investigating how user-driven service development approach is manifested on the level of the developer authorities. In addition to this, we examine the kinds of orientations from which the developer authorities are listening to the message composed of the municipal residents' voices. When naming these orientations, two interrelated factors have been taken into consideration; firstly, what kinds of expert roles (e.g., that of a doctor) the public authorities manifest when listening to and discussing the message formed of the voices, and, secondly, what kinds of meanings are assigned to the municipal residents as service developers (e.g., as feedback givers) by the public authorities, and what kinds of orientations there seem to be in the background, when the public authorities listen to the message.

14.4.2.1 About Defining the Approach

In the case study the developer authorities used officially approved modes of thinking when defining what user-oriented service development means in the social and health care sector. This was manifested e.g. as the use of administrative jargon from the field of customer-driven service development, as can be seen from the following quotes: "The customer point of view is taken into consideration in all development" and "Customer centricity is the key". Another manifestation of the use of official modes of thinking in the definition process were references to structural models used in official documents, such as "Customer centricity is located in the upper third of the upturned triangle". In this example the triangle referred to the profit card model currently in use.

In addition to this, service users seemed to include not only the end-users, i.e., the municipal residents, but also the organisations and organisational staff that seemed to be 'part' of the new or renewed service offered to the customer. This research material could not be used to find out the form of this 'participation', but a general statement can be made that taking these parties into account and labelling them as 'users' seemed to be based on an idea of the services as a process. This view can be seen to be linked to von Hippel's (2005: 3) definition of users, in which users "are firms or individual consumers that expect to benefit from *using* a product or a service. In contrast, manufacturers expect to benefit from *selling* a product or a service". For instance, a doctor at an emergency department can use and utilise the outpatient care laboratory services when treating a patient, or refer the patient to specialised care when necessary.

One method the authorities used in order to develop the services in a user-driven manner was either to assume the role of the customer or examine the customer's point of view from a distance, as one participant put it when discussing the development of technology-based services: "We have always thought about how the customer sees and experiences things". To imagine and to let the knowledge of the user be based upon feelings and imaginations represents the lowest level of user involvement (Nordic Innovation Centre 2007).

Besides making these assumptions, the authorities seemed to feel that the authentic voice of the municipal residents can be allowed to be heard, but they did not want to commit to having it affect service development. Even though their basic attitude towards listening to the voice of the customer was positive, the significance of the voice for service development was belittled for several reasons. Some of these reasons were related to the fact that customers were seen to be ignorant about production-related factors pertaining to the services or about their own needs. Some, on the other hand, were related to an assumption from the authorities' part that the voices' message only contained demands, as can be seen from the following:

> "We listen to the customers' opinions about service development, but the public sector is not an automat that makes every wish come true." (Female, has worked in her position for 4.5 years)

Interestingly, the voice of the customer was mainly seen to express wishes, opinions, or demands – in other words, not real needs. Needs were seen as an important factor guiding service development, but the definition of the needs was to be based on the authorities' expertise, or at the very least to be done under the management of authorities. One participant wrote that experts are in the role of agents, because the customers cannot define their needs in customer-centred service development by themselves. "Our aim is not a satisfied customer, but the good of the customer", was the comment of one participant.

14.4.2.2 Orientations of Listening to the Message of the Voice

Based on the case study material, we can point out two orientations from the developer authorities' part relating to the reception of the message compiled from the users' voices, namely, the administrative authority orientation and the profession orientation. These orientations are not manifested as personal attitudes, but the same person can utilise both orientations and different shades thereof flexibly in the different phases of the processing of the message.

When the developer authorities assume the administrative authority orientation, they are concerned whether administrative rules and guidelines are observed. Taking the message of the voices of just a small group into consideration in the development of outpatient care services raises concerns about the equal treatment of all municipal residents.

This orientation is also manifested in the authorities' disposition towards the function of the participation of the residents and the significance of the information conveyed in the message. The developer authorities seemed to feel that the function of the municipal residents' participation seemed to be giving feedback. This was in accordance with the residents' own view of their role. The interpretation of the information in the message was markedly service system-centred. The information that emerged was used to assess the successfulness of the already completed developments in the outpatient care service process. Shifting the perspective actually towards the service users was less important. The view in which the contents of the message were seen as a way to increase the authorities' knowledge and understanding of the municipal residents as service users – i.e., as a source of customer understanding – was not given prominence when the message was interpreted.

The profession orientation is more prominent in the reception of the message than the administrative authority orientation mentioned above, especially from the point of view how effective the message was among the developer authorities. Two partially intertwined factors can be pointed out in this orientation. These are the authorities' background in the natural sciences (more extensively than in the social sciences) and their professional substance expertise. Including only a small amount of service users in the service process development was problematic when contrasted with the developer authorities' background in the natural sciences and the quantitative research tradition of the natural sciences. The credibility of the message was questioned, because the information was not acquired according to the scope and generalisation principles used especially in medical research and research in the field of nursing science. In addition to this, positioning the information based on the customers' experience among the information sources of the dominant development paradigm was problematic as well.

In the case study the developer authorities were given the chance to familiarise themselves with the contents of the message beforehand, and the message was also discussed in more detail in the meeting between the developer group and the researchers. Interestingly, some parts of the information in the message were ignored by the developers, while others raised even very intensive discussion. What these parts had in common was that they were focused on the work of doctors and medical expertise in general. For instance, the service users' view of a good visit to the doctor was contradictory with that of the doctors themselves. The messages that raised the most controversy were disregarded as potential informational raw material for the service development. For example, when discussing the aforementioned 'good visit to the doctor', the users' wish for a comprehensive good treatment of the patient including, e.g., eye contact was ignored, because from the medical point of view the visit to the doctor is functionally about making a diagnosis, a problem which has more of a technical solution rather than a social one. Another interesting issue was that medical expertise was given more weight than expertise in the field of nursing science when discussing the usability and effectiveness of the message.

In sum it can be stated that based on the research results the municipal residents seemed to be forming the 'voice of the users' from four different positions. These were:

- I as a taxpayer;
- I as a customer;
- I as an expert on illnesses and care;
- I as expert on the service system and process.

In contrast, two different orientations could be identified from the part of the public authorities when they were listening to the message, namely:

- The administrative authority orientation, and
- The profession orientation.

14.5 Challenges of User-Involvement in the Public Sector

The purpose of this study is to find out what kind of challenges there are in a public sector service innovation approach that seeks to involve the municipal residents. In this case study, conducted in the field of social and health care services, we approach the recognition of the challenges from the points of view of both municipal residents and authorities.

In the previous section (14.4), we have examined how the municipal residents are disposed towards the possibility of giving and producing the user voice in the context of a renewal of public sector services, and how enabling of and listening to the public sector users' voice is manifested on the level of the authorities. In the following paragraphs we seek to answer our research question by analysing the research results from the two points of view (municipal resident and authority). The following six challenges indicate the problematic issues pertaining to a user involvement innovation approach in the public sector.

14.5.1 Challenge 1

According to the results, the service users find it important to have the possibility to participate in the development of public services. It can be said that – in the light of their different roles ('I as a taxpayer' and 'I as a customer'), expertise and knowledge – the possibility of the municipal residents to participate even comes across as a right.

The results also show that the users' disposition towards their possibility to participate in service development is enthusiastic and responsible. This can be seen to be due to, for instance, the fact that the users felt that they had adequate information and experience enabling them to state their views on the contents and production methods of the services they need to support their own well-being. These resources become evident in the 'I as an expert on illnesses and care' and 'I as an expert on the service system and process' positions used in the production of the user's voice.

The roles of service funders, controllers, customers and experts on their own well-being experienced by the municipal residents reinforce the necessity of a collaborative service development model also in innovation activities pertaining to public services. As the opportunity arises, at least some of the municipal residents are willing to do their bit in the service renewal process. In addition to this, unlike in for example the general public's letters and writings in newspapers or the Internet, the municipal residents' voice contains different tones if some effort is put into the discovery of those tones and voice-enriching methods are utilised.

From the municipal resident point of view the challenge of practical service innovation activities that involve the municipal residents is whether there will be enough occasions of low-level involvement for the residents – in other matters than those provided in legislation as well. In addition to this, there is the question whether these arenas for participation enable the creation of a methodologically fruitful customer-driven development information repository?

14.5.2 Challenge 2

Besides being enthusiastic, however, the municipal residents also seemed to harbour some doubts concerning the significance of the participation. One critical factor affecting the residents' motivation to participate was trust – trust in the significance of their participation and input in the service development work conducted by the authorities.

The service users' participation motivator was a desire to give the service developers some feedback. This attitude might suggest that, as pertains to the development of public services, the municipal residents see themselves as objects. In other words, they are the target for which the public authorities develop the services. When considering the principles of innovation activities that involve the service user, this view is problematic. Involvement, as defined in this study, refers precisely to facing the municipal residents as development partners and resources, not the objects of development activities.

In the municipal resident point of view the challenge is how the municipal residents' trust in the significance of their participation can be increased and how a public service development role based on the subject status and partnership of the municipal residents can be reinforced.

14.5.3 Challenges 3 and 4

The authorities – and their development practices in particular – have a central role in facing the two challenges presented above. The development practices can be seen as manifestations of the dominant public service development paradigms of the time. In this case study, user-driven service development was recognised as a principle governing service development, and moves towards realising it in practice had been made for example by assuming the position of the customer in the development work. According to our results, despite these measures, a clear challenge for the authorities seems to be to realise this approach in practice, at least to the extent that is meant by the term in the context of this study.

According to the results, the authorities have adopted user-driven service development as a term. However, it is defined more strongly through administrative language and administrative models than as a dynamic activity in itself. The position of the municipal residents and the service users in the development work is somewhat awkward. The residents must be heard, but the authorities perceive problems in the active involvement of residents and the usability of the information they bring in.

Since the authorities are looking at the matter from the perspective of their own administrative culture, they are concerned about the equal treatment of all municipal residents, because only a relatively small group of service users gets the possibility to participate in the production of the voice. In addition to this, when viewed from within the research tradition of the natural sciences, the information produced with the voice of the customer method is perceived as questionable in the customer segment in question due to issues relating to scope and generalisability. In addition to this, being able to read and understand the customer's experience would seem to require more practice.

According to the results, the authorities' view of the municipal residents does not seem to be altogether fruitful as pertains to the practical realisation of service renewal activities that include the residents. To use an extreme example, the municipal residents are seen as greedy individuals wanting various new services. In addition to this, it can be seen that the interests of the service users and the service producers are not constructively close to each other. The reason to this is the users' inability to assess their service needs and the prerequisites relating to service production. Because of, among other things, these characteristics associated with the municipal residents, the authorities' expert knowledge is seen as crucial for service development.

From the point of view of the practical implementation of the service innovation activities that involve the municipal residents, a challenge is for the developer authorities to be prepared and willing to expand their information-producing practices in service development towards collaborative information production. This expansion is also connected to the view of the municipal residents as experts on their own experiences, as well as the municipal residents' general and social and health care sector-specific development resources.

These challenges to the public authorities are connected to the previously mentioned challenges that emerged in the municipal resident-related points of view. What is central is *whether the authorities are willing to utilise the available development resources, and how prepared the authorities are to implement and establish these practices as part of their development culture.*

14.5.4 Challenge 5

According to the analysis of this case study, the service user point of view is generally taken into consideration in service development. However, the methods used, which are mainly based on assumptions, are not among the most fruitful forms of inclusion. Using other methods, like for example the storytelling method, would enable and allow the municipal residents gain a subject status in the public service development activities. In addition to this, those other methods would be a more useful way of receiving positive feedback than feedback questionnaires, which could be a motivating factor in service innovation activities when seen from a human resource point of view.

The results also suggested that the authorities did not see it their duty to give weight to the voice of the municipal residents in service development. Interestingly, it could even be inferred that the public authorities feared that they would lose their power to the residents, were the residents' voice given a stronger status than that of an opinion or a wish. One reason for this seemed to be a certain view of the municipal resident-involving methodology. The authorities displayed a dichotomous attitude towards the importance of the voice in development activities: either the voice is allowed to dictate and formulate the entire end result of the work, or it is given only marginal importance. The latter seemed to be the safer option. Moreover, the authorities did not seem to connect the service user-involving service renewal methodology with the concept of the interpretative nature of service renewal processes (see Lester and Piore 2004) and with how the information gained from the users can be converted to a form that supports service development (see, e.g., Hennala and Melkas 2010).

The authorities will face the challenge of understanding the different forms of the service innovation approach that involves the users, and of gaining the ability to assess when the involvement of the users is practical and what other factors (research results, experiential knowledge, production-related and economic principles, legislation) are essential – and to what extent – in the development of a new or renewed service. In addition to this, new involvement methods should boldly be tried in user-driven service development, and, when necessary, methodological competence should be complemented by networking with other experts. Networking would also facilitate time management.

14.5.5 Challenge 6

In the light of the results of this study, the reception of the message compiled from the users' voices would seem to be defined by not only the administrative authority orientation, but the profession orientation as well. This orientation became clearly evident when discussing the contents and rejecting the influence of the message when the matters pertained to medicine. We can come to the cautious conclusion that a medical point of view was emphasised in all discussions of the service users' message. This study did not investigate how professional hierarchy, which is still strongly present in the health care sector, affected the discussion.

From the authorities' point of view the challenge is to make visible and discern the substance expertise based on profession and the customer's expertise when listening to and discussing the users' message. Listening to the points of view of different professions in order to support multi-voiced development work would seem to require attention as well.

14.6 Summary and Conclusion

Users are increasingly seen as subjects of innovation processes instead of serviced customers within them. The framework of practice-based innovation emphasises the roles of 'shop-floor' service producers and service users in the creation of innovations in the context of public sector service renewal. The focus of this approach is on Mode 2b innovation production (see "Epilogue" in this book). This chapter is an attempt to give some new perspectives to the Mode 2 innovation processes that include the users' voice in them.

The purpose of this study was to increase understanding of a service renewal approach based on the voice of the service users in the public sector. From the voice-of-the-customer perspective, the citizens act as the giver and producer of the voice and the authorities as the enabler, listener and exploiter of the voice.

The present case study has increased understanding of the possible issues that may slow down or even totally hinder the involvement of users and, hence, also their voice production in public sector innovation activities. In addition to this, the study indicates that there might also be problematic features related to listening to the service users' voice and having the authorities utilise it.

When analysing the case study results it should be noted, particularly as pertains to the voice giver and producer (municipal resident), that the results may be excessively positive. For instance, municipal residents in the study felt that it was very important for them to be given the possibility to participate in the service renewal, but in practice the storytelling sessions failed to involve the sought-for number of patients. The same observation was made in another study based on service users' voice (Hennala et al. 2011), in which senior citizens were asked to participate in the renewal of the service concept of a well-being centre. Those involved in the development were motivated and found the issue important, but recruiting participants is not easy.

In the analysis of the developers' results we must emphasise that this was a relatively limited case study. Therefore it could prove interesting to conduct further studies and examine how much the emergent six challenges of user involvement innovation approach can be said to apply to the public social and health care sector in general, and how the situation changes over time.

Emphasising the service users' point of view in service development increasingly strongly will transform views on the nature of service innovation activities in the public sector. It can be expected that in the future the demands to include the service users in service renewal activities will increase, if the public service system continues to adopt the market-based operational logic and the societal trend where the service users' demands for services increase continues. This can become evident as, e.g., different public service production models, such as public–private partnerships, challenging public service producers to compete for customers (Hennala et al. 2008), and when the service users make service providers compete for users according to quality criteria that are significant to the users themselves. A participatory expertise must be established beside the substance expertise of the authorities. Simultaneously, the municipal residents' ability to participate in the constructive development of services affecting them must be reinforced as well. The two poles should be positively and continuously affecting one another in a loop that feeds itself.

As a main contribution of the present study six visible challenges for public sector user-involvement were outlined. In a nutshell they are as follows:

- Challenge 1: To create user-driven easy-access participation arenas for municipal residents, based on methodological solutions that enrich customer-oriented service innovation.
- Challenge 2: To increase the trust of the municipal residents in the significance of their participation and to reinforce their service development role based on their subject status and partnership.
- Challenge 3: The authorities' will and ability to produce information collaboratively.
- Challenge 4: The authorities' will and ability to understand the role of the service users as a valuable resource for service renewal.
- Challenge 5: To understand and assess the different forms of participatory innovation activities and their suitable (or unsuitable) uses in different situations.
- Challenge 6: To identify 'the customer's voice message', to recognise the customer's expertise, and to differentiate between the substance expertise-based profession and the developer's expertise.

References

Alam, I. (2002). An exploratory investigation of user involvement in new service development. *Journal of the Academy of Marketing Science, 30*(3), 250–261.

Alam, I. (2006). Process of customer interaction in new service development. In B. Edvardsson (Ed.), *Involving customers in new service development* (pp. 15–32). London: Imperial College Press.

Ali-Yrkkö, J., Honkanen, P., Hyysalo, S., & Uotinen, J. (2006). *Globalisaatio, innovaatio ja kansalaisuus – ProActiivisia näkökulmia innovaatiopolitiikkaan [Globalisation, innovation and citizenship – ProActive perspectives into innovation policy]*. Helsinki: Kauppa- ja teollisuusministeriö. (In Finnish.)

Anttiroiko, A.-P., Haveri, A., Karhu, V., Ryynänen, A., & Siitonen, P. (2003). *Kuntien toiminta, johtaminen ja hallintasuhteet [Operations, management and managerial relations of municipalities]*. Kunnallisalan tutkimuksia. Tampere: Tampereen yliopisto. (In Finnish.)

Blazevic, V., & Lievens, A. (2008). Managing innovation through customer coproduced knowledge in electronic services: An exploratory study. *Journal of the Academy of Marketing Science, 36*, 138–151.

Buur, J., & Matthews, B. (2008). Participatory innovation. *Journal of Innovation Management, 12*(3), 255–273.

Carr, S. (2007). Participation, power, conflict and change: Theorizing dynamics of service user participation in the social care system of England and Wales. *Critical Social Policy, 27*(2), 266–276.

Cavaye, A. L. M. (1995). User participation in system development revisited. *Information & Management, 28*, 311–323.

CIS. (2004). *Innovation in Europe – Results for the EU, Iceland and Norway*. Luxembourg: Office for Official Publications of the European Communities. ftp://ftp.cordis.europa.eu/pub/innovation-smes/docs/results_from_cis3_for_eu_iceland_norway.pdf Accessed 13 June 2010.

Edvardsson, B., Gustafsson, A., Kristensson, P., Magnusson, P., & Matthing, J. (2006). Introduction. In B. Edvardsson (Ed.), *Involving customers in new service development* (pp. 15–32). London: Imperial College Press.

Fuglsang, L. (2008). Capturing the benefits of open innovation in public innovation: A case study. *International Journal of Services Technology and Management, 9*(3-4), 234–248.

Gales, L., & Mansour-Cole, D. (1995). User involvement in innovation projects: Toward an information processing model. *Journal of Engineering and Technology Management, 12*, 77–109.

Granovetter, M. (1973). The strength of weak ties. *The American Journal of Sociology, 78*, 1360–1380.

Grabher, G., Ibert, O., & Flohr, S. (2008). The neglected king: The customer in the new knowledge ecology of innovation. *Economic Geography, 84*(3), 253–280.

Hartley, J. (2006). *Innovation and its contribution to improvement. A review for policymakers, policy advisers, managers and researchers*. London: Department for Communities and Local Government.

Hasu, M., & Saari, E. (2009). Julkisen sektorin innovaatiotoiminta ja sen johtaminen kansallisena haasteena [Innovation activities in the public sector and their management as a national challenge]. In V. Valovirta & J. Hyvönen (Eds.), *Julkisen sektorin innovaatiot ja innovaatiotoiminta* (pp. 112–132). Esiselvitys Sektoritutkimuksen neuvottelukunnalle. Osaaminen, työ ja hyvinvointi 11/2009. Helsinki: Sektoritutkimuksen neuvottelukunta. (In Finnish.)

Hennala, L., Linna, P., & Pekkarinen, S. (2008). Julkisen sektorin innovaatiotoiminnasta [On innovation activities in the public sector]. In V. Harmaakorpi & H. Melkas (Eds.), *Innovaatiopolitiikkaa järjestelmien välimaastossa* (pp. 93–108). Acta series no. 200. Helsinki: Suomen Kuntaliitto and Lappeenrannan teknillinen yliopisto. (In Finnish.)

Hennala, L., & Melkas, H. (2010). Surffailua, taiji-voimistelua ja karjalanpiirakoita – Ikäihmisten monitahoinen ääni hyvinvointipalvelujen innovaatiotoiminnan voimavarana [Surfing on the web, tai chi training and Karelian pies – The multifaceted voice of ageing people as a resource of innovation activities in well-being services]. *Gerontologia, 3*, 231–248. (In Finnish.)

Hennala, L., Melkas, H., & Pekkarinen, S. (2011). Customers as innovators in senior service markets: An examination of innovation potential and characteristics. *International Journal of Service Science, Management, Engineering, and Technology, 2*(1), 30–51.

Jaeger, B. (2009). *User-driven innovation in the public service delivery*. Paper presented at the 2009 Annual Conference of European Group for Public Administration, Malta. http://www.egpa2009.com/documents/psg1/Jaeger.pdf. Accessed 18 August 2010.

Janssen, K. L., & Dankbaar, B. (2008). Proactive involvement of consumers in innovation: selecting appropriate techniques. *International Journal of Innovation Management, 12*(3), 511–541.

Jeppesen, L., & Molin, M. (2003). Consumers as co-developers: Learning and innovation outside the firm. *Technology Analysis & Strategic Management, 15*(3), 363–383.

Kristensson, P., Gustafsson, A., & Archer, T. (2004). Harnessing the creative potential among users. *Journal of Product Innovation Management, 21*, 4–14.

Kristensson, P., & Magnusson, P. R. (2010). Tuning user's innovativeness during ideation. *Creativity and Innovation Management, 19*(2), 147–159.

Lester, R. K., & Piore, M. J. (2004). *Innovation – the missing dimension*. Cambridge, MA: Harvard University Press.

Magnusson, P. R. (2003). Benefits of involving users in service innovation. *European Journal of Innovation Management, 6*(4), 228–238.

Mokka, R., & Neuvonen, A. (2006). *Yksilön ääni. Hyvinvointivaltio yhteisöjen ajalla [An individual's voice. The welfare state in the era of communities]*. Report 69. Helsinki: Sitra (the Finnish Innovation Fund). (In Finnish.)

Nambisan, S., & Baron, R. A. (2009). Virtual customer environments: Testing a model of voluntary participation in value co-creation activities. *Journal of Product Innovation Management, 26*, 388–406.

Nordic Council of Ministers. (2006). *Understanding user-driven innovation*. TemaNord 2006:522. Copenhagen: Nordic Council of Ministers.

Nordic Innovation Centre. (2007). *User-driven innovation: When the user makes the difference*. Oslo: Nordic Innovation Centre. http://www.nordicinnovation.net/_img/user-driven_innovation_2007_web.pdf. Accessed 30 April 2010.

Osborne, S., & Brown, K. (2005). *Managing change and innovation in public service organizations*. London & New York: Routledge.

Pekkarinen, S., Tura, T., Hennala, L., & Harmaakorpi, V. (2006). Innovativeness as a challenge and an opportunity for the public sector. Paper presented at 'Innovation Pressure', International ProACT Conference, 15–17 March, Tampere, Finland.

Prahalad, C. K., & Ramaswamy, V. (2004). Co-creation experiences: The next practice in value creation. *Journal of Interactive Marketing, 18*(3), 5–14.

Pässilä, A., & Oikarinen, T. (forthcoming). Research-based theatre as a facilitator for organizational learning. In A. P. Meusburger, & A. Berthoin-Antal & E. Wunder (Eds.), *Knowledge in organizations – Learning organizations*. Vol. 6. of the series Learning, Knowledge and Space. Berlin: Springer.

Raasch, C., Herstatt, C., & Lock, P. (2008). The dynamics of user innovation: Drivers and impediments of innovation activities. *International Journal of Innovation Management, 12*(3), 377–398.

Toivonen, M. (2004). *Expertise as business. Long-term development and future prospects of knowledge-intensive business services (KIBS)*. Dissertation. Espoo: Helsinki University of Technology. http://lib.tkk.fi/Diss/2004/isbn9512273152/isbn9512273152.pdf. Accessed 4 January 2010.

Von Hippel, E. (2005). *Democratizing innovation*. Cambridge, MA: MIT Press. http://web.mit.edu/evhippel/www/democ1.htm. Accessed 4 January 2010.

[illegible]

Chapter 15
The Increasing Use of Dramaturgy in Regional Innovation Practice

Philip Cooke

Abstract This chapter reports on advances in regional innovation practice. Regional innovation has become a maturing field of economic governance. Regions have become more prominent actors in the innovation field in the past decade. Innovation is widely seen by supranational, national and regional governance bodies and agencies as a mainspring of improved regional economic performance and wellbeing. Leading regional innovation practitioners are increasingly being understood as catalysts of innovation, a development in their earlier role as being supporter or partner in innovation essentially conducted by others. One technique this chapter devotes attention to where regional 'orchestration' of innovation occurs is the use of narrative, drama and non-scientific laboratory experimentation to open business and community minds to the constructed regional advantages of innovation. The theoretical context is 'post-cluster' hence platform-minded and using matrix models to induce innovation through stimulating cross-cluster 'transversality'.

15.1 Introduction

Much regional development agency responsibility is administrative, fund-seeking or recruiting foreign direct investment to raise regional innovation. Most European regional development agencies have sector or cluster policies, but only a little over half of the 340 or so have responsibility for innovation (Cooke 2010). Very few have either informal or, less, formal methodologies aimed at getting cluster incumbents to cross-pollinate or cross-fertilise their knowledges. There are very few Venn diagrams showing overlapping inter-cluster horizontal knowledge dynamics. But there are some that practice such 'transversality' and a few others

P. Cooke
Centre for Advanced Studies, Cardiff University, Cardiff, UK
e-mail: cookepn@cardiff.ac.uk

H. Melkas and V. Harmaakorpi (eds.),
Practice-Based Innovation: Insights, Applications and Policy Implications,
DOI 10.1007/978-3-642-21723-4_15, © Springer-Verlag Berlin Heidelberg 2012

that understand it and aspire to evolve appropriate methods to take advantage of cross-sectoral knowledge dynamics and open up the 'white spaces' as they are sometimes referred to, or what Burt (1992) refers to as 'structural holes', where innovation opportunities are at their most tempting and potentially abundant.

This of course chimes perfectly with Schumpeter's (1975) concept of innovation as recombinations of knowledge. In Andersen's (2011) recent invocation of Schumpeter's regional innovation theory, it is made clear that Schumpeter radically separated the role of innovator and entrepreneur. The innovator recombines knowledge, utilising mainly ideas and knowledge that already exist, while the entrepreneur separately implements the innovation, utilising skills of risk management, market attraction and arbitrage. In its modern guise, the regional innovation agency steps closer to the 'innovator' role and moves away from performing no significant entrepreneurial role, which nevertheless remains mainly the province of the firm. Nevertheless, catalytic innovation is far more than mere 'support' or such equivalent supply-side lame ducks as 'promote', 'encourage', 'assist' or 'partner' firms in their quest for innovation. Supply-side economics was the neo-liberal mantra, adopted by regional, national, multilateral and supranational agencies everywhere in the form of 'spatial specialisation' policies like clusters and clustering. Now, demand-driven innovation is back to pick up the pieces left by the failures of supply-side neo-liberalism. Anne Golden has articulated the innovation policy dilemma as follows: 'Should reliance be placed on urban specialisation – as Harvard professor Michael Porter claims? ....Or should we trust in diversity – as the late urban expert Jane Jacobs believed?' (Golden 2009). This chapter looks beyond Porter but stops short of observing full-blooded Jacobian diversity, satisficing on the evolutionary economic geography of 'related variety' (Frenken et al. 2007).

The chapter is constructed as follows. It begins in the next section (15.2) with a theoretical overview of some of the key concepts introduced already, which will feature prominently in the empirical cases of advanced regional innovation practice outlined. Hence there is a discussion of the core evolutionary economic geography concept of 'related variety' which underpins the whole analysis. But more time and space are devoted to policy concepts like knowledge recombination methods, 'white spaces' and 'structural holes', platforms, matrix methodologies, dramaturgy and transversality. An attempt is made diagrammatically to show how they fit together as elements of a demand-driven innovation model. In the second main section (15.3) of the chapter, theory will be juxtaposed with regional innovation policy practice of the more advanced kind. Three representative models will be discussed and those utilising dramaturgy will be highlighted. Other dramaturgy 'learning regions' will be alluded to. Reasons for this 'cultural turn' in practice will be adduced. Amongst the principal contributory elements, it will be argued are: the rise of creative industries as progenitors of services innovation; the need for innovation 'orchestrators' to move beyond mere advertising into the 'experience economy'; and the sheer cognitive added-value that can arise from using dramaturgy in translational knowledge activities. After that, the chapter ends with concluding remarks.

15.2 A Conceptual Framework for Advanced Regional Innovation Policy Practice

'There is nothing so practical as a good theory' argued Kurt Lewin, the father of social psychology (Lewin 1951) while Keynes had a different take on the same idea: 'Practical men, who believe themselves to be quite exempt from any intellectual influences, are usually the slaves of some defunct economist' (Keynes 1973). Both display profound understanding of the framing process, whereby predominant ideas pervade consciousness to the extent that, never having read the relevant text, a practitioner may absorb theory and utilise it for good or ill. It is possible that this is beginning to occur in relation to the core evolutionary economic geography concept of *related variety*. This is by no means a defunct economic concept, it is too young to have been fully tested yet, but the fact that advanced practitioners of regional innovation policy are behaving as though they had read, for example, Frenken et al. (2007) suggests it may be good theory.

15.2.1 Spatial Process Concepts

15.2.1.1 Related Variety

The basic idea of *related variety* is that economic development is caused by interactions between elements in regional economies displaying relatedness in technology or industry focus and has the following powerful theoretical implications. First, applying the notion of related variety has led to new insights in the externalities literature. In particular it suggests, with empirical support, that these 'spillovers' as they are now known, occur in two ways: by spinoff firms and labour mobility (see Wenting 2008). Second, it has provided additional insights to the question whether or not extra-regional linkages matter for regional growth. Third, relatedness is now also investigated utilising social network analysis techniques. Fourth, the notion of relatedness enriches the literature on labour mobility, which is often regarded as one of the key mechanisms through which knowledge diffuses. Fifth, relatedness may also show its relevance through entrepreneurship dynamics. Experienced entrepreneurs (those that have acquired knowledge in related industries), as opposed to spinoff companies, may play a crucial role in the regional diversification process. Insights available from evolutionary economic geography in relation to regional economic growth have thus been outlined; in the following, they are further elaborated. First, applying the notion of related variety has led to new insights in the externalities literature. Empirical studies tend to show it is not so much regional specialisation or regional diversification (Jacobs 1969) regarding externalities that induce knowledge spillovers and enhance regional growth, but a regional economy that encompasses related activities in terms of competences (i.e. regions well endowed with related variety). Second, it has provided additional insights to the question whether or not extra-regional linkages matter for regional growth. Adopting a relatedness framework, empirical studies on trade patterns tend to show that it is not inflows of knowledge per se that matter for regional growth, but inflows of knowledge that are related (not similar) to the existing knowledge base of regions. Related flows concern new knowledge that can be understood and exploited and, thus, be transformed in regional growth.

Third, relatedness is now also investigated in network analysis because tracing movements of knowledgeable persons can reveal relatedness in action. For instance, studies show that collaborative research projects tend to create more new knowledge when they consist of agents that bring in complementary competences. Fourth, the notion of relatedness refreshes recognition of the importance to regional growth of labour mobility. This too is often regarded as one of the key mechanisms through which knowledge diffuses. Recent studies show that neither inflows nor outflows of labour are properly assessed if not also considering how these knowledge flows match the already existing knowledge base of firms and regions. Fifth, relatedness may also show its relevance through entrepreneurship dynamics of the kind indicated by Schumpeter (1975). Experienced entrepreneurs (those that have acquired knowledge in related industries), as opposed to spinoff companies, may play a crucial role in the regional diversification process. More generally speaking, longitudinal studies show that long-term development of regions depends on their ability to diversify into new sectors while building on their current knowledge base. However, at least two problems exist methodologically in application of related variety to economic processes. What might be called simple related variety may predict regional innovation potential by *ex ante* analysis of regional industrial composition. Hence similar industries, like varieties of engineering, may reveal actual evidence of high innovation and high spinoff and labour mobility factors assisting it. For example, a UK study shows engineering to be the most innovative industry category, ahead of highly innovative categories like: other manufacturing; creative industries; and knowledge-intensive business services (KIBS; Chapain et al. 2010). But there is a strong element of unpredictability regarding innovative cross-overs between unrelated industries. This is extremely hard, to the point of impossibility, to predict and may only be observable *ex post* as something akin to 'revealed related variety'. Equally so, the issue of related variety outside the region or even country probably also only allows for *ex post* understanding. In these respects, fascinating as it is to see advanced innovation emerging from co-operations between, for example, otherwise unrelated agricultural and automotive research, as occurs regarding biofuels, the practical implications of this for regional policy are unclear but may become less so as the chapter proceeds.

15.2.1.2 Recombination

The second key process, rather than practice, concept concerns 'recombinant methods' of innovation, as proposed by Schumpeter. How does this occur? Veblen held the view that the ultimate source of innovation lay in dis-satisfaction with the status quo on the part of the innovator. This is highly likely to be user-driven in the sense that the innovator her or himself may experience the initial dis-satisfaction. Or s/he may learn of dis-satisfaction by a third party such as a doctor with respect to a drug's efficacy. Or finally a quite widespread segment of the community, for example, concerning pollution or environmental degradation may display dis-satisfaction. Dis-satisfaction reveals prospects of a 'lead market' that can be exploited entrepreneurially if the innovator comes up with a workable improvement.

In the history of innovation, numerous world class novelties arose unexpectedly when a different innovation target was being aimed at. Post-it notes are the classic 'useful innovation' of this kind. As a 'repositionable' or 'releasable adhesive' it began as a solution in search of a problem. Initially conceived of as a better kind of bookmark, consumer demand after a marketing trial showed its real value lay as a visual reminder listing device. In Cooke and Morgan (1998) the example is given of Borg Warner the US steering gear firm needing to cut cost by stamping rather than milling components, which led them to innovate an advanced stamping machine that created more demand than the product it made. Prosthetic joints became more efficient when discoveries made in aerospace ceramics were applied to inadequate pre-existing solutions. Very many other innovations are explicable in terms of these cross-over effects in contexts of dis-satisfaction with existing combinations of knowledge. Proximity to the initial user may be advantageous but many such cross-overs occur at trade exhibitions (Maskell et al. 2006).

15.2.1.3 Structural Holes/White Spaces

This brings us to the third of our spatial process concepts, the identification of 'white spaces' or 'structural holes' where innovation opportunities and potential are argued to lie according to this approach (Burt 1992). He contrasts two types of interactions that may create the social capital underpinning innovation opportunities. The first is that it occurs among strongly interconnected network members, a contrast made earlier by Granovetter (1973) with his notion of strong network ties, which may in practice turn out to be too inward-looking and specialised to avoid 'lock-in'. The weight of past relations and interactions among actors and institutions under a strong ties regime can be innovation inhibiting. The second possibility is that social capital for innovation may be better served where the network has actors who can broker interconnections between disconnected interests. This is cognate to the 'weak ties' argument advanced in Granovetter (1973) where 'lock-in' is avoided because the burden of obligations, favour exchanges and traditionally strong and regular social capital interactions is missing. This leaves greater scope for new interactions that might be desirable for innovation to ensue. This is consistent with the innovation mechanism embodied in the concept of 'recombinant methods' discussed above.

Two interesting and important theoretical insights are then discussed in Burt (1992). First he demonstrates the widely-assumed prevalence of sociological and social psychological research insights that intra-group knowledge circulation predominates over inter-group knowledge circulation. This is shown for work groups, divisions of firms and, indeed, whole industries where no actor is simultaneously aware of innovation opportunities resting with other actor networks. In passing, the perspective on collective socio-technical processes known as actor-network theory (ANT) can be criticised for its failure to recognise the weaknesses of strongly interconnected networks. This lacuna is particularly clear in the expressed view that network agents or 'actants' are 'foundationally indeterminate', with 'no a priori substance' but only receive their nature from the networks in which they associate (Latour 1999). So, although heterogeneous, such networks are highly determined, even over-determined. This seems to go too far, since it begs the question of how knowledge for recombination 'breaks out' of its network 'silo'. While agreeing that 'silos' are inefficient, whether in business divisions or industrial clusters, not least for the time they waste in diffusing knowledge, even Burt (1992) recognises that it eventually reaches everyone of consequence to it.

The second important observation is that while it may seem that innovation inhibiting tight networks need to be replaced in their entirety by innovation promoting looser networks with good inter-network antennae, both have a key role to play. Thus, as we have seen, the presence of inter-network brokerage functions is seen to enhance efficiencies of time and space. However, Burt also contends that closed networks can be critical to recognising structural holes. This, if it can be sustained, is a highly practical argument that – from a cluster perspective – means policy makers should continue to build cluster strengths while brokering inter-cluster platform linkages that fill the white spaces or structural holes with innovations. However, this insight needs probing before that useful practical judgement can be settled upon. How are structural holes identified from tight networks? Burt (2000) draws a distinction between structural holes outside the network and those inside it. The latter exerts a negative effect on the network's ability to reach consensus. It is also empirically frequently the case that some hierarchy prevails as well as good internal network communication and cohesion. These constraints allow the tight network to reach conclusions about the need to seek innovation elsewhere than inside the network or specialisation setting. Hence strong cluster organisations are valuable but limited in their horizons, while looser ones tend on all criteria analysed in Burt (2000) to be superior.

15.2.1.4 Platforms

This leads to the final process concept, edging towards a practice concept, which is the notion of 'platform'. This is represented in spatial process terms by the following diagram (Fig. 15.1). We have mapped a set of four conceptual relations that connect 'relatedness' as a condition of industry co-existence that is potentially value-adding for innovation strategy, to 'recombination' as the means by which innovation results in the commercialisation of new knowledge. Recombination involves crossing 'structural holes' among networks and filling 'white spaces' between clusters. If the network or cluster itself is internally well-integrated and

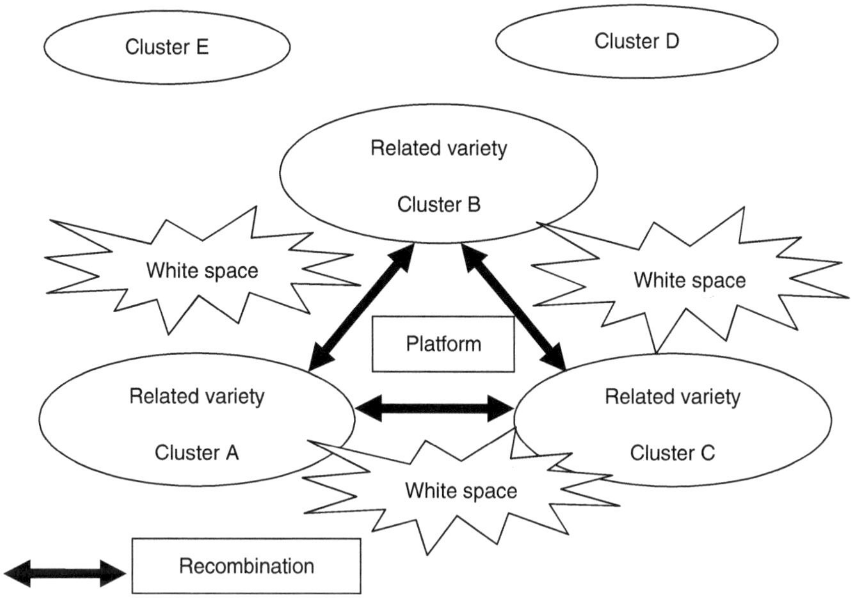

Fig. 15.1 Regional inter-cluster innovation platform model

led, this enhances efficiencies, leading to recognition of the need for network brokership. Hence well-functioning tight clusters will be more likely to make moves to evolve platforms, but they will do it more slowly and hesitantly than looser networks having many brokers already. Poorly organised, internally porous clusters will, in all probability, be ineffective and fail.

The platform is thus the result of successful, repetitive moves by orchestrators to integrate disconnected clusters across structural holes, filling up 'white spaces' with innovations that come from Jacobian knowledge spillovers. The diagram shows recognition that recombination of knowledge among some clusters may be unrecognised, unexploited or impossible. Indeed, it clarifies the distinction between structural holes, which are, in principle, present between all clusters and 'white spaces' which are those selected to be filled up. Recall, once again, that the key actors are, on the one hand, *innovators*, who envision the recombination and the subsequent innovation and, on the other, *entrepreneurs*, who implement innovations commercially. They may overlap, but much evidence from the practice of venture capitalists towards, for example, academic innovators, suggests that entrepreneurial knowledge is required to manage an innovative firm (Kenney and Florida 2000).

15.2.2 Policy Theory Concepts

We now come to the second part of this chapter, which is more interested in practice and concerns policy theory concepts dealing with how, practically, platform policy is conducted. This has already been elaborated for the Finnish case by Harmaakorpi and Melkas (2005) and Harmaakorpi (2006) as the regional development platform method. This is seminal work in the field of regional innovation systems theory (see, for example, Cooke 2008). It is empirically based in a study of regional resilience in the Finnish furniture industry region of Lahti (on 'resilience' see Bristow 2010; Simmie and Martin 2010). However, this is not the only type of regional innovation context; indeed platform policy in support of exploiting white spaces has been more evident in relatively prosperous European regions, although it is also practised in some peripheral regions. Hence, there is a need for a conceptualisation of this practice approach as it is deployed in different regional settings. Nevertheless, returning to the Finnish original, heralding a Mark 3 regional innovation system approach that moves beyond mainly 'institutional' (Mark 1) or mainly 'entrepreneurial' (Mark 2) models (see Cooke et al. 2004) here we see the early processes and institutions of dramaturgy appearing in regional innovation policy practice.

15.2.2.1 Dramaturgy

Dramaturgy is a branch of the theatrical arts which involves bringing productions to life on the stage. The role of a dramaturg in a production can be quite varied, depending on their training 'school' and the philosophy of the theatre the dramaturg is working for. Different national theatre traditions also have differing beliefs on the role of dramaturgy in theatre production. In Harmaakorpi's exposition of the 'toolkit' utilised in recombination stimulation in Lahti (Fig. 15.2), among numerous innovations involving more than one Living Laboratory, is Innolink. This is part of the organisational process of integrating regional platform policy approaches with

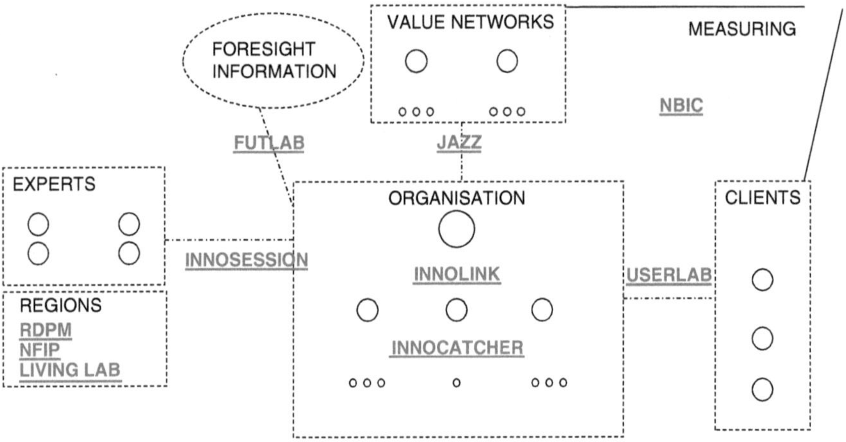

Fig. 15.2 The Lahti regional innovation system platform policy toolkit (source: Harmaakorpi 2009)

technological and other experts with value networks and client firms. One of the means of identifying possible future developmental trajectories and giving them regional relevance is the use of creative methods. One of the most interesting and successful of these has been the use of theatrical drama to connect ideas to a systematic analysis of knowledge, resources and the potentialities and possible pitfalls of co-operation. Scripts are written by professional dramatists in the form of short sketches articulating mentalities, blockages, barriers and change requirements in contexts where the drama involves simulations of real world experiences of firms and institutions confronting the need for innovation. The short dramas or sketches are then acted out to an audience of potential users of the ideas and solutions presented. The cognitive plateau achieved collectively by users then feeds into the succeeding phases of the innovation process, such as Jazz, where improvisation on any emergent opportunity themes is encouraged, Innocatcher, where potentially creative notions and concepts are articulated, captured and integrated, and Userlab, a Living Lab where testing and trialling of ideas occurs and process outcomes are simulated. We shall see later in this chapter how Living Labs are used elsewhere as test-beds or what might be termed 'rehearsals for reality' in relation to innovation.

The work of Pässilä and Oikarinen (2010a, b; Pässilä et al. 2011, this volume) deserves mention here for the innovation of Research Based Theatre (RBT) they have developed. This approach uses live, scripted mini-dramas to explore different 'frames' of meaning projected by actors in, for example, business innovation and change management workshops. Accordingly, it combines innovation management studies with art and cultural studies to 'get behind' the 'presentation of self in everyday life' that is acted out by the powerful and the powerless. In process terms, it is accordingly highly revelatory of hidden biases in attitude and perception. However, it may be argued that it is weaker on the content side in that not all actors in a firm or organisation would have the necessary professional expertise to deliberate on, say, a biotechnology innovation. Moreover, it would be difficult but not impossible to equip non-professionals using techniques of deliberative democracy or polling and it would be interesting to observe RBT in such a context.[1]

15.2.2.2 Orchestration

For such an innovation policy process to be enacted requires organisation and management. This is for three reasons: first, the users of innovation involved in

[1] In conversation, we agree that mine is a more institutional approach than the symbolic interactionism at the individual level these experts practice. Although their 'Boalian' perspective facilitates conflict analysis of a collective kind, it may not yet be capable of illuminating the 'dark matter' of power relations, which are ultimately more institutional, often gaining from wide social legitimacy, than individual.

the quest for innovation may be diverse in sectoral terms, although they probably consist of sectoral sub-blocks rather than having wholly individualised capabilities. Second, the central task of dramaturgy is to open up the prospects for cross-pollination of knowledge and capabilities in as harmonious a manner as possible. Finally, there is a need for the special skills, sometimes called ‘hybrid’ skills, of an animator or facilitator who is cognizant of relatively deep elements of actor technological and business processes, how these may possess potential from the viewpoint of transversality, and what the methods are that can assist these to be fulfilled. Other kinds of hybrid knowledge are also pertinent, regarding, for example, multi-level economic governance and relevant programmes, as well as a network of contacts in relevant scientific, technological and creative research fields that may be included as process advisers. Such skills are not particularly those of a Kotter-style ‘leadership’ regime (Kotter 1996). The critique is well-known of this type of change management model for its authoritarianism, narrowness of focus and short-termism. But, equally, an overly inclusive ‘learning discourse’ model would probably be too slow, recursive and diffuse (Argyris and Schön 1978). Under these circumstances an intervening ‘orchestration’ model, neither strongly hierarchical nor overly networked, is to be preferred (Wallin 2006, 2010).

Of the four kinds of orchestrator identified by Wallin (2006), namely: promoter; architect; auctioneer and conductor, the roles of promoter and conductor are most relevant in collective action rather than individual firm contexts. The promoter mobilises the network and the conductor provides it with a narrative that connects individual and collective concerns to contextual opportunities. In the former sense the promoter mobilises interest in a series of events and the conductor facilitates the means to meet audience or customer demand. The orchestration process thus defines the value-creation priorities of business networks and their resources. It specifies relationships with stakeholders, such as customers, employees, unions, suppliers, competitors, and other internal groups. It takes account of contexts that may create conflicting interests, or impose risks to firm networks. It seeks to identify innovative means of creating value. The orchestration process model is itself subject to continual review as a response to actual and possible changes in perceived business conditions. In regional innovation policy settings, the ideal promoter and conductor of proceedings is the specialist innovation agency, whose interest may still be in profitability for itself but only in relation to selling its specific skills rather than competing with its business customers. Such an agency may house the infrastructural knowledge discussed above to assist promotion and the transversality knowledge to stimulate innovation. As will be shown in the empirical part of the chapter, orchestration can also usefully be assisted by narrative if not dramaturgy in both its promotion and conducting roles.

15.2.2.3 Transversality

In discussing the conducting aspect of orchestration policy, transversality has been mentioned more than once. It is important now to say what it is and why it is important. Recall that a driver of this whole policy perspective is the evolutionary economic geography insight regarding the extra value that accrues to urban or regional economies displaying related variety (Frenken et al. 2007; Boschma and Frenken 2003). However, two important things were noted in the section discussing the concept. The first is that the mere existence of related variety does not automatically mean that it will be exploited. Market failure may prevail due to asymmetric information resulting in knowledge spillovers and mutual absorptive capacity being under-used for well-known reasons of satisficing first articulated by Simon (1957). Firms may not have the resources, time or priority to look outwards at narrow transversality as much as they ideally should. Here, as in other situations, market failures warrant third-party interventions that, in this instance for a fee, bring opportunities for potential innovation gains to their attention. Second, it was noted that transversality in the broader sense can also be a source of useful interactions among firms leading to innovation. Where such search and selection behaviour is prioritised by the firm in one cluster or sector seeking to fill a 'white space' by connecting to firms in wholly other clusters or sectors, they may be successful, even within their region. However, achieving 'revealed related variety' in this way is even more demanding, but its potential is sufficient to warrant orchestration of potentially valuable articulations as far as possible within the region and reasonable outside it. As will be shown in the empirical part of this chapter to follow, here is where 'story-telling' skills are utilised to the utmost to create a 'red thread' of narrative that might appeal to the broadest possible innovation-seeking audience.

The means by which this can first, be promoted, then, second, orchestrated are in their infancy and being experimented with contemporaneously. The platform model in Fig. 15.2 is as comprehensive as any, but could be said to be rather vague on precise methodologies and instruments for ensuring successful transversality interactions. Of interest in the accounts to follow is the absence of a single 'cookie-cutter' model of how to do it. The few models that exist are rather region specific but by no means incapable of being typologised, as we shall see. It is because of the importance for new regional innovation policy to be as professional as possible about innovation instruments that this aspect is given some prominence. For the first time the regional innovation agency takes on a catalytic role in the innovation process. The interaction of the key concepts of relatedness, recombination, structural holes/white spaces, platforms, dramaturgy, orchestration and transversality endow the third party brokerage capability with a special resonance compared to the past. Earlier models of regional innovation have focused on supply-side measures like attracting foreign investment, designing science parks and incubators or building clusters as main means of promoting regional innovation. Now innovation is moving towards demand side drivers, with a focus on lead markets, public procurement and consumer preference for ethical production and consumption. Helping to identify revealed related variety opportunities to meet demand-driven innovation is the key new challenge faced by regional innovation policy and its practitioners.

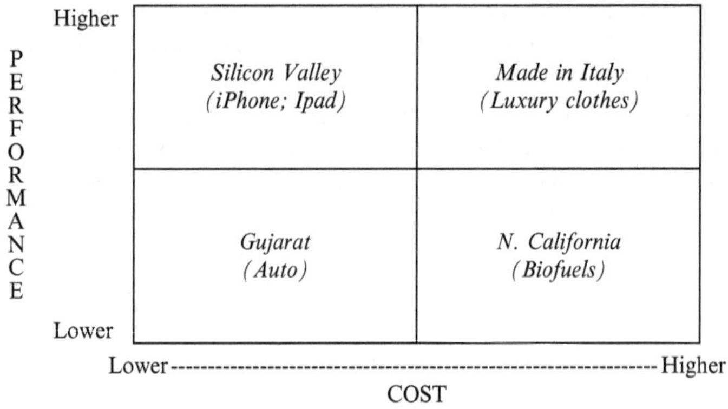

Fig. 15.3 Regional innovation quadrants

15.2.2.4 Matrix Methods

One of the simplest yet most effective instruments for facilitating transversality and its associated innovation opportunities is 'the matrix' (see Eriksson et al. 2010). This is an elaboration of the technique of the 2×2 quadrant or double entry tabulation that social scientists and management theorist find useful as heuristic devices in both teaching and research. Where concepts interact, illumination, clarification or even innovation may be achieved as shown in Fig. 15.3. The interest in this illustrative quadrant is that it shows innovation is found in many and diverse locations, at different positions in global markets, and with different drivers in terms of relations between cost and performance. One of the generic findings from empirical studies of innovation is that much of it is driven by the twin contradictory forces of cost reduction and quality improvement. Higher performance at lower cost is the usual cause of innovation, its source of increased productivity and its key contribution to increased competitiveness. These are useful and important messages to convey quickly to students, policy makers, customers and suppliers. Thus, Silicon Valley is a highly innovative region – home of disruptive innovations (Christensen 1997) – such as the Apple ICT innovations over the years. By contrast, Italy is home to many industrial districts that, to survive, have tended to keep moving up-market with higher consumer cost but increasingly high quality output. This is 'sustaining' innovation as Christensen (1997) calls it in contrast to price-lowering innovation more associated with the disruptive kind. New market segments such as the middle-class Indian (or for that matter Chinese) market have invoked price lowering product innovation in the indigenous car industry, with Tata Nano production centred in Gujarat province. Finally, northern California, partly the Bay Area, partly Silicon Valley, is home to new entrepreneurs producing biofuels in pursuit of renewable automotive and other energy. However, it is more expensive and displays lower performance than traditional hydrocarbon fuels even though it commands a niche consumer market of ideologically green consumers. Thus we see the power of matrix methods of thinking.

However, the use to which matrix thinking has been put in innovation practice is thus far twofold. First, some regional innovation practitioners have compared their industry mix along the 'y' axis with their toolbox of innovation support instruments along the 'x' axis and categorised cases accordingly. Thus, for example, firms in one sector might tend to be supported by access to technical university laboratories, while firms in another might be recipients of clustering policy, as appropriate. An alternative matrix approach has been to adopt a policy perspective, for example 'sustainable development' and identify industries appropriate to different aspects of that perspective. Thus green construction might be included on the production side, organic food on the consumption side and green logistics at the intersection of the two. Finally, regional industries have been compared with technologies known to be utilised by one industry or cluster but not by others. Innovation policy here is to have innovations developed in one industry demonstrated to other industries in workshops, exhibitions and conferences. The key aim is innovation transfer. The orchestrating agency then brokers conversations, dialogues and meetings among interested potential innovation adapters and adopters, and assists as needed with extra expertise, for example from university or private laboratory research establishments. In each of these applications, a narrative is evolved, whether from a sustainable development viewpoint or a healthier living perspective that endows the promotion and conducting of the innovation process an attractive storyline 'hook'.

15.3 Three Kinds of Dramaturgy in Regional Innovation Practice

It is relatively easy by now to see the rationales underlying the use of dramaturgy in regional innovation discourse. There are three clear roles it can be observed to play. First, modern innovation facilitation, in which the third-party broker or agency takes on a more active role, requires greater awareness of the need for successful branding. This is imperative where the agency is reliant on fees to sustain itself and must demonstrate a professional and knowledgeable way of attracting client attention against competing calls on customer time. In Gilmore and Pine (2007) this process is known as *placemaking*, a term coined by architect Jon Jerde:

> 'Companies that embrace placemaking understand a fundamental dictum for contending with authenticity.... the best way to generate demand for any offering – whether a commodity, good, service, other experience, or even transformation – is for potential (and current) customers to experience that offering in a place so engaging that they can't help but pay attention....'
>
> (Gilmore and Pine 2007: 149)

In other words, clients filter what to prioritise according to the relevance and *authenticity* of what they have or expect to have experienced. The fakery involved in advertising, which is both ubiquitous and oversold, nowadays has little purchase on client attention. The 'place' can be permanent or temporary, real or virtual but it must associate cognitively with an authentic, lived experience. Drama is, in the literal sense, unreal but also a heightened form of reality in which important experiences can be captured, juxtaposed and resolved with efficiency, effectiveness and the opportunity for customer participation.

Second, as Harmaakorpi (2006) showed from experience of reactive innovation change management, scenario-designed dramaturgy was intended to and actually helped change basic assumptions of key actors, raising awareness of the importance of institutions. Old industrial policy mentalities and measures were transcended. A previous perception was that absence of clusters and dynamic lead industries or firms such as Nokia doomed the region. This view was changed by application of policy platform discourse concerning transversality. Accordingly, the power of transversal networks led to a revaluation of the region's own capabilities. What problems had to be overcome? Foremost, 'knowledge' access, flow and integration asymmetries were overwhelming. Hence, a basic kind of 'policy pedagogy' had to be formulated and conducted. Second, creative, imaginative thinking about the future lacked the specificity to enable platform-envisioning to evolve. Extra time had to be devoted to diffusing visualisation skills, dramaturgy assisting significantly in bridging cognitive divides. Shifting mindsets from individualistic to more collective action perspectives took longer and was more difficult than expected. Third, even the main change management texts deal only with the internalised world of the single firm. In the *externalised* world of the policy platform far greater knowledge depth and complexity must be handled to enhance 'visualisation' and 'potentialisation' for the region. Finally, this new process proceeded raggedly and in time-consuming ways, but a key lesson was that the novelty of terms and concepts makes it initially difficult for outsiders to understand that clear communication is vital. Dramaturgy, with its heightened and concentrated sense of reality can assist in explaining and clarifying learning discourse.

Third, more and more innovation support has inevitably to involve service-sector innovation since service employment and businesses are the majority in virtually all economies. This has distinctive innovation characteristics. Some of it is largely dormant from an innovation or creativity viewpoint. But some parts of service industry, notably the creative industries, and knowledge-intensive business services, are measurably close to levels of innovation in manufacturing, at least in the UK (Chapain et al. 2010). Such businesses are either themselves creators or users of dramaturgy in their core business models. Accordingly they are acutely aware of the authenticity and creativity dimensions of the innovation services they may be purchasing. This implies placemaking with, as Gilmore and Pine again suggest *flagship* status. This means the innovation agency having a portfolio of experience-oriented services on offer in the form of permanent demonstration space or specific types of innovation service availability. It also means having access to appropriate enterprise and innovation support personnel that understand and utilise dramaturgy where appropriate and together make the agency's premises a site to be trusted to provide value-adding innovation and creativity-optimising opportunities. In what follows, three examples of the utilisation of varieties of dramaturgy in regional innovation policy will be showcased, with a further two noted in the concluding section.

15.3.1 Transversality in Pulp, Paper & Graphics

Sweden provides good examples of distinctive ways of building regional innovation platforms within a regional innovation system format that is based on transversality and uses elements of dramaturgy. This is practised most completely in Region Värmland. The region displays strong transversality regarding added-value use of its numerous existing clusters. At regional level orchestration includes a 'global megatrends' analysis, identifying key trends of relevance to the region's industry and society in which each cluster organises action plans in response. Areas like 'innovative environments (living environment)', 'entrepreneurship', 'clustering', user-driven innovation, competences and information and communications have underpinned cluster priorities. Thereafter the region has identified 'innovation platforms' as they call them, to stage interactions at related variety cluster interfaces. Clusters include The Packaging Arena; The Paper Province; Compare ICT Cluster; Food Valley and Graphics Valley (Fig. 15.4). Each cluster has been exposed to a scenario analysis process and the results further informed the regional development programme. The 'innovation platforms' include: The Packaging

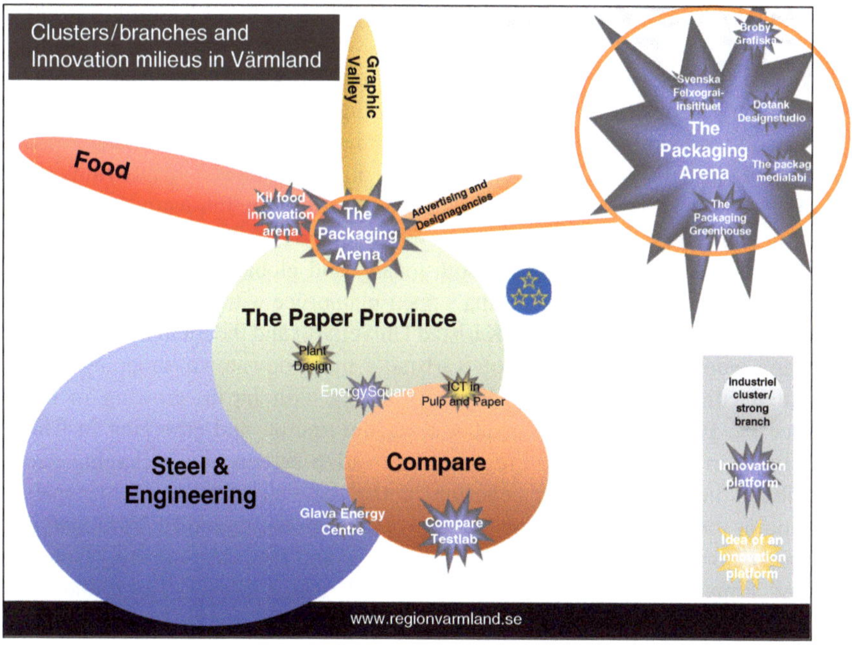

Fig. 15.4 The Värmland regional innovation system & its clusters (source: Region Värmland)

Greenhouse; The Packaging Media Lab; Do-tank Design Studio, Broby Cross-Media College; Flexography Institute; Food Innovation Arena; ICT in Pulp & Paper; Compare Testlab; Energy Square; Glava Energy Centre and others. Of importance here is the manner in which the inter-locking nature of Värmland's clusters is designed to take advantage of the high degree of knowledge and innovation interactively transferring from university research (notably Karlstad University) to industry in the region and beyond it. Värmland's Regional Development Programme was approved in 2008 during which year the Glava Energy Centre and Kil Innovative Food Arena initiatives were established, followed in 2009 by Karlstad University setting up a Care & Wellness Centre and in 2010 by the establishing of ten regionally funded cluster professorships in Karlstad University.

In this segment of the account attention is focused on the cultural tradition of storytelling and narrative that is expressed in platform developments based on transversality in Värmland. The region boasts the first female Nobel literature laureate, Selma Lagerlöf, who was nominated in 1909. Her books built on the storytelling traditions of Värmland, still celebrated in the annual literature festival near Karlstad, and this has been a 'red thread' in aspects of regional development strategy. Accordingly, evolving from a basic pulp and paper industry, the region expressed its identity through securing advances in word-related and graphics technologies, as expressed in the development of the Swedish Flexography Institute and Broby Cross-Media College in the region. The Cross-Media College utilises the region's transversality in 'white spaces' pedagogy as will be seen. It is rurally located in what is now branded 'Graphics Valley' in Sunne, alongside the Swedish Flexography Institute and some 20 spinoff businesses such as *Flexigraph, Exonera* etc. Although it was originally a school for graphic media such as printing, packaging design, and graphic research, it is currently also a national centre for photography, scriptwriting, film-making and digital media arts. These skills are marketed worldwide but are also utilised in orchestration policy. Thus the Packaging Media Lab calls on graphics, photography, digital media and scriptwriting skills in creating living laboratory test-bed demonstrations of innovative packaging, printing and store layouts for its various retailing clients.

Cluster interaction is important both locally and globally. This is evident, for example, within the Packaging Arena's open-plan office with the Packaging Media Lab, Service Research Centre (of Karlstad University) and the IncuPac incubator in the same complex in the heart of Karlstad's main downtown shopping centre. Communications abroad are already very strong with Japan, growing in India, whence the Packaging Arena agent is a regular visitor, and emerging in China. As a key part of the Värmland regional innovation system, the Packaging Arena contributes immensely to the cohesion of the regional economy. It occupies a position at the cross-roads of the leading clusters in the region of Värmland and is closely involved with orchestrating initiatives and facilities, notably 'Graphics Valley' firms who supply global clients like *Nestlé*, *Unilever*, *Tesco* and large Swedish brands like *Löfbergs Lila* coffee. Among the 'white spaces' currently being explored are those, on the one hand, concerned with cross-fertilisation links to other regional clusters including those linked to Agro-food, Steel & Engineering and ICT inside Värmland and, on the other, extra-regional clusters such as 'Printed Electronics' at Norrköping in Ostragotland region. Among matrix elements being innovated are 'green packaging'; flexible logistic bar-code graphics; and releasable adhesives.

Värmland is a peripheral, rural region ranking seventeenth out of Sweden's 21 regions in regional GDP. Successful platform orchestration, stressing authenticity of regional tradition in innovative ways and recombining creative and technological knowledge dramaturgically has nevertheless raised the regional game. Multi-level governance has been important to regional evolution. As a poor region, Värmland suffers from 'excellence' criteria in the allocation of extra national scientific research funds. However, funds to support its transversal platforms have been accessed from outside the region. Tillväxtverket (TVV; formerly Nutek) supports The Packaging Arena through its cluster programme and, as conduit for EU Structural Funds to Sweden's regions, TVV also channels regional development funds into Värmland. Accordingly, the region and university jointly funded the ten new cluster chairs in response to being overlooked in the national contest for research excellence funds. This further strengthens efforts in the quest for innovative 'white spaces' but stresses the importance of building on regional identity to secure national and international competitiveness with varieties of public innovation support funding.

15.3.2 Transversality & Orchestration in Cluster Renewal: Marche Region, Italy

Marche region in Italy has branded itself with the grand narrative of being a classic, creative Italian industrial district economy. As farming declined from the 1950s and 1960s entrepreneurship and innovation grew in small-firm networks concentrating on agro-food processing; footwear and leather; textiles; white goods; and furniture. Like Värmland it is seen as peripheral, yet it has above-average regional GDP for Italy and sees itself as a quality economy, promoted with the 'Made in Marche' brand. However, in the depths of the global financial crisis of late 2009 the common perception of policy and representative economic actors with respect to the Marche regional economy was that it had been deeply, possibly fatally, undermined by the crisis. Many small firms had gone to the wall; even larger firms in the agro-food industry had been bankrupted. Globalisation had, in any case, made trading conditions for Marche firms considerably more difficult, not least because most were in traditional industries like footwear and leather, textiles, furniture and food. Only the substantial electro-mechanical industry, boasting such globally-recognised names as *Indesit-Hotpoint* and *Ariston* white goods production were at least somewhat future-oriented in their business planning. In this respect, the engagement of the economic governance and innovation system elements responsible for the new *Domotics* (domestic robotics) platform was notable. This included

Fig. 15.5 Green domotics in Marche's LEAF community design

an innovative connection to a 'green domotics' model of the LEAF green construction and eco-community initiatives promoted by Marche firm *Loccioni* (Fig. 15.5).

Here is an exemplar of the innovative potential of the region when 'related variety' is harnessed in a somewhat transversal manner, exploiting for potentially substantial market opportunity the existing creative assets of the regional economy. Another exemplar of modernity in industrial application of knowledge spillovers from related variety of industry is observable in the rise of a new Nautical Cluster combining the creative talents of the domotics, furniture and leather industries in boat building and services. These illustrations show clear recognition of the importance of transversality as a survival strategy in a region beset by crisis in its foundational business model.

The key economic development orchestrator is *Sviluppo in Marche* (SVIM). This is a small agency that interprets innovation strategy guided by the Laws passed by the regional government. These give 'rules of the game' for different policy areas and SVIM forms its sense of priorities and activities according to its interpretation of these. Hence, for example, most of the time of the 30 staff of SVIM is spent on the arduous task of tendering mostly successful bids for EU Structural and Framework Funds. The efficiency of Marche Region in disbursing such funds means that in 2009 it was the first Italian region in actually ensuring allocation of funds to recipient organisations and partners. Such funds as these form the largest part of the SVIM budget. In 2008 this was reflected in a global budget for projects of all kinds, numbering 150 approved, of €35 million. The annual budget of the organisation, by contrast, was of the order of €3–5 million. Seeking such funding is guided by five fields of activity: Innovation; Research; Energy; Credit Access; and Internationalisation. However, this focus had occasioned closure of the 'Localisation (FDI) of Companies' and 'Sustainable Environment' functions leaving Environment embedded – usefully – but secondarily in the Innovation function.

There is a widespread perception in Marche that the clusters that have served the region well for decades are in need of renewal. But the global financial crisis of 2008–9 had a more profound effect on Marche firms than previous downturns. This is almost certainly because of the fact that it was principally a crisis of the banks and specifically a crisis which froze credit, especially to small family businesses. Here is revealed, perhaps for the first time a negative aspect of the Marche model. It was clearly built on the assumption that small, regular amounts of credit would be available through high trust, reputational links among small firms and regional or local banks. While that assumption prevailed the system could function in a flexible but specialised way within the distinctive clusters. But when even local banks, knowing their creditors for decades, experienced something akin to a collective nervous breakdown in late 2008, credit dried up, firms started to go to the wall and pressure on the regional government to respond with an extensive system of credit guarantees became irresistible.

The 'creativity' narrative is widely utilised both in branding the region's assets and its products. Some quality niche firms like *Tod's* or the *Prada* and *Ferragamo* manufacturers located in the region are better-known internationally than Marche and some regional branding is reflected back from these firms. Clearly, Marche's competitive specialisation in furniture, shoes, clothing and leatherwear depended significantly on a high degree of design skills in the regional community. As in the case of the marine, especially yacht-designing, cluster signs are that transversality policy is well-advanced. There is also an indication of the emergence of creative industry, especially in animated films and other content. Ancona is a regular host city for the International Animation Exhibition, not least because the city and Marche more widely possess a tradition, competence and identity in the animated cartoons field. These skills are utilised in the cityscape where cartoons are used to enliven the dreary export vegetable-oil containers in the harbour. One of the leading producers is highlighted in Fig. 15.6.

15.3.3 Regional Innovation Orchestration of Transversality: Bayern Innovativ

An exemplary model of the whole panoply of new innovation concepts is found in Bayern (Bavaria) Germany as summarised below. This account is focused upon the platform-building activities of *Bayern Innovativ*, a governance agency for regional innovation based in Nürnberg. Since its establishment in 1995, the agency has identified key industries among many that were beneficiaries of cluster policy paid

Fig. 15.6 Animation: an emergent creative cluster in Marche

for by Bavaria's resource windfall when it sold its share in the regional energy supplier. Five of these clusters are managed on contract by this private innovation agency. Technically the clusters were cross-tabulated against key technologies to find the inter-disciplinary and inter-industry innovation potentials of 'related variety' in the regional economy. Many innovations (on average 10%) have ensued from the over 1,000 per year 'conversations' facilitated between neighbouring sectors concerning technological co-operations, applications and resulting innovations. A small part of the new platform thinking involves recognition of the importance of enhancing sustainable development as part of a new green vision concerning renewable energy and clean technologies.

How does *Bayern Innovativ*'s proactive regional innovation policy work? Fig. 15.7 gives an indication showing matrix management of potential innovation opportunities at intersections between regional industries. These are points where conversations among distinct and by no means obviously neighbouring business sectors are facilitated.

Accordingly, where these dialogues facilitate personal discussion between experts and customers, sustainable cooperation networks may be developed. Examples of the roughly 10% of marketed innovations arising from these co-operations include:

Laser technology adapted to beam nanoscale droplets onto microarrays for rapid bioanalysis
Mechatronic systems for car engine management that have been transferred to bus steering systems
Portable fuel cells that have been applied in automotive electronics

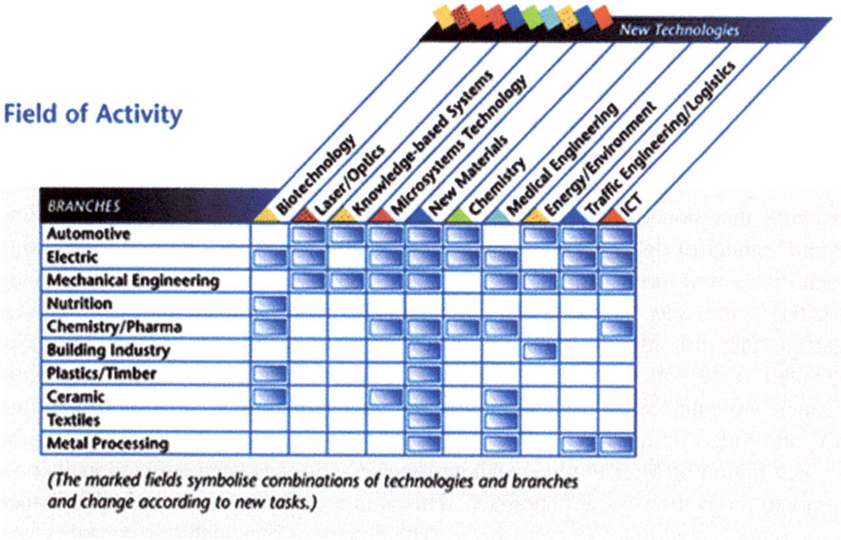

Fig. 15.7 *Bayern Innovativ*: technology platforms (source: Bayern Innovativ – http://www.bayern-innovativ.de/2009)

Plastic injection moulding processes from button manufacturing which have been implemented in automotive plastic components

A logistics and transport company that has secured a contract with one of the world's largest Internet suppliers

A technical textile producer won a contract in medical engineering.

Hence, *Bayern Innovativ* initiates business-driven project co-operations across disciplines and branches, taking into account the latest results from the scientific community. Over the past decade the agency has forged new pathways and created a portfolio of cooperation platforms and networks that have generated an extended, sustainable network structure. Both the platforms and the networks are in demand at regional, national and international levels. Accordingly, firms are 'orchestrated' to explore 'white spaces' between related industries and recombine knowledge for innovation. Interestingly, as in some of the other transversality types discussed, a strong creative element underpins platform management. Thus the agency sees its role not as 'leadership' but 'orchestration' of regional innovation interests (see also Wallin 2006). Each event BI organises – some 400 annually – has a narrative 'red thread' running through it designed to attract participants. For instance, an 'Innovative Materials' event aimed at vehicle and new materials manufacturers held in Augsburg in early 2010 played on the theme of 'Light Living & Leisure' to connect engineering with wider societal issues, attracting 450 participants and key presentations on fibre composites, aluminium and new alloys from the likes of Audi, BMW, Daimler, Opel and Porsche, alongside materials specialists like Huber, Voith and ThyssenKrupp.

15.4 Conclusions

It is worth noting that, although there are relatively few regional innovation agencies engaging with these new 'orchestration', recombination and 'white spaces' quests to enhance regional innovation potential, a few others that are beginning to work in this way deserve mention. Thus Skåne is a second Swedish example that pioneered the 'white spaces' concept and has established a 'sounding board' model of deliberation among regional innovation stakeholders concerning identification of likely candidate white spaces. This is echoed in ways its 'Moving Media' cluster has shown strong transversality among new media like computer games, the film industry and film-based tourism. Moreover, policy has been devoted from both national economic development and innovation promotion agencies to enhance creative and innovative infrastructures in relation to the film, TV and video drama industry evolving in the region. A further example is the French region of Midi-Pyrénées where transversality has been adopted as its new regional innovation system approach. This is in a context of economic dominance by a major global industry (aerospace). The processes by which transversality was arrived at as a policy development exercise was orchestrated by the regional governance system. These are open-ended but a platform of transversality among agro-food, healthcare biotechnology and aerospace exists in which transversality is to be imagined and implemented by firms selected as high-potential innovators, many acting as supply-chain hubs or '*firmes pivots*'. This strategy aims to secure regional prosperity by enhancing platforms for innovation and their transversality.

Finally, the introduction to the chapter promised a conceptual framework, generalising the kind of concepts and practices described into a frame of reference for academics and policy-makers alike to test out in other contexts to determine the extent to which dramaturgy plays any role in evolving regional innovation policy formulation processes. In Fig. 15.8 a relatively straightforward attempt is made to illustrate connectivity among groups of concepts and practices using a regional innovation systems framework. This shows the full working of the model between the process sub-system in the upper level and the policy and practice sub-system at the lower level. Two-way interactions occur between and within the sub-systems. This model achieves three important things of relevance to modern regional innovation policy. First, it shows a high degree of overlap between spatial processes of both related variety and revealed related variety and policy mechanisms to further exploit these for regional innovation purposes. Second, it recognises in its incorporation of dramaturgy in the policy toolbox, that regional innovation policy-makers should raise their game in terms of popular appeal to economic actors and utilisation of dramaturgy in the economic actor – policy-maker interface. Finally, it demonstrates the new demand-side role of the regional innovation policy system as a catalyst and orchestrator of regional innovation opportunity, rather than a mere provider of supply-side support as has been the case for most of the neoliberal era since about 1980.

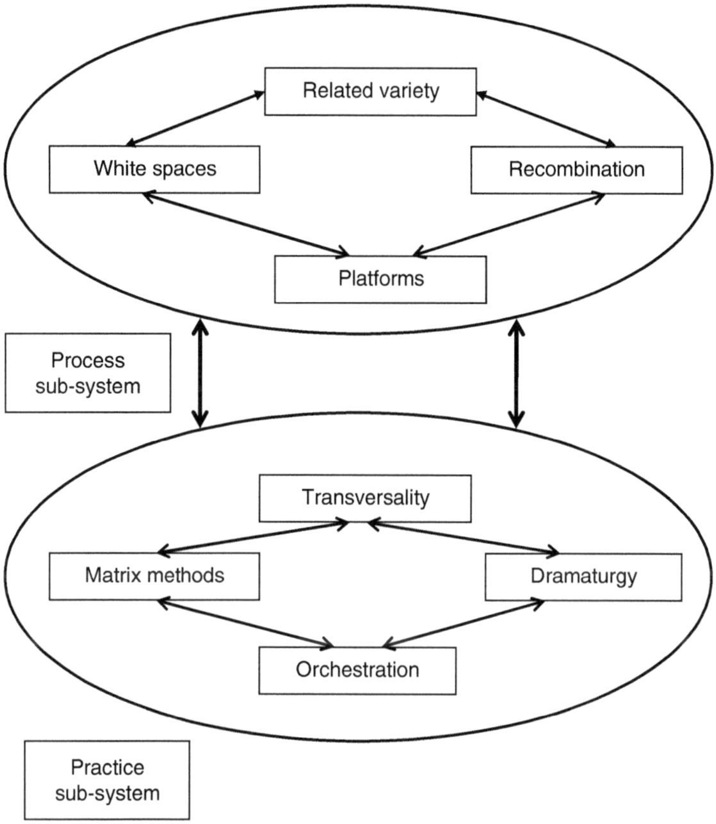

Fig. 15.8 Transversality: key process & practice conceptual framework

Acknowledgements I developed these insights in the following contexts. Much of the first half of the chapter was stimulated by my engagement in the EU 6th Framework RTD project whose acronym was EURODITE and the VINNOVA policy evaluation exercise 'The Matrix' throughout 2009. The sponsors and colleagues are thanked. Case material came respectively from commissions to evaluate Tillväxtverket clusters in Sweden in early 2010; from the OECD evaluation mission to Marche region in late 2009; and in connection with our InterReg ReThe project in early 2010. All sponsors and respondents are thanked for their support.

References

Andersen, E. (2011). Schumpeter and regional innovation. In P. Cooke et al. (Eds.), *The handbook of regional innovation and growth*. Cheltenham: Edward Elgar.

Argyris, C., & Schön, D. (1978). *Organizational learning: A theory of action perspective*. Reading: Addison-Wesley.

Boschma, R., & Frenken, K. (2003). Evolutionary economics and industry location. *International Review of Regional Research, 23*, 183–194.

Bristow, G. (2010). Resilient regions: Re-'place'ing regional competitiveness. *Cambridge Journal of Regions, Economy and Society, 3*, 153–167.
Burt, R. (1992). *Structural holes: The social structure of competition*. Cambridge, MA: Harvard University Press.
Burt, R. (2000). The network structure of social capital. In I. Sutton & R. Staw (Eds.), *Research in organizational behaviour*. Greenwich: JAI Press.
Chapain, C., Cooke, P., De Propris, L., MacNeill, S., & Mateos-Garcia, J. (2010). *Creative clusters and innovation: Putting creativity on the map*. London: NESTA.
Christensen, C. (1997). *The innovator's dilemma*. Boston: Harvard Business School Press.
Cooke, P. (2004). Introduction: Regional innovation systems – an evolutionary approach. In P. Cooke, M. Heidenreich, & H. Braczyk (Eds.), *Regional innovation systems*. London: Routledge.
Cooke, P. (2008). Regional innovation systems: Origin of the species. *International Journal of Technological Learning, Innovation & Development, 1*, 393–409.
Cooke, P. (2010). Matrix policy: rationales and good examples. In A. Eriksson, P. Cooke, J. Wallin (Eds.), *The matrix: Post-cluster innovation policy*. Stockholm: Vinnova.
Cooke, P., & Morgan, K. (1998). *The associational economy*. Oxford, UK: Oxford University Press.
Frenken, K., van Oort, F., & Verburg, T. (2007). Related variety, unrelated variety and regional economic growth. *Regional Studies, 41*, 685–697.
Gilmore, J., & Pine, J. (2007). *Authenticity*. Boston: Harvard Business School Press.
Golden, A. (2009). Foreword. In D. Wolfe (Ed.), *21st century cities in Canada: the geography of innovation*. Ottawa: The Conference Board of Canada.
Granovetter, M. (1973). The strength of weak ties. *The American Journal of Sociology, 78*, 1360–1380.
Harmaakorpi, V. (2006). Regional development platform method as a tool for regional innovation policy. *European Planning Studies, 14*, 1093–1112.
Harmaakorpi, V. (2009). Platforms to develop regional capabilities. Presentation to Vinnova Policy Review Workshop, February 24, Stockholm, Sweden.
Harmaakorpi, V., & Melkas, H. (2005). Knowledge management in regional innovation networks: The case of Lahti, Finland. *European Planning Studies, 13*(5), 641–659.
Jacobs, J. (1969). *The economy of cities*. New York: Vintage.
Kenney, M., & Florida, R. (2000). Venture capital in Silicon Valley: fuelling new firm formation. In M. Kenney (Ed.), *Understanding Silicon Valley: Anatomy of an entrepreneurial region*. Stanford: Stanford University Press.
Keynes, J. (1973). *The general theory of employment, interest & money*. London: Macmillan.
Kotter, J. (1996). *Leading change: Eight steps to transform your organisation*. Boston: Harvard Business School Press.
Latour, B. (1999). On recalling ANT. In J. Law & J. Hassard (Eds.), *Actor network theory and after*. Oxford: Blackwell.
Lewin, K. (1951). In D. Cartwright (Ed.), *Field theory in social science: Selected theoretical papers*. New York: Harper & Row.
Maskell, P., Bathelt, H., & Malmberg, A. (2006). Building global knowledge pipelines: The role of temporary clusters. *European Planning Studies, 14*, 997–1014.
Pässilä, A., & Oikarinen, T. (2010a). Research-based theatre as a facilitator for organisational learning. In P. Meusberger & A. Berthoin-Antal (Eds.), *Knowledge in organizations – learning organizations*. Berlin: Springer.
Pässilä, A., & Oikarinen, T. (2010b). *Bridging organizational learning culture and innovative dialoguing by storytelling*. Lahti: Lappenranta University of Technology, Lahti School of Innovation (mimeo).
Pässilä, A., Oikarinen, T., & Vince, R. (2011). The role of reflection, reflection on roles: Practice-based innovation through theatre-based learning. In H. Melkas & V. Harmaakorpi (Eds.), *Practice-based innovation: Insights, applications and policy implications*. Berlin: Springer.

Schumpeter, J. (1975). *Capitalism, socialism & democracy*. New York: Harper.
Simmie, J., & Martin, R. (2010). The economic resilience of regions: Towards an evolutionary approach. *Cambridge Journal of Regions, Economy and Society, 3*, 27–43.
Simon, H. (1957). *Models of man: Social and rational*. London: Wiley.
Wallin, J. (2006). *Business orchestration*. London: Wiley.
Wallin, J. (2010). Business orchestration for regional competitiveness. In A. Eriksson et al. (Eds.), *The matrix: Post-cluster innovation policy*. Stockholm: Vinnova.
Wenting, R. (2008). *The evolution of a creative industry: The industrial dynamics of fashion design*. Utrecht: University of Utrecht Press.

Part III
Case Studies and Policy Implications

Chapter 16
Service Innovation and Service Design in the German Printing Industry

Christina Cramer and Christiane Hipp

Abstract Service innovation has received less attention in science over the last years compared to product innovation, although the importance of service innovation is steadily growing. In fact, for the printing industry service innovation and the service design approach have received almost no attention at all. Because of the current crisis in the printing industry that is due to digitalising, there is considerable pressure to find new business opportunities. Service innovation and design can perhaps help to overcome the crisis, so that companies can remain competitive in a dynamic business climate. Therefore, this chapter analyses cross-industry benchmarks in the field of service innovation and service design. The use of the Circle of Service Design is shown in order to highlight possible opportunities for the printing industry to develop better and optimised services in the future.

16.1 Introduction

Innovation research has traditionally been focused on the manufacturing sector and the early phases of innovation management, including ideation and creativity. However, service innovation and design in the service industries as well as services in manufacturing have received less attention.

C. Cramer
Brandenburg University of Technology, Cottbus; Druckhaus Cramer, Münster, Germany

C. Hipp (✉)
Department of Economics and Business Sciences, Brandenburg University of Technology, Cottbus, Germany
e-mail: hipp@tu-cottbus.de

H. Melkas and V. Harmaakorpi (eds.), *Practice-Based Innovation: Insights, Applications and Policy Implications*,
DOI 10.1007/978-3-642-21723-4_16, © Springer-Verlag Berlin Heidelberg 2012

Miles describes service innovations as having a “Cinderella status, being neglected and marginal” (Miles 2000, 2005), although the future belongs to value creation models based on service and innovation (e.g., Franke and Piller 2004; Morelli 2006). As a consequence of digitalisation and lower circulation, among other factors, companies in the printing industry are faced with problems such as low demand, decreasing sales, and even bankruptcy. This chapter addresses service innovations as a way to deal with technical change and competitive advantage in the printing industry. In Germany the printing industry needs new business models, but there is no research area in ‘service innovation for the printing industry’ in practice. The questions for this chapter are: What can the printing industry learn from successful service innovators like McDonalds and how can service innovation improve the business situation in the printing industry?

This chapter focuses on service innovation and especially service design as one aspect of service innovation. The chapter has three main objectives: firstly, to review service innovation in the context of current research in order to determine what are the new aspects or methods from a scientific point of view; second, to analyse service innovation in practice in order to find out how companies deal with service innovation and service design; and third, to identify benchmarks in the field of service innovation and service design, so that the printing industry can learn from successful innovators.

16.2 Entering the Printing Industry

The printing industry is a small sector in Germany. The industry is comprised of printing companies (offset, gravure, letterpress, screen-printing etc.), prepress – and premedia companies, flexography (stamps, engravers), and print finishing or bookbinding enterprises. Ninety-five percent of the companies have fewer than hundred employees. Figure 16.1 shows an overview of companies and employees in the industry.

Most of the owners of companies in the industry manage their companies themselves; there are only a few large companies with employed management. The industry contributes only 2.3% of the gross production value and employs 2.8% of the labour force of the manufacturing industry. However, the importance of the printing industry for the economy and society is much larger than these figures suggest. The distribution of information is not only one of the most important cultural tasks for a society; it is also a guarantee for its stability. The printed media (newspapers, magazines, and books) are information carriers in the culture’s political, social and individual life. Although today there are substitute products like online news and e-papers, the internet, Google, and so on, printed information will continue to be a very important data medium in the future (Heine and Hauser 2002).

The growth of the printing industry is influenced by different factors such as technical development, consumer demand and behaviour. The industry as a whole has

Number of employees	Companies			Socially insured employees		
	Number	Share (%)	Change in 2000–2009 (%)	Number	Share (%)	Change in 2000–2009 (%)
1-9	7.482	69.86	-23.6	23.820	13.84	-25.5
10-19	1.405	13.12	-26.1	19.206	11.16	-25.2
20-49	1.136	10.61	-19.0	34.365	19.96	-19.5
50-99	402	3.75	-15.4	27.490	15.97	-14.3
100-499	267	2.49	-16.3	52.277	30.36	-14.2
500-999	15	0.14	-34.8	10.079	5.85	-35.0
1000 or more	3	0.03	-70.0	4.933	2.87	-64.8
Sum	10.710	100.00	-23.1	172.170	100.00	-22.8

Fig. 16.1 Companies and employees in the German printing and media industry (source: Deutscher Drucker No. 24/23.7.2009, Calculations BVDM [Bundesverband Druck und Medien/ The German Printing and Media Industries Federation])

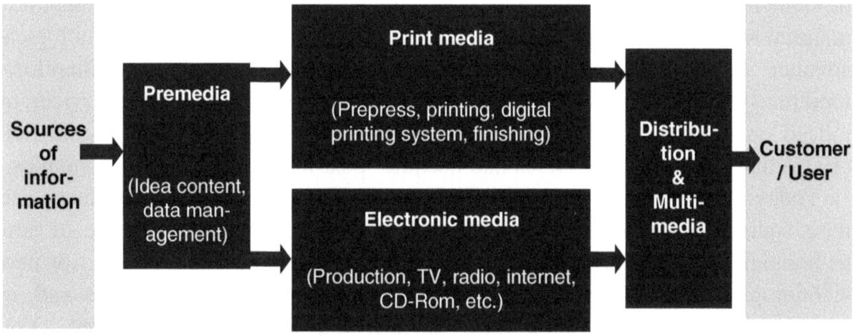

Fig. 16.2 Structure of the multimedia-system including the print media (source: based on Kipphan 2001: 8)

generally been undergoing enormous technological and structural changes over the last few years, resulting in an industry-wide crisis in terms of demand (Andjelic 2008; Cummings and Le Maire 2008). The growth in ‘multimedia’ documents including collections of photos, music and videos, etc. has been phenomenal in recent years. Because multimedia includes the printing industry; this leads to a growth of printed information and is evidence for the importance of printed information in the future (Print Media Academy 2008). Figure 16.2 shows the structure of the multimedia-system including print media (based on Kipphan 2001).

Figure 16.2 shows that multimedia implies the use of several types of information. Human communication is multimedia as humans transmit information in different ways (Evanschitzky and Iyer 2007; Cummings 2008). The fact that printing processes are controlled electronically leads to consistent high quality and better productivity and shows the progressive convergence of the digital and print world (Biagi 2007; Cummings 2008). In summary, the printing industry is undergoing cyclical and structural changes; it is getting more competitive and complex, whereas the information and communication sector is expected to provide a positive impulse for the whole printing industry. The challenge for the printing, publishing and manufacturing companies is to see these changes as an opportunity to find new products, services, processes, and business models, rather than to see them only as the threat of over-capacity, excessive supply, and ruin of the printed world (Schremper 2009).

16.3 Service Innovation and Service Design in Science

Adam Smith noted very early (1776) that "the labour of the menial servant. . .does not fix or realise itself in any particular subject or vendible commodity. His services generally perish in the very instant of their performance, and seldom leave any trace or value behind them. . ." So the idea, which goes back to Adam Smith, is that it is material wealth that counts and that it is technology which determines technological advance. This idea has dominated economic thoughts for two centuries. Therefore, until the 1980s very little research had been undertaken on innovation in services, or indeed on services in general, despite the observation by Fuchs (1968) and Bell (1973) that advanced economies had become 'post-industrial' service economies.

Today, innovation research still concentrates on new technology, products, and their sources. To this day, much innovation study concerns the source of new technologies and services. In essence, there is a focus on the generation of new technologies and services, rather than their diffusion and use. Services and, in particular, service innovation and its diffusion have been neglected (Von Tunzelmann and Acha 2005). Only a few authors have emphasised the importance of non-technological innovation in the economy (e.g., Tidd and Hull 2003).

The science of services poses the challenging question of how 'service' can be defined. There is a variety of definitions. One helpful definition comes from TEKES (2010, www.tekes.fi), Finland's research agency:

> Service innovation is a new or significantly improved service concept that is taken into practice. It can be for example a new customer interaction channel, a distribution system or a technological concept or a combination of them. A service innovation always includes replicable elements that can be identified and systematically reproduced in other cases or environments.

It is important to recognise that some services have a high cognitive and informational component. Services like consultancy fall into this category, that of 'knowledge-intensive services'. These services constitute the best people and laboratories for exploring the possible specificities of innovation in services, because the quality of those services is dependent on the expertise of human beings' behaviour and interaction (Gallouj 2002).

Typical characteristics of services are:

- The interactive role of the customer
- A challenge in defining and measuring output
- Knowledge and capital intensity

The properties of service innovation, also known as 'IHIP' taxonomy, include (Zeithaml and Bitner 2000; Green et al. 2001; Gallouj 2002):

- **I**ntangibility
- **H**eterogeneity
- **I**nteractivity
- **P**erishability

What these studies all have in common is that they concentrate on 'interactivity or the interactive role of the customer' as one characteristic of a service. This interactivity is also an important factor that concerns service design; in their research on experience design, Pullman and Gross (2004) found that one of the key elements for success is creating opportunities for customers to interact with each other in order to gain entry into a community.

Services or service innovations are often contrasted to goods and manufacturing, although increasingly, firms today do not consider themselves to be solely 'services' or solely 'manufacturing', but see their role as that of providing solutions for customers that involve a combination of products and services. Thus, the differences between service firms and manufacturing firms have been blurred because they all act as the 'problem solver' for their clients. Concerning the increase of the service dimension in the overall economy, Green et al. (2001) pointed out that "both the white-collar share of industrial workforces and the service share of industrial firms output are generally growing, together with the growth of specialised service firms. . .."

Innovation studies should recognise this blending of service and manufacturing and focus on the interactions and complementarities between these two forms (Van Ark et al. 2003), rather than only on services in the traditional service sector. The Institute of Innovation Research (IOIR 2003) considered the links between the knowledge-intensive services sector and the science base and concluded that "most services remained poorly linked into technological innovation systems". Evidence from the Community Innovation Statistics (CIS 2006) confirmed that firms across sectors innovate in both services and goods. The analysis in Fig. 16.3 of innovations in goods and services in different sectors makes clear that it is not only service firms that innovate in services.

A recent study of service innovation in Ireland presented a three-fold typology of service innovation: (1) business model innovation, (2) process / system innovation, and (3) service product innovation (Forfás 2006). The business model innovation includes complete or partial changes in how money is earned. Process/system innovations are changes in the way information is exchanged between the service

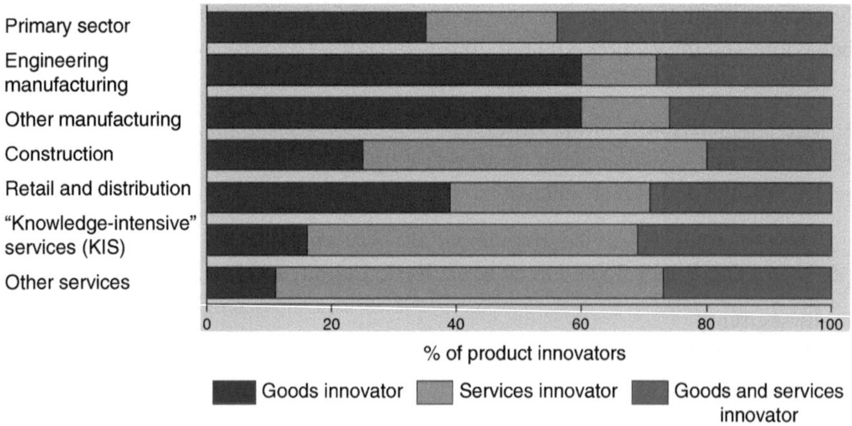

Fig. 16.3 Analysis of goods and services innovations (source: based on Community Innovation Survey CIS4)

provider and the customer. Service product innovations are very similar to the traditional way of dealing with innovation in the manufacturing sector in that they involve the introduction of new services. However, even though the interdependence between manufacturing and services is growing, and sometimes their borders are blurring, the specific nature of services requires distinct support and intervention models and policies (Pilat and Wölfl 2005). Therefore, this chapter focuses on service product innovations in the printing industry.

The second term in the title of this chapter, 'service design', is broad, has diverse meanings (Stacey et al. 2002), and is frequently equated with service engineering (Veryzer 2005). Service design lacks a strong sense of research direction; only a limited number of academic publications in the field of 'service design' are extant (Sangiorgi 2009), e.g. the book "Service Design" written by Birgit Mager and Michael Gais in 2009. While product design exists in the realm of the tangible products – good examples for successful product design are Apple, Phillips or Bang & Olufsen – service design deals with the intangible and the conceptual issues and involves a high level of rethinking and user engagement (Mager and Gais 2009). Everyone in a company, whether it is a Business-to-Customer or a Business-to-Business market, is responsible for the design of the organisation's services, even though most employees would never think of themselves as designers (The Guardian 2010). In this way, the quality of a company's service innovation is closely connected to the ability of its employees (Marsh 2010).

Service design as an area of research began in the early 1990s (Manzini 1993; Erlhoff et al. 1997). The service design approach is only part of the overall service innovation practice and needs to be connected with existing economic and organisational approaches. Service design defines a legitimate area of intervention with specific design tools and methods. Service design tools are both an evolution of existing marketing, management, and design tools and the result of a dedicated

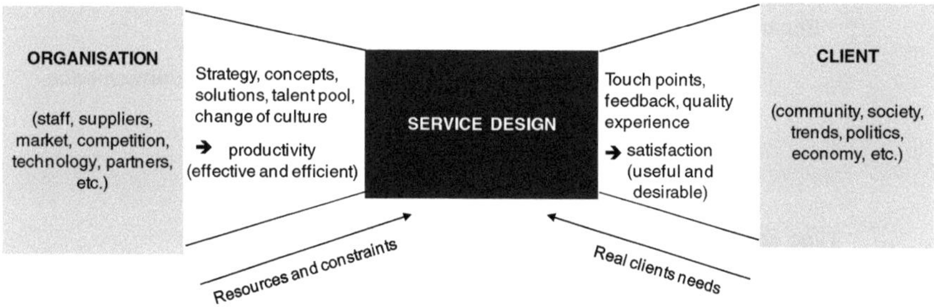

Fig. 16.4 Service design as an interface (source: based on Moritz 2005: 152f)

design approach, but the service design approach is a very different way of thinking to deliver value for companies and clients (Moritz 2005). Service design can contribute to the modification of the commodified and existing services and in the development of innovative service offerings (Maffei et al. 2005). In conclusion, services are seen as products, which can and must be planned and delivered systematically (Evanschitzky and Iyer 2007).

Figure 16.4 shows service design as an interface itself; it creates a win-win-situation for organisations and clients and helps to develop and implement successful service concepts. This kind of working together aims at making people aware and able to use their creativity to deal with change and co-develop innovative solutions (Sangiorgi 2009).

16.4 Importance of Service Innovation in Practice

Many companies make changes to their service offerings, but only a few succeed in creating 'market-creating service innovation', service innovations that generate new markets or reshape existing ones (Berry et al. 2006) and promise much greater upside potential than incremental innovations. Google or eBay are good examples of 'market-creating service innovation'. A radical market-creating service innovation can lead to a competitive advantage more effectively than an incremental change to a service can. Customers perceive market-creating service innovations as offering new benefits of sufficient appeal that their behaviour is influenced, as is the behaviour of the company's competitors. Berry et al. defined four types of these market-creating service innovations (see Fig. 16.5).

Concluding innovations can be characterised on two dimensions: whether they offer new core benefits or a new way of delivering a core benefit, and whether the service must be consumed where and when it is produced or can be consumed separately from its production. (See practical examples in Fig. 16.5.) Often the knowledge of the customer, the 'demand side' of the equation, is particularly important since many services are simultaneously created and consumed at the

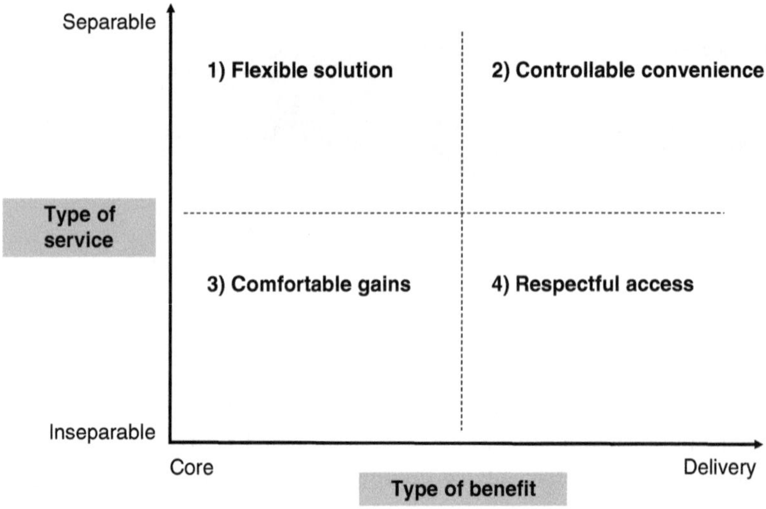

Fig. 16.5 The four types of market-creating service innovations (source: based on Berry et al. 2006: 59)

same time, but all four types of service innovation need to be built on the basis of customers' needs in order to create added value for the customer.

When businesses undergo a paradigm shift from a business that focuses on the exchange of goods to a business in which service is fundamental to economic exchange (Vargo and Lusch 2004), an improved understanding of service innovation complements the established knowledge of manufacturing innovation. In the great majority of both manufacturing and service industries, only a minority of firms use universities as a source of information for innovation, but different types of research are appropriate. For example, experiential services use 'empathic research', trend watching, and so on, which may be best gathered through the private sector, and this is where service design as a new approach is needed.

New forms of added value are the result of integrating product and services. The combination of product and service holds a key function in service innovations, but the challenge is to combine them in a way that the hybrid generates added value for the customer and cannot be imitated easily by competitors (Bienzeisler 2008). But how are market-creating service innovations developed? Is it the best way for companies to use universities for market research, their clients to get ideas, or just combine product and services into a hybrid, but what is the solution for the printing industry?

Perhaps the new and strategic approach of 'service design' can be of help to the industry. In an increasingly complex and fast-moving media world, a new set of methods is needed to address the challenges. Service design combines old and new tools to provide a new way of thinking. Design thinking helps to build up ideas and provides the ability to create solutions that matter and that deliver value (Zwiers 2009). Service design improves and creates innovative services with regard to the customers' needs; it is a process that identifies where, when, and how organisations can make the services they provide more valuable for their customers. Service designers work in a holistic way in which the functional and emotional aspects of a service – the storyboards, the actors, and stage settings – are considered in the design process (King and Mager 2009).

To clarify this holistic approach to innovation, the next section identifies benchmarks in the field of service innovation – and especially service design – as a contribution to cross-industry analysis.

16.5 Benchmarks: Service Design

User-centred service design begins holistically by understanding customers' needs. Service design aims to support various modes of use, not just specific activities, and then matches customers' interactions to the right channels. The goal of service design is to determine which aspects of the service add real value for customers and which are less critical or even lower the customers' interest. This section analyses institutions and companies that work in this way in order to highlight key lessons learned for application to the printing industry.

During research and analysis of benchmarks for service design, two key institutions are mentioned repeatedly: Köln International School of Design (KISD), whose first professor for service design, Birgit Mager, was founder of research centre Sedes and co-founder of the first service design network (SDN); and IDEO, the benchmark organisation for global design consultancy. The projects by Mager, the service design specialist, and IDEO are benchmarks in the field of service design. Figure 16.6 shows the core areas of these institutions.

These institutions and the way they approach their work provide good and detailed examples in the field of service design. Since 2000 other service design agencies, such as Engine, live/work and Transformator, have been founded. They all pursue the idea that services need design (e.g., Mager and Gais 2009) and refer to the overall approach as 'design thinking': an approach to problem-solving that uses design methodologies to tap into opportunity. These methods, which include observation, prototyping, building, and storytelling, can be applied by a wide range of people to many organisational challenges. Service design practice defines and builds customer experiences in a variety of industries by using tools to create a tangible and actionable vision of the desired result (Moed and Samalionis 2009).

McDonald's, Siemens Betriebskrankenkasse (SBK) and Volkswagen AG provide examples of service design projects that can help to deepen the research and analysis of benchmarks:

1. McDonald's
2. Siemens Betriebskrankenkasse (SBK) (Siemens health insurance fund)
3. Volkswagen AG

Institution	KISD	IDEO
Definition	Institute at Köln International School of Design since 1995; focused on service design (Prof. Dr. Birgit Mager)	Best and global design consultancy; was formed in 1991 by a merger of three design firms (David Kelley, Bill Moggridge, Mike Nuttall)
Philosophy	Academic orientation: sharing results of research and analysis of service design with others (international network, workshops, lectures, consultancy projects, etc.) and developing innovations	More commercial than academic orientation: helping hundreds of companies across the globe understand their customers and design their services (interactive projects)
Work examples	McDonald's, Siemens Betriebskrankenkasse SBK, VW AG, Motorola, etc.	McDonald's, Ford, Toyota, 3M, Microsoft, BBVA, Air New Zealand, Bank of America, etc.

Fig. 16.6 Service design institutions

1. Service design at McDonald's: About three years ago in 2006 McDonald's set up a first customer experience group of designers inside the company to work with the corporate development group in designing services by taking a holistic approach to the design of McDonald's services. In 1999 McDonald's had opened an innovation centre in Chicago where new products are tested and get designed (Weil 2009). At the Service Design Conference (sdn) conference in 2008 Dennis Weil (Vice President, innovation and concept development at McDonald's) pointed out that finding new ways to use design methods to drive businesses is an essential part of successful innovation management (sdn 2008; see sdn video at http://conference08.service-design-network.org/media/videos/video10.html). Therefore, he likes to push the boundaries of integrating design into innovation and works together with the service design specialists IDEO and KISD, Prof. Mager.

   KISD project: The 15-students-team 'service lab' of Prof. Mager created one room at the university looking like a McDonald's store to experience the service system. It was a 6 weeks project to create new and improved existing services. The result was a technology-based service where customers can get information on products and ingredients by using screens offered in the McDonald's store; customers can also order online and get their meals very quickly, upholding the 'service in 60 seconds' legend at Mc Donald's. The service lab also created a new hygiene service because eating with hands needs special hygiene. Small pills as a product innovation can be used to clean hands before eating, which at the same time is a great service innovation as well.

   The 'innovation lab' at the global innovation centre of McDonald's set up another project called 'future services at McDonald's'. Three service designers, thinking and acting like entrepreneurs, and IDEO San Francisco worked together on this project. They designed new services, e.g. with new media and technology (ordering food by using the mobile phone) which were tested at the innovation centre in Chicago. Until today McDonald's built up different prototyping and testing centres all over the world to improve the service system; one testing store is situated in Munich to check the store experience continuously from the customers' point of view. In Kirchheim, experts and test customers try new food, new order systems, new lighting, and new furniture in order to improve the McDonald's stores and get new ideas from the test takers.

2. Service design at SBK: The 'Live Healthy' project, with which KISD helped SBK, researched why some people live unhealthy lives and found that people often require concrete and practical support as well as useful information to reach their health-related goals. The process also needs to be entertaining and easy to implement in everyday life. SBK used service design to set up an online community where people exchange their experiences with activity and healthy eating. Another new service was an activity sensor to record daily movement and to provide motivational feedback. Customer satisfaction can be achieved only through a positive emotional experience and through service design that is immersed in the customers' worlds and wishes (Wenning 2010). Other projects, like the 'pregnancy' project, were successful as well (Mager and Gais 2009) in using the service design approach to go on a 'customer journey' and see the world through the eyes of clients (Funk 2007).
3. Service design at VW: Like the printing industry, the automobile industry has suffered from the economic downtown and faces an unstable future. Therefore, true innovation is needed to 'rethink the car'. Cars should do more for people than simply carry them around. A few years ago VW began using the service design approach to develop new cars and services like car-on-demand-services, communication, entertainment and even a car-earning-system in which one's car earns money while parked. VW corporate service designer Felix Sommerville-Scharf is sure that service design can save manufacturers from falling into the 'commodity trap' in which products compete on price alone because, in his view, it is better to extend the value chain and make customers enthusiastic than to suffer from price wars (Pearce 2010).

The lessons learned from benchmarks in service design indicate that it is important for an organisation to set up an environment for designing services. Figure 16.7 shows the four steps to implementing a service design approach.

It is important to fix the service design approach in the service strategy as an overall approach of a company. Innovation in service largely depends on attitude and changes in the way people interact and think. Therefore, the environment has to be open-minded toward service creation and design, and everyone in the organisation needs to feel responsible for the company's services. The four steps to a service design approach can help to create such a service design environment.

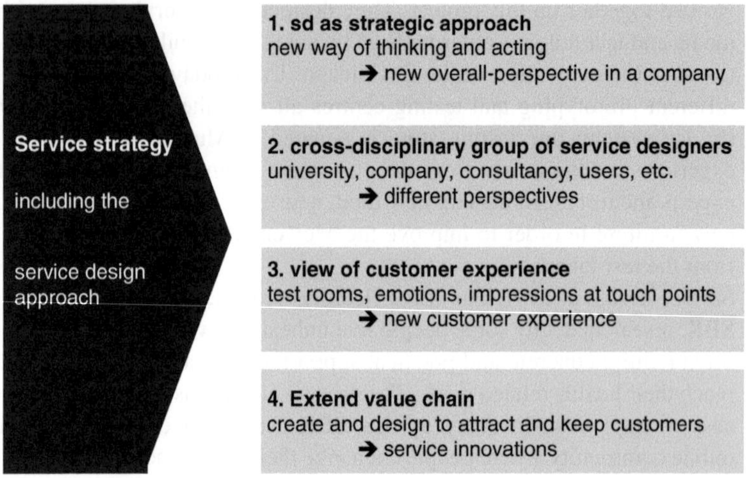

Fig. 16.7 Four steps to a service design (sd) approach

The service design project itself has to be planned in a process such that the strategic approach can be transferred into the operational practice. The four 'RDSI steps' can help in setting up a service design project (Brown and Wyatt 2010):

1. **R**esearch & analysis (discovering, experiences)
2. **D**evelopment (conception, designing)
3. **S**pecification (building, testing)
4. **I**mplementing (documenting, evaluation)

Following the RDSI steps can help in the creation of practice-based service innovations with a focus on the customers' view and using the language of the consumer. The company can decide whether to integrate technical experts like a design consultancy, universities or users in the service design process or not. Good designers are good storytellers and a good story makes ideas accessible and brings added value to the customer.

Moritz, one of Mager's students, defined six service design steps, called the '6 SD categories': understanding, thinking, generating, filtering, explaining and realising. Depending on the service design project, the steps can be used in parts, as a whole, or in several iterations. In each step one needs special skills, such as the ability to define targets, to develop hypothesis, to identify key insights, and to brainstorm experience (Moritz 2005).

The printing industry can use both the strategic and the operational approach to service design to ensure its future. While there is no reason to think these aspects could not be transferred to the printing industry, it will be difficult to change people's minds from thinking in terms of 'products or technology' to thinking in terms of 'service innovation through service design'. In this way, a Business-to-Business-service could be more complex than a Business-to-Customer-service, but this problem is recognised by the service design approach and can be solved. Summing up, the printing industry needs new service concepts in the future because an organisation that provides services needs to address innovation, development, and research in a different way than one that is concerned only with products; this is where service design comes in.

16.6 The Circle of Service Design in the Printing Industry

The authors' recommendation for the printing industry is to use the 'Circle of Service Design' to develop market-creating service innovations because the steps can help to develop better and optimised services in future (Fig. 16.8).

As stated in Fig. 16.4 above, service design is an interface and connects organisation and clients to affiliate the inside and outside view. Figure 16.7 showed the four steps to a service design approach. The Circle of Service Design keeps these aspects in mind and deepens the concepts of service design by using the RDSI-steps to transfer the approach into practice. The Circle shows that a good service design approach will never end because continuous market-creating service innovations are needed for the professional and effective innovation management.

A practical example of how to use the Circle of Service Design shows how the Circle can help transfer the concept of service design into practice. Imagine that a big manufacturer of printing machines is in trouble because of lack of demand, sales, and revenue. Because of over-capacity in the printing industry, no printing company is likely to need machines in the next several months or even years. What should the company do?

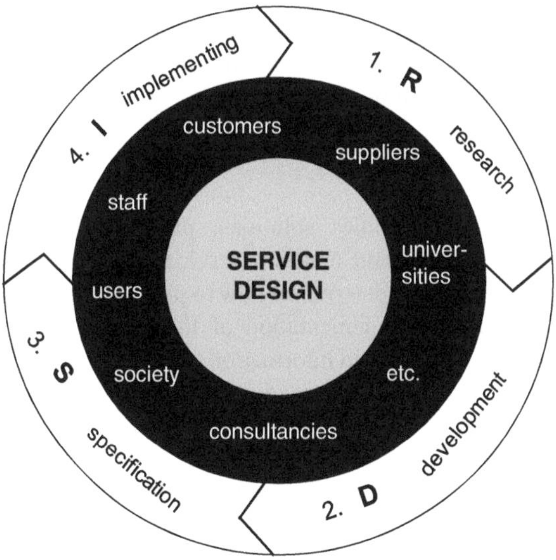

Fig. 16.8 The 'Circle of Service Design', own illustration

Using the Circle of Service Design in the printing industry: A mixed team of people from different backgrounds (economies, industries, branches, etc.) is set up for a 2-day-SD-workshop (this is not fixed and can also be 3, 5 or 10 days or alternatively an online-task force can be set up) to solve the problem described above. The group members should have different experiences with services and service design since the service design approach uses a wide range of tools, methods and techniques.

Step 1: Research (day 1)

Step 1 contributes researching the clients' conscious needs: The task force researches the needs of printing companies and their future demand. What do they need, if they do not need machines for the foreseeable future? What kinds of services might they desire?

This step ensures that the findings are linked with the reality and that they are relevant to the problem. Thorough market research done in concert with universities before the workshop can help to deliver insights on the needs of customers. Including clients in the workshop team can also help reveal the needs of the market.

➔ To understand the printing companies' needs, the 'customer's journey' must be understood

Step 2: Development (day 1)

Step 2 entails identifying criteria, turning complex data into insights, setting a strategy and objectives, and developing a SD action plan to create a new idea and service. In other words, what should be developed to meet the customer's needs?

This step focuses on a topic; in the example, one topic for a consultancy service offered by the printing machine manufacturer might be 'the future of the printing industry', which could help printing companies get ideas about the future of the media world, its products and possibilities. This makes sense because printing companies are often very small and do not have the capacity and knowledge (trained employees and competences) to analyse the market on their own. A big manufacturer usually has a big research and development department, where such information can be prepared more easily.

➔ To produce ideas that are true to the customer's needs based on the insights and strategy

Step 3: Specification (day 2)

In step 3 the workshop specifies solutions, prototypes or processes; writes guidelines or business plans; and makes a service happen using quality service delivery. In other words, step 3 determines how to provide the new service in detail.

Step 3 prepares for the implementation of the defined service offer. In the example, the specified solution is an information service on the future media market concerning which products, contents, media formats, and so on will be needed in the market in 5–10 years. The manufacturer of printing machines can create a platform using the web 2.0 to offer the solution and to build the new service, which is a market-creating service innovation, a controllable convenience service like a Google service offer (Fig. 16.5). However, instead of being used by everyone, it would be a special website with a required user-ID and password and, based on the level of service a printing company had bought (e.g., a 'standard service' that offers primary information, data and key facts on the media market or a 'premium service' that addresses individual problems of the printing company and offers specialised data, the company will pay for the service and get the required information online via the new service platform).

➔ To take the service to the market; all details are checked finally to provide the new service which ensures the best possible service performance

Step 4: Implementing (day 2)

In step 4 the workshop determines how to implement the defined services, including continuous improvement. This step includes consideration for the fact that a service is never perfect but can always be improved. Systems are complex and the environment is dynamic, so it is necessary to maintain and re-develop the service offer. Consequently, step 4 is not the end of the SD Circle but a new beginning. In our example, the step to be implemented could be a new feature on the website, such as a discussion forum for all 'premium service' users. The feature would enable users to communicate and discuss a topic, like young people via Facebook or Skype talk about their experiences (Shepherd and Davis 2009). A printing company could use the forum to discuss issues with other printing companies or even with the manufacturer. The purpose of this step, then, is to implement and improve the developed service.

Using the Circle of Service Design (e.g., in a 2-day workshop) could lead to a successful solution to a problem in the printing industry because the Circle includes and considers the dynamic environment in the media sector. The industry practitioners who attend the workshop are part of the target group, as well as supporters in co-operation from around the globe. The different backgrounds of the attendants will lead to new ideas for services in the printing industry.

➔ To build and improve the developed service

16.7 Summary and Future Prospects

Service design projects are likely to be a crucial part of the future because they foster behavioural change. A design approach should be systemic and include as many touch points as possible in order to ensure the adoption of new actions and services (Papeschi 2010). Effective service design brings together the right people at the right time to create opportunities. It solves problems by allowing companies to focus on what they do best and to achieve more by working together. Therefore, the future lies in well designed services.

Although all companies that provide services need to design their services, only a little research on service design has been done so far. This chapter shows that service design is a concept that can help companies develop and deliver great services. Service design projects improve factors like ease of use, satisfaction, loyalty and efficiency across areas such as environments, communications and products, without ever forgetting the people who deliver the service.

Service innovation – especially service design – is a strategic approach for companies, a kind of expertise that has not yet been analysed for the German printing industry. The communication sector as a whole is subject to change as electronic publishing, multimedia, digital data transfer, among other things, create new rules in the media markets. The printing industry has increasingly become a media service provider and information processor that not only produces information in the form of print products but also makes it available on other media devices like e-readers, CD-ROM databases, and online services. In the future, the sole print product of today, like a book or a newspaper, will be linked with electronic media to a much greater degree than it is today, offering a significant opportunity to use the service design approach to create value for the future of print media.

The German printing industry and its companies need service innovations to overcome the current crisis in demand and to remain competitive and relevant. By using the Circle of Service Design as a practical approach, the industry can get new ideas of service innovations. This approach identifies new commercial opportunities for implementation to attract new and keep existing customers. Printing companies – especially manufacturers of machines – in the industry have to be open-minded towards the new concept of service design, even though change can sometimes be expensive and risky. But the example given in this chapter showed a situation in which a total change in thinking of the business base would seem to be needed and effectually. However, the service design's approach to integrating the customers' view from the beginning by describing a customers' journey through a service and plotting how they might become aware, join, use and pay for a service (Mager and Gais 2009) seems to be a promising course of action for an industry hungry for new ideas.

> "There is an urgent need for professional exchange in order to better and faster develop the field of service design." (Birgit Mager, co-founder of the first service design network, sdn)

References

Andjelic, A. (2008). Transformations in the media industry: Customization and branding as strategic choices for media firms. In C. Dal Zotto & H. van Kranenburg (Eds.), *Management and innovation in the media industry* (pp. 109–129). Cheltenham: Edward Elgar.

Bell, D. (1973). *The coming of post-industrial society: A venture in social forecasting*. New York: Basic Books.

Berry, L., Shankar, V., Parish, J., Cadwallader, S., & Dotzel, T. (2006). Creating new markets through service innovation. *MIT Sloan Management Review, 47*(2), 56–63.

Biagi, S. (2007). *Media impact: An introduction to mass media*. Belmont, CA: Thomson Wadsworth.

Bienzeisler, B. (2008). Business transformation: New organizational and business models. In D. Spath & W. Ganz (Eds.), *The future of services: Trends and perspectives* (pp. 247–264). München: Hanser.

Brown, T., & Wyatt, J. (2010). Design thinking for social innovation. *Stanford Social Innovation Review*, (Winter), 29–35.

Community Innovation Statistics. (2006). *Fourth Community Innovation Survey (CIS 4) and European Innovation Scoreboard (EIS)*. Luxembourg: Community Innovation Survey.
Cummings, T. & Le Maire, B. (2008). *Print media distribution, a look at infrastructure, systems and trends*. Printing Industry Center Series 3. Rochester: Rit Cary Graphic Arts Press.
Erlhoff, M., Mager, B., & Manzini, E. (1997). *Dienstleistung braucht Design, professioneller Produkt- und Marktauftritt für Serviceanbieter*. Cologne: Hermann Luchterhand Verlag GmbH. (In German.)
Evanschitzky, H., & Iyer, G. (2007). E-services, opportunities and threats. *Journal of Value Chain Management, 1*(1/2), 177–196.
Forfás. (2006). *Services innovation in Ireland – options for innovation policy*. Dublin: Forfás.
Franke, N., & Piller, F. (2004). Value creation by toolkits for user innovation and design: The case of the watch market. *The Journal of Product Innovation Management, 21*, 401–415.
Fuchs, V. (1968). *The service economy*. Cambridge, MA: National Bureau of Economic Research.
Funck, A. (2007). Von Obdachlosen lernen. Dienstleistungen, Service-Designer haben überraschende Wege gefunden, sie intelligenter und menschenfreundlicher zu machen. *Brand Eins, Magazine, 05*.2007, 112–116. (In German).
Gallouj, F. (2002). *Innovation in the service economy: The new wealth of nations*. Cheltenham, UK: Edward Elgar.
Green, L., Howells, J., & Miles, I. (2001). *Services and innovation: Dynamics of service innovation in the European Union*. Manchester: University of Manchester & UMIST. Prest/Cri.
(The) Guardian (2010). *Service Design*. Media Guardian. In association with service design network. 15 March 2010.
Heine, J., & Hauser, T. (2002). Dienstleistungsstrategie am Beispiel eines global präsenten Druckmaschinenherstellers – Ansätze für das Thema "Qualitätsmanagement". In G. Lay & P. Jung Erceg (Eds.), *Produktbegleitende Dienstleistungen – Konzepte und Beispiele erfolgreicher Strategieentwicklung* (pp. 69–79). Berlin, Heidelberg: Springer. (In German.)
Institute of Innovation Research. (2003). *Knowing how, knowing whom: A study of the links between the knowledge intensive service sector and the science base (IOIR)*. Manchester: University of Manchester/UMIST.
King, O., & Mager, B. (2009). Methods and processes of service design. *Touchpoint, The Journal of Service Design, 1*(1), 20–29. sdn, Köln.
Kipphan, H. (2001). *Handbook of print media: technologies and production methods*. Berlin: Springer.
Maffei, S., Mager, B., & Sangiorgi, D. (2005). *Innovation through service design. From research and theory to a network of practice – a users driven perspective*. Helsinki: University of Art and Design. 22–24 September 2005.
Mager, B., & Gais, M. (2009). *Service design, design studieren*. Paderborn: Fink.
Manzini, E. (1993). Il Design dei Servizi. La pregettazione del prodott-servizio. *Design Management, 7*, Giugno, 7–12. (In Italian.)
Marsh, N. (2010). In celebration of "silent designers". In *The Guardian: Service Design*, 15 March 2010, 1 ff. Media Guardian. In association with service design network.
Miles, I. (2000). Services innovation: Coming of age in the knowledge-based economy. *International Journal of Innovation Management, 4*(4), 371–389.
Miles, I. (2005). Innovation in services. In J. Fagerberg, D. Mowery, & R. Nelson (Eds.), *Oxford handbook of innovation* (pp. 433–458). Oxford: Oxford University Press.
Moed, J., & Samalionis, F. (2009). Design's odd couple. *Touchpoint, The Journal of Service Design, 1*(1), 30–37. sdn, Köln.
Morelli, N. (2006). Developing new product service system (PSS): methodologies and operational tools. *Journal of Cleaner Production, 14*(17), 1495–1501.
Moritz, S. (2005). *Service design. Practical access to an evolving field*. London: KISD.
Papeschi, F. (2010). Rethinking behaviours: Beyond persuasion, questions and ideas on long-lasting change, environment and service design. *Touchpoint, The Journal of Service Design, 2*(1), 46–53. sdn, Köln.

Pearce, F. (2010). What can you do on four wheels? Service design car industry. In *The Guardian: Service Design,* 15 March 2010, 10 ff. Media Guardian. In association with service design network.

Pilat, D., & Wölfl, A. (2005). *Measuring the interaction between manufacturing and services. STI working paper*. Paris: OECD.

Print Media Academy (2008). Die Stärken von Print in der Marken – und Unternehmenskommunikation. In Print Media Academy, Heidelberger Druckmaschinen AG und Hochschule der Medien/ Institut für Werbung und Marktkommunikation (Hrsg.), *Projektstudie WS 2007/2008*. Stuttgart: Print Media Academy. (In German.)

Pullman, M., & Gross, M. (2004). Ability of experience design elements to elicit emotions and loyalty behaviors. *Decision Sciences, 35*(3), 551.

Sangiorgi, D. (2009). Building up a framework for service design research. 8th European Academy of Design Conference, 1st–3rd April 2009, The Robert Gordon University, Aberdeen, Scotland.

Schremper, R. (2009). Wachstum und Innovation in der Medienindustrie. In A. Picot & S. Doeblin (Eds.), *Innovationsführerschaft durch Open Innovation – Chancen für die Telekommunikations-, IT- und Medienindustrie* (pp. 93–104). Berlin, Heidelberg: Springer. (In German.)

Shepherd, F., & Davis, J. (2009). The innovation architectures of Facebook. In A. Ozok & P. Zaphiris (Eds.), *Online communities and social computing*. Berlin, Heidelberg: Springer.

Smith, A. (1776). *An Inquiry into the Nature and Causes of the Wealth of Nations*, 1976 Edition. Edited by R. H. Campbell, & A. S. Skinner with W. B. Todd. Volume I, Book II, Chapter III, Of the accumulation of capital, or of productive and unproductive labour (p. 330). Oxford: Oxford University Press.

Stacey, M., Eckert, C., Earl, C., Bucciarelli, L., & Clarkson, P. (2002). *A comparative programme for design research*. Paper presented at the Common Ground Conference, 5–7 September, London, UK.

TEKES (2010). *A broad-based view on innovation, Pioneers of service business 2006–2013*. Helsinki: Tekes, The Finnish Funding Agency for Technology and Innovation. www.tekes.fi. Accessed 10 June 2010.

Tidd, J., & Hull, F. (Eds.). (2003). *Service innovation – organizational responses to technological opportunities and market imperatives* (Series on Technology Management, no. 9.). London: Imperial College Press.

Van Ark, B., Broersma, L., & den Hertog, P. (2003). *Service Innovation, Performance and Policy: A Review*. Research Series, no. 6. Directorate-General for Innovation, Ministry of Economic Affairs.

Vargo, S., & Lusch, R. (2004). Evolving to a new dominant logic for marketing. *Journal of Marketing, 68*, 1–23.

Veryzer, R. (2005). The roles of marketing and industrial design in discontinuous new product development. *The Journal of Product Innovation Management, 22*, 22–41.

Von Tunzelmann, N., & Acha, V. (2005). Innovation in low tech industries. In J. Fagerberg, D. Mowery, & R. Nelson (Eds.), *The Oxford handbook of innovation* (pp. 407–432). Oxford: Oxford University Press.

Weil, D. (2009). Service design at McDonald's, an interview with Denis Weil. *Touchpoint, The Journal of Service Design, 1*(1), 48–49. sdn, Köln.

Wenning, C. (2010). Innovative thinking: The experts' view. In *The Guardian: Service Design*, 15 March 2010. Media Guardian. In association with service design network.

Zeithaml, V., & Bitner, J. (2000). *Services marketing – Integrating customer focus across the firm*. Boston: Irwin/McGraw-Hill.

Zwiers, M. (2009). Dutch design. Time for a new definition. *Touchpoint, The Journal of Service Design, 1*(1), 16–19. sdn, Köln.

Chapter 17
Innovation, Cities and Place: An Empirical Study of the Knowledge System in Vancouver and Its Place on the Pacific Rim

Brian Wixted and J. Adam Holbrook

Abstract As a generalisation the innovation systems literature has downplayed the overall physical geographic setting of particular places and the connections between cities in national or international urban systems. This chapter examines the innovation history of a few of Vancouver's peculiarly human capital intensive clusters noting how the isolation with no nearby cities and its connectedness (a Pacific gateway point) appear to have shaped its trajectories. This analysis begins to make a few sketches of how place and innovation can come together emerging from a 10 year study of innovation clusters and cities in Canada.

17.1 Introduction

Virtually all innovation studies have been carried out in industrialised countries, and most of these studies have been carried out in Europe and eastern North America. These studies have induced governments to focus their innovation policies on promoting innovation in what is perceived to be high value sectors – i.e. high technology manufacturing, and in places where there are intensive agglomerations of industrial activity. One feature of densely populated mega-regions is that agglomerations of industrialised activity (which in Europe can stretch across national boundaries or in the USA across state borders) is that they have lines of communication within themselves and among the different agglomerations that radiate out in all directions. Even port cities in this context have lines of communication that link them to several agglomerations in their hinterland.

B. Wixted (✉) • J. A. Holbrook
Centre for Policy Research on Science and Technology, Simon Fraser University, Vancouver, Canada
e-mail: brian_wixted@sfu.ca

H. Melkas and V. Harmaakorpi (eds.),
Practice-Based Innovation: Insights, Applications and Policy Implications,
DOI 10.1007/978-3-642-21723-4_17,

But there is another model of development; one that is typically found in nations which have a more recent history of settlement and industrialisation, or which have unusual geophysical characteristics that have in turn fashioned a different geography of industrialisation. In these cases development can take place in a linear fashion as industry and thus innovation follows a transportation route. Western Canada is an example of this where development proceeded west along the trans-continental railway line as settlers moved in from Europe and eastern Canada. There are similar models in Australia (sea trade along its east coast) and with some differences, the western coast of the US and Chile (mountain barriers).

Do policies that promote innovation in what could be described as a two-dimensional economic space work in a one-dimensional situation? Or should there be a different set of innovation policies that take into account the transportation and communication constraints created by a single corridor? What are the implications of this unique geo-economic situation on innovation in these one-dimensional corridors, and what unique policies might be required?

The work for this chapter stems from 10 years of research on the Canadian innovation system which has investigated in comparative fashion the development of clusters and city centric innovation systems. This chapter has evolved out of thinking about the Vancouver example with an awareness of the greater project's results. Detailed specific analysis of Vancouver's innovation system has been published elsewhere (Wixted and Holbrook, forthcoming) but that paper left open the question of how being on the Pacific Rim has influenced the history of Vancouver's innovation system. Drawing upon a wide range of existing research and our own work we try to move one step closer to a framework for discussing the emerging development of the Pacific innovation structures. A conclusion of this research is that while geography matters policy needs to be active. Innovation policies and the development of innovation systems, regardless of scale, must be pursued at the city level, perhaps even eclipsing those policies developed by national or sub-national governments. The evidence presented here also supports the inheritance view of clustering (Klepper 2001) and begins to suggest that policies that support early career stage professionals gaining business experience not just academic education may be quite important.

17.2 The Dimensional Space of 'Innovation Systems'

A huge literature now exists on the concept of innovation systems, with the focus of the studies gradually narrowing from nations (Lundvall 1992; Nelson 1993; Freeman 1995; Edquist 1997; Sharif 2006), to regions (Cooke 1998), clusters (OECD 1999, 2001) and now to cities (Florida 2002). But this narrowing of the spatial focus often does so by continuing to ignore the larger scale phenomena. From a policy perspective then, it has become increasingly unclear what impact the different level of analysis should have.

One aspect of the innovation systems literature which has been rather conclusively shown is that little of it problematises system boundaries (Wixted 2009). This leads to an initial contention of this chapter, that the population and economic landscape which form the milieu of academic research has influenced the research agenda to a greater extent than is generally acknowledged.

While Fig. 17.1 (below) highlights both the mega-population regions of Western Europe, Eastern United States, East Asia and South Asia, it also reveals some other information of significance. Canada, Australia, the western United States, Chile and even western Russia have regional structures that are rather linear and in some cases best described as an archipelago of populations. The latter designation can best be applied to Australia and Canada (east and west of Ontario-Quebec).

Much of the innovation systems literature emanates from Europe where populations centres form densely packed regions with major competing centres on all four points of the compass and in relatively close proximity for many key cities. Although there has been a both a geographical turn (Martin 1999) and a relational turn (Yeung 2005) in economic geography; the first emphasising place and history, while the second emphasising the relationships between actors in a particular locality – many of these locations are within what can be described as a two dimensional system. Yeung, for example identifies the relational turn stating:

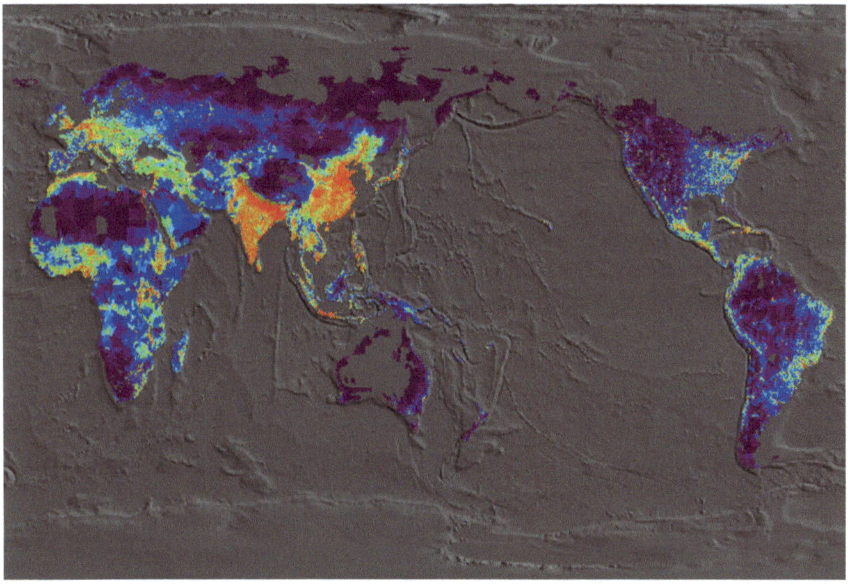

Fig. 17.1 Adjusted world population counts 2000 (source: Center for International Earth Science Information Network [CIESIN], Columbia University; and Centro Internacional de Agricultura Tropical [CIAT] 2005. Gridded Population of the World Version 3 [GPWv3]: Population Grids. Palisades, NY: Socioeconomic Data and Applications Center [SEDAC], Columbia University. Available at http://sedac.ciesin.columbia.edu/gpw)

> Recent theoretical and empirical work in economic geography has experienced what might be termed a 'relational turn' that focuses primarily on the ways in which socio-spatial relations of economic actors are intertwined with processes of economic change at various geographical scales. (2005: 109).

Earlier, in what is now argued by some to be an 'outdated' (see Barnes 2003) analytical emphasis in economic geography prevalent in the 1950s and 1960s was that of central places in two dimensional space. The emphasis was on the development of 'central places' (cities) and their hinterlands (see for example Christaller 1966). The key question was why did places develop as central and larger and some remained small villages? Christaller put it in these terms:

> Gradmann has called the chief profession of a town, namely, 'to be centre of its rural surroundings and mediator of local commerce with the outside world'. As one might think, this chief profession affects the small country towns which are really exceptions, being nothing more than the centres of their rural surroundings. But it also affects in the same way the larger towns, not only in respect to their immediate vicinities, but also in regard to their places in systems of many smaller towns. (1966: 16).

While Krugman and others have returned to these older ideas the innovation system scholars have never adopted the perspective to question the interrelationships and interdependencies between regions, preferencing a focus on the history, local context and actor dynamics of specific places (cities, regions, cluster, nations). Thus, a large percentage of cluster studies for example, ignore the spatial distribution of clusters or the interconnections between them (as is argued by Oinas and Malecki 2002; Bunnell and Coe 2001; and Wixted 2009). However, as the context for most of the current studies is the densely populated mega-regions of the Western Europe and the Eastern United States, it would not be too much of a stretch of the imagination to suggest that the two dimensionality of economic space forms the unconscious presuppositions of the research questions. The individuality of place is an interesting question if there are many different localities with a given geographic distance than 'could have' (but didn't) given rise to high rates of innovation. But such a context as noted above does not fit the geographical context of everywhere. If across large geographical distances there were only a few major centres to start with, then the focus needs to be different.

One hundred years ago the geographic positions of Australian and Canadian city innovation systems were not too different from their modern realities as both were systems of ports of a fashion. Of course each system has different historical, physical and contingent factors. Australian cities, being set within an island nation with poor water supply in its interior, are all seaports providing an administrative and service bridge between the high value produce inland and the world's markets. Canada's cities, being landlocked, were different. They were established at intervals, often at major river crossings, along the first Canadian transcontinental railway, the Canadian Pacific railway (CPR), but in each case they were there to aid the trans-shipment of commodities. Until the arrival of jet aircraft in the mid to

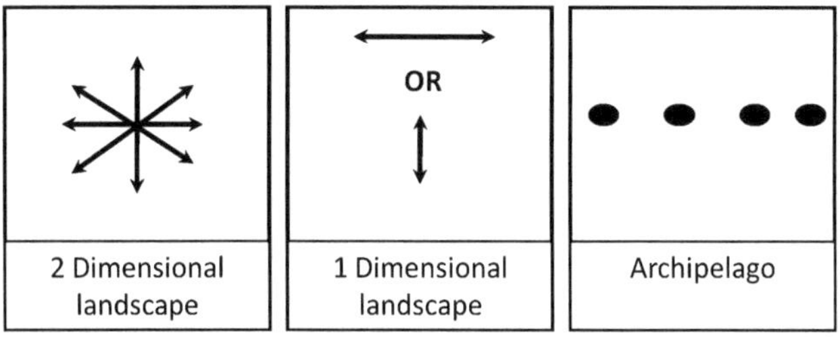

Fig. 17.2 Different dimensions of city system configurations

late 1950s, Vancouver[1] was primarily accessible by sea or rail as even road access was difficult. So instead of one model of geography we should have at least three (Fig. 17.2).

There is little reference in the work on innovation systems to states where populations are either linear or broken up. We could use words such as; periphery, gateways and others but each of these has complications. 'Periphery' or semi periphery (Wallerstein 1976; Boreham et al. 1989) are used in reference to places that are distant from the international economic centres (i.e. Western Europe and the Eastern USA) but this can emphasise distance from the centre while not illuminating the urban structure dimensions. It also de-emphasises the network bridging role that such places can play. 'Gateways' is also somewhat problematic as it is traditionally emphases transport, particularly multi-modal (sea, rail, truck) trans-shipment rather than innovation (Rodrigue et al. 2009). In some cases 'entrepôt' (Phillips 2002) is useful where foreign investment drives local activity, particularly for foreign sales, but again as we shall show for Vancouver this doesn't capture the essence of its innovation functions.

The problem with our language of 'innovation systems' is that we want to think of them as a unified structures and forget that a higher level notion of systems requires better ideas on sub-level configurations. The results of the Innovation Systems Research Network over the years could be used to describe the innovation functions of various cities. Saskatoon is an entrepôt for biotechnology (Phillips 2002), Calgary as a technology services hub for the oil and gas production assemblage (Ryan et al. 2011) and Vancouver as a location for high human capital intensive activities from movies to fuel cells. However, we can go further and describe the degree to which cities are linked to a broader suite of regional

[1] Although Seattle is relatively near physically, there has been little interaction between the two cities. Initially it was because both were competing in the same resource businesses. Seattle branched out into high-tech as a result of WWII and Cold War investments by the US military which did not spill over into Canada. Finally, the result of heightened border security on the part of the US has led to a throttling of transactions between the two cities – e.g. Microsoft has establishment a base in greater in Vancouver to connect with Canada and easier immigration. As a result Vancouver has had to look eastwards to the rest of Canada rather than south (with the exception of course, of the film industry).

economic activities. Saskatoon is internationally connected in a few resource-based industries, Calgary is highly integrated with its sub-national regional production system based on the petroleum industry, but Vancouver is increasingly operating as a separate economy dependent upon a series of sub-economies that include; legitimate economic activity, as well as the black economy (drugs trafficking and cash transactions) and what is locally known as the 'red' economy (Chinese investment in property).[2]

However, within the framework of the existing literature, it is not straightforward to sketch hypotheses on the expected innovative and technological or knowledge intensive industries that might be found in particular cities. The literature of the Marshall-Arrow-Romer[3] versus Jacobs (MAR-Jacobs) debate helps little as very few papers take any account of the macro-geographic urban system setting, apart from perhaps city size. In recent years innovation systems research has begun to take a more sophisticated approach to the multi-scale nature of geography; incorporating national, regional and city levels (with cluster studies stretching across multiple layers). However, apart from a focus on the internationalisation of R&D centres (OECD 2008) there has been too little work exploring the multi-spatial linkages across these scales and spaces. Carlson (2006) for example states:

> ...in view of the fact that most studies of innovation systems focus on national innovation systems, it is not surprising that little direct evidence is found that innovation systems are becoming global. The main focus in this literature is on institutions at the national level. But national institutions may influence innovation systems at regional, sectoral or technological levels differently. However, at these lower levels there has been little work done with a view toward internationalisation of systems (as distinct from corporate innovative activity). Also, not all institutions are national. For large firms, national institutions may be most important, while for small and new firms, subnational institutions may also be important (2006: 65).

This is beginning to change with, for example, Wixted (2009) mapping the international flows of components in a range of complex products for the period between 1970 and 2000. However, that study only focused on the interactions between national clusters and using large scale multi-country input–output modelling techniques. The book argued that while some industries are truly global (aerospace and automobiles) others are more restricted to continental (and smaller) trading areas.

In a more recent paper Wixted and Bliemel (in preparation) has looked at trade flows for 20 classes of manufactured products. Based on the year 2000 and

[2] There are a number of terms that can be used in this matter; underground (see BC Stats 1994), black (Thomas 1999) or hidden (Frey and Pommerehne 1984). Drugs have been estimated to be worth nearly 3% of the provincial economy (Easton 2004), and cash also around 3% in 1994 (BC Stats 1994), and though these would be intersecting sets they are not duplicates.

[3] Simply put the MAR argument is that innovation benefits from environments that specialized (typified by industrial towns), while Jacobs argues is that cities in all their industrial and cultural diversity are better for innovation (see for example Glaser et al. 1992; Wolfe and Bramwell 2008; Beaudry and Schiffauerova 2009).

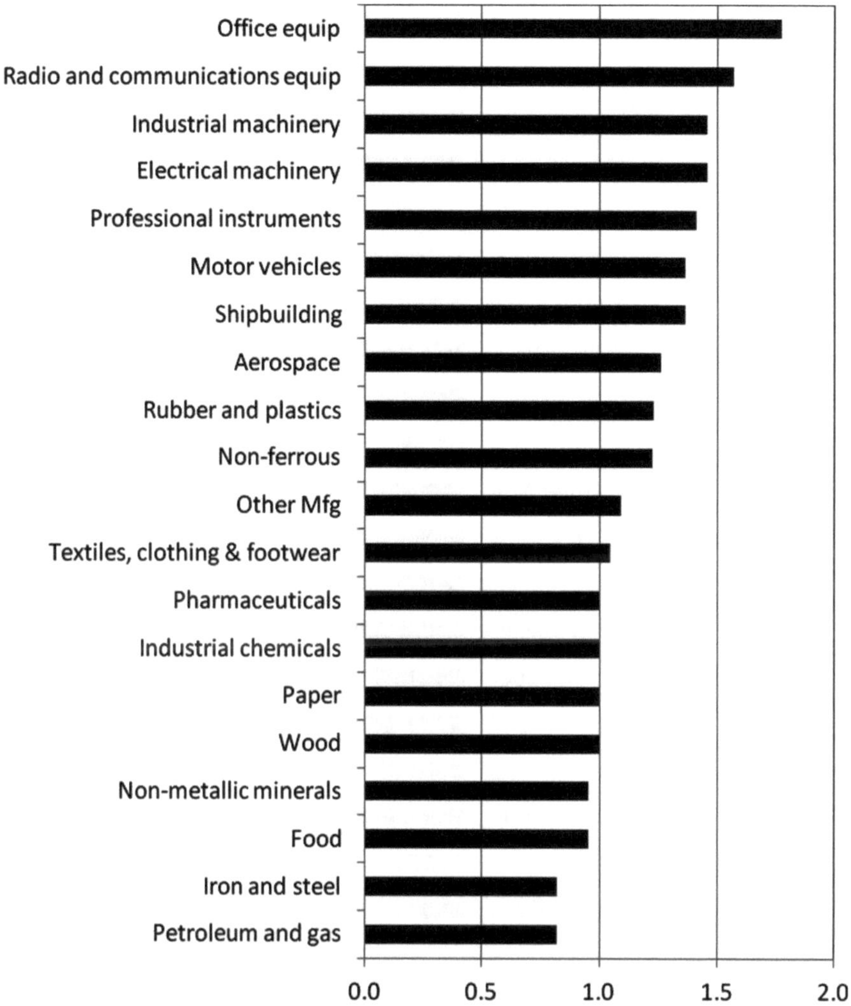

Fig. 17.3 Significant trade links by industry as ratio to the number of economies in the model (source: Based on work for Wixted and Bliemel [in preparation])

significant trade flows, although there are no clear breaks there appear to be three distinguishable sets of industries. Some are truly globally connected while others are only barely connected (Fig. 17.3). Interestingly, trans-Pacific connectivity is far greater than trans-Atlantic connectivity (Fig. 17.4).

Except for industries such as aerospace, professional instruments, radio and communications equipment and office equipment, Pacific trade crossing are of much more importance than trans-Atlantic ones. This is somewhat surprising as the 'buzz' is between Europe and Eastern North America as witnessed through either air line flight data or internet traffic flows (Harrison 2007).

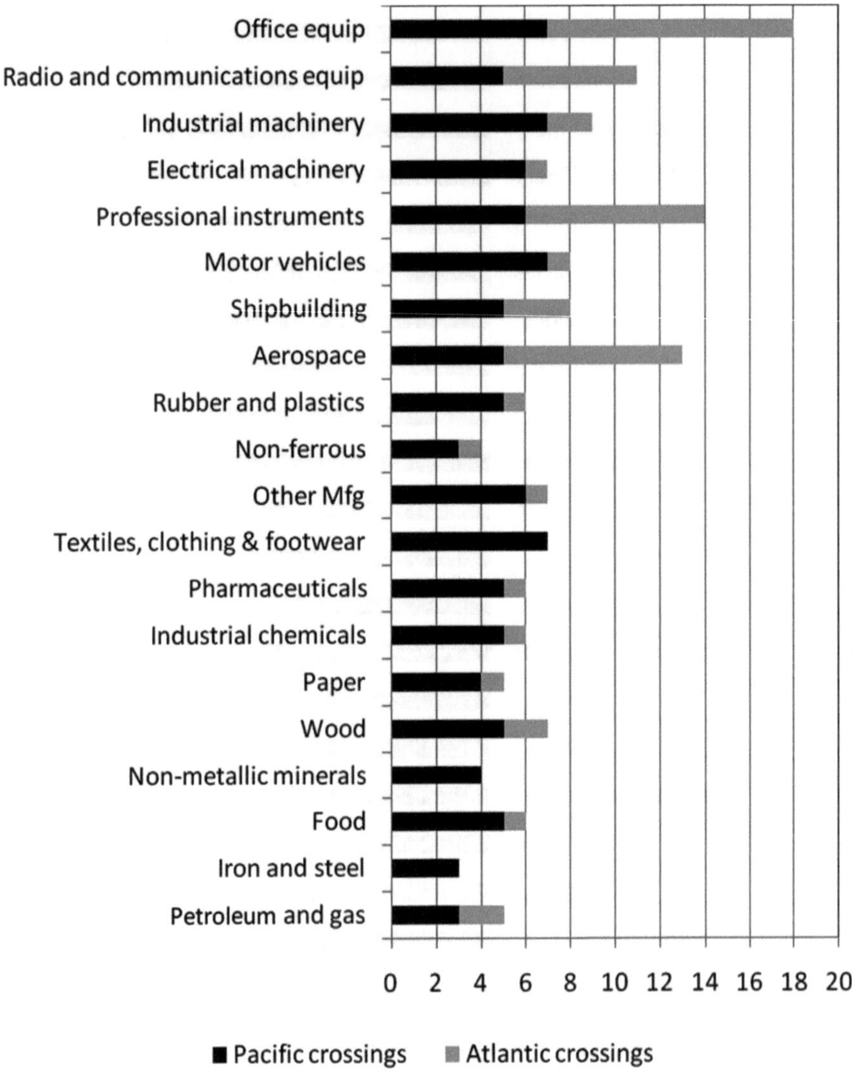

Fig. 17.4 Trans-Pacific and Trans-Atlantic significant trade links by industry (Notes: Europe to Asia links are excluded from the analysis but this is not a large number. Source: Based on work for Wixted and Bliemel [in preparation])

17.3 The Canadian National Innovation System

The national innovation systems (NIS) literature emerged from evolutionary economic theory and joined with more mainstream science, technology and innovation policy analysis. Analyses of regional innovation systems (RIS) by contrast, came initially from regional science and economic geography, as well as institutional economists and sociologists (see discussion in Holbrook and Wolfe 2000). But what is a region? What are the characteristics one should look for, in order to talk about regional or even local systems of innovation (cities) instead of a national approach? Regions are often defined in terms of shared normative interest (cultural areas), economic specificity (mono-production systems) and administrative homogeneity (governance areas). To these must be added other criteria, such as a non-specific size (except that of being part of a nation state); identifiable cultural or industrial mix; an ability to be distinguished from other regions in terms of these criteria; and, possession of some combination of internal cohesion characteristics (Cooke 1998).

For federations, the national innovation systems (NIS) concept should be understood as being a set of more complex relationships than that of a centrally-administered nation, since there are often provincial/state level institutions and actors paralleling national level institutions and actors, with some policies or powers under provincial control, and others under federal control. Canada is one of the few true economic and social (as well as political) federations in the world. In the OECD, only Australia, the US and Germany come close to the unique structure and socio-economic features that exist in Canada. Thus, unlike most nations in the OECD, and other parts of the industrialised world, the Canadian NIS is different. A key element of the Canadian federation is the allocation of most economic powers to the national government and the assignment of social responsibilities – particularly health and education – to the provinces.

In Canada, the NIS is regionalised; the Ontario/Quebec economy is not the same as the BC or Prairie regions. Indeed, Canada is a country of metropolitan 'islands': Vancouver, Calgary, Toronto, Montreal, etc. National statistics are heavily weighted by the industrial activity that takes place in the Windsor – Quebec City corridor. Thus for Canadian policy makers there is an important question: what are the boundaries of a functional system of innovation and what determines its viability? How small or large is a region?

In economic terms Canada is a thin ribbon of settlement, mainly within 200 km of the US border (Fig. 17.5). In eastern Canada, the major cities are closely linked to the eastern US economy, and thus are the northern periphery of that large two-dimensional economy. Western Canada is quite different. The major cities are strung out along the Canadian transcontinental railway system like pearls on a necklace. Indeed the railway was for all of them a major factor in their establishment and development in the late nineteenth and early twentieth century. These cities are separated by hundreds of kilometres of sparsely settled land – farms, ranches and forests.

Holbrook and Wolfe (2000) have argued, at least in the case of Canada, in order to understand the NIS, one must first understand the RIS. Is the Canadian NIS the sum of a number of RIS, based either on economic regions or provincial boundaries? In the Canadian context this sum is distorted by the wide variation in sizes of the regional systems – national level data (and the ensuing analyses) of the Canadian system of innovation are heavily biased by the economic activities occurring in the two major industrialised provinces, Ontario and Quebec. In most developed nations, innovation, science and technology policies are formulated by

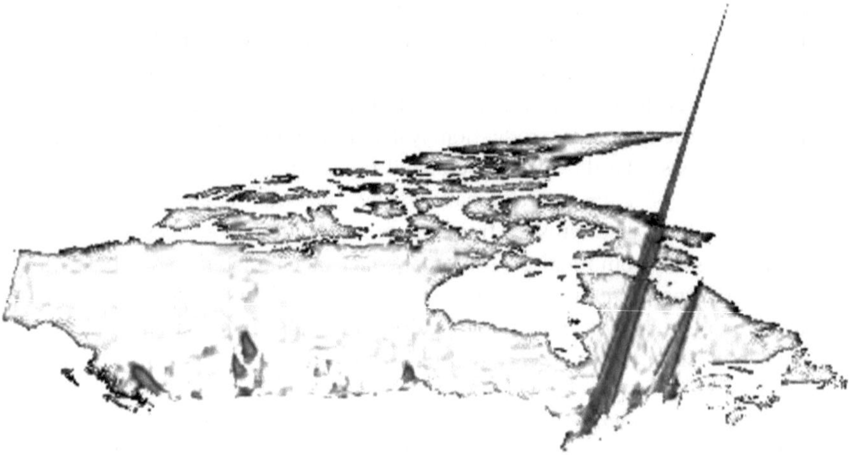

Fig. 17.5 Canadian GDP by location (Source: Nordhaus 2006)

the central government, yet most innovation takes place locally. Thus nation-wide innovation policies may not affect each region equally, and could conceivably be counterproductive.

In May 2001 a network of researchers drawn from the five regions of Canada: Atlantic Canada, Québec, Ontario, and Western Canada launched a project 'Innovation Systems and Economic Development: The Role of Local and Regional Clusters in Canada', that examined the impact and importance of cluster-driven innovation in Canada (Holbrook and Wolfe 2005). This research network, the Innovation Systems Research Network (ISRN), investigated how local networks of firms and supporting infrastructure of institutions, businesses and people in communities across Canada interact to spark economic growth. When the analyses of the Canadian NIS went a step 'down' from the national level to a regional one, social issues emerged more clearly, and the importance of communication and interaction were highlighted. The innovation systems studies of the flows of knowledge, and how knowledge is created, have realised the importance of the interaction between the different actors (particularly users and producers of knowledge and innovation – see e.g. DeBresson 1996), mutual trust (Maskell and Lorenzen 2004 on the benefits of clustering) and of course proximity needed to facilitate these flows.

While previous studies of innovation systems and economic development have been limited to individual regions, this project analysed how the growth of clusters contributes to economic growth and development within a number of regions across Canada. Research focused on more than 25 clusters across the five regions in newly emerging knowledge-intensive areas (i.e.: information and communication technologies [ICT], wireless, new media and biomedical) as well as in more traditional sectors (i.e.: manufacturing, wood products, food and beverage, automotive and steel). Studies on IT, biotech, new media, and wireless clusters were carried out in at least three separate regions, to get an understanding of the regional differences across the country. Recognising that, in Canada, the manufacturing

Table 17.1 Canadian local (metropolitan) innovation systems

Business innovation dimension	Governance structure		
	Grassroots	Network	Dirigiste
Localist	St. John, New Brunswick St. John's, Newfoundland & Labrador	Halifax	Québec city
Interactive	Saskatoon Winnipeg	Calgary Edmonton Victoria	
Globalised	Ottawa	Toronto Vancouver	Montréal

Source: Holbrook 2006

sector is concentrated in eastern Canada, studies in Ontario/Quebec also included automobiles, aerospace, and food products among others.

Arguably, in Canada, some provincial boundaries, such as those between Saskatchewan and Manitoba, or among the Maritime Provinces (the localist business innovation provinces), are artificial in terms of innovation systems. Canadian RIS can extend beyond provincial boundaries, or in some cases, such as the Ottawa and greater Toronto RIS, be contained within one province. Indeed in the more successful regions, from the point of view of innovativeness the RIS can be subdivided into local systems of innovation (LIS), which are usually based in individual cities. Table 17.1 gives a possible taxonomy of local systems of innovation in Canada. The structure is derived from Cooke (1992 and Cooke et al. 2004). Grassroots is defined as where the cluster governance characteristics are self-generated. Networked governance infers that multiple tiers of actors come together particular geographic points to act on the innovation system. Finally, dirigiste is defined by actors primarily above and beyond the local being the principal motivators of change.

17.4 The Study Methodology

Why should Vancouver be a major centre of innovation in Canada? More particularly, why should it be host to a number of human-capital intensive clusters? The study of Vancouver was based on an examination of human capital based economic activities that are important to the region that are mostly a little unexpected in a city so far from other major population centres. The five activities chosen were:

- Fuel cells (predominantly based on hydrogen technologies) utilising interviews and published papers based on 19 in-depth interviews;
- Bio-pharma (a range of firms creating human health oriented biotechnologies or traditional pharmaceuticals) based on 31 interviews;
- Motion pictures based on extant literature;
- New media (mainly the electronic games sector) based on 73 interviews; and
- Wireless technologies based on 36 interviews from a previous project and extant literature.

Table 17.2 Cluster interviews

Cluster	ISRN 1	ISRN II
Fuel cells		19
New media	21	10
Wireless	72 (Alberta & British Columbia)	1
Bio-pharma	32	4
Miscellaneous		13
Total interviews	125	47

The second wave of ISRN research focused on the MAR-Jacobs debate which is couched in terms of intra-city spillovers either within or across industries, but we can use the material to investigate the role of Vancouver on the Pacific Rim (Table 17.2). We used the survey responses to test the perceptions of industry participants on the source of important inputs. Agglomeration processes worked out in the evolution of centripetal vs. centrifugal forces: the logic here is that the start up time of clusters can show whether there are centripetal cost pressures at work which have delayed the start up times of clusters. For example, a technology industry may have started up somewhere else in the 1980s but only got going in BC in the late 1990s. This would be a primary indicator that industry evolution was expanding the geographic frontiers and creating a space for new players.

More traditional indicators of innovation such as R&D expenditures and results from conventional (OECD-type) innovation surveys were not used. There are important reasons for not weighting such information highly for Vancouver. Data on R&D expenditures is not published on a city basis by Statistics Canada, thus data at the provincial level would have to be used. More importantly, there is an unavoidable built-in bias towards central Canada in Canadian innovation surveys due to the way the samples have to be constructed.

Vancouver in particular has very little manufacturing compared to the national average and as our analysis (below) reveals Vancouver's enterprise demography is weighted more towards smaller businesses.

17.5 Analysis

Vancouver is mid sized city (2.5 million population) on the western fringe of the North American continent. Beckstead and Brown (2004) reports that Vancouver has an industrial diversity very similar to Toronto although having a significantly smaller population. The key question to be addressed by this section is whether the clusters started early and grew more or less autonomously via say the inheritance model of cluster development or whether their development has been greatly influenced by external events or even transplanted from elsewhere.

It is not sufficient to examine the MAR-Jacobs questions of 'spill-ins' (as one might call the benefits flowing into particular activities from other firms in the same city or the diversity of the city itself), but also the labour markets for these human

Table 17.3 Industry start-up timeline

Cluster	Vancouver/BC origins	Global origins
New media	1980s	1980s (California)
Bio/pharma	1981	1976
Fuel cells	1980s (Ballard Power)	1950s (space program) / re-emerged 1980s
Wireless communications	Late 1930s	1930s
Motion pictures	1980s	LA – 1900s (1920s – beginning of the boom – Storper & Christopherson 1987)

Source: Wixted and Holbrook (forthcoming)

capital intensive clusters and the start-up conditions (Table 17.3). For more detailed descriptions and analysis of the following cases please refer to Wixted and Holbrook (forthcoming).

17.5.1 Fuel Cells

Of all the human capital intensive activities, fuel cells is the smallest and most specialised in terms of spillovers. Although it has benefits from R&D capital from outside the region, it is significantly an undiversified cluster as it relies on local R&D facilities and access to near-local talent at the University of Victoria (Holbrook et al. 2010).

17.5.2 Bio/Pharma

The prime characteristic of the bio/pharma cluster in Vancouver is one largely of IP rent seeking. For example it doesn't have the critical mass of important centres east and south but companies do have strategies for managing the construction of value in Vancouver. In response to the question 'how does your firm benefit from being located in this particular urban region with its mix of firms/institutions' one business leader responded:

> "To a limited extent the company does benefit from ideas generated in the local milieu and the skilled people that reside here. But all these things are comparative. Compared to being in Saskatchewan, in Saskatoon, I would think probably yes. Compared to being in the Bay Area, absolutely no. We're here because some groundbreaking science was done here, and the scientists who created those innovations decided it was worthwhile trying to commercialise them, hence we're here. We get enormous benefit from continuing to work with them. In the community there's a reasonable level of managerial expertise. It's not terribly high; there's not a lot of people, but there are a number. And there are some people to go to for various sorts of help. There is a pool of skilled scientific staff, but not an enormous number because we have a relatively small biotech community here. So, yes, there are advantages compared to being out of the loop completely."

One interesting insight came from another business executive:

> "We mainly collaborate with Australia, New Zealand, California, Montreal and North Carolina."

So in bio-pharma the specific location of Vancouver has helped it survive but has been a barrier to its development beyond a certain size.

17.5.3 Wireless

The wireless cluster has had a long and varied path in Vancouver relying on investment from outside the region as well as, for a time, an innovative, provincially owned telephone company. Today the cluster has close ties with the new media cluster. This closeness is highlighted by the very recent amalgamation of the two industry associations WinBC (for Wireless) and New Media BC into DigiBC, an organisation representing approximately 22,000 employees and 1,300 companies.

17.5.4 Motion Pictures

What is interesting about the story of Vancouver and motion pictures is there has been a very traditional split between foreign film development in Vancouver and Canadian film industry activities in Toronto (Coe 2001). This trend has been weakening in more recent years. The dependency of Vancouver on external interaction is seen by the need to continually reassess the tax credit status of foreign films (primarily out of the USA). Coe for example, highlights the closeness of Vancouver and Hollywood products:

> As a location, Vancouver was extremely well placed to benefit from the opportunities being created by the on-going vertical disintegration of the Hollywood studios. The city is close to Los Angeles (2.5 h flying time), provides an enviable 'west coast' quality of life, is in the same time-zone thereby allowing easy co-ordination of activities between the two centres, has a mild climate which allows all-year-round filming and offers a large range of different scenic locations within 1–2 h drive of central Vancouver (2001: 1760).

In 2009 this industry in Vancouver experienced significant economic turbulence during the global financial crisis, but seems to have nevertheless attracted significant investment.

17.5.5 New Media

The new media specialisation and diversity pattern is summed up excellently by the following quote by a manager in a major new media producer in response to a question regarding the major benefits of Vancouver.

> "Vancouver is a beautiful city – a draw for recruits. Sea to sky [a reference to the seaside highway between Vancouver and Whistler]. Has a lot of amenities. Great place to make video games because there are lots of companies that make video games. Very incestuous here . . . Hollywood North helps. Strong film and television. Cross fertilisation, we use those professionals. Great educational institutions that produce high caliber potential hires. On the Pacific Rim gateway including to the east. . . .. It is cheaper to develop here than in the US. It was the case, anyway for many years."

In this one quote one can see that the advantage for Vancouver is simultaneously local, diverse (interactions with motion picture industry) and translocation (given references to being on the Pacific Rim as well being a gateway to the East – Toronto and Montreal). The reference to cross-fertilisation is important. This is an explicit acknowledgement of the value of clustering among both competitors and similar, but not necessarily competitor industries (i.e. new media and motion pictures).

In 2009 this industry in Vancouver also experienced significance economic turbulence during the global financial crisis. However, based on the significant capital sunk costs in the region and the history of startup entrepreneurship it is expected that this cluster will show resilience to the turmoil.

17.5.6 Summary

We have summarised our finding on the local industry lifecycle position as it is today relative to the global situation (Fig. 17.6).

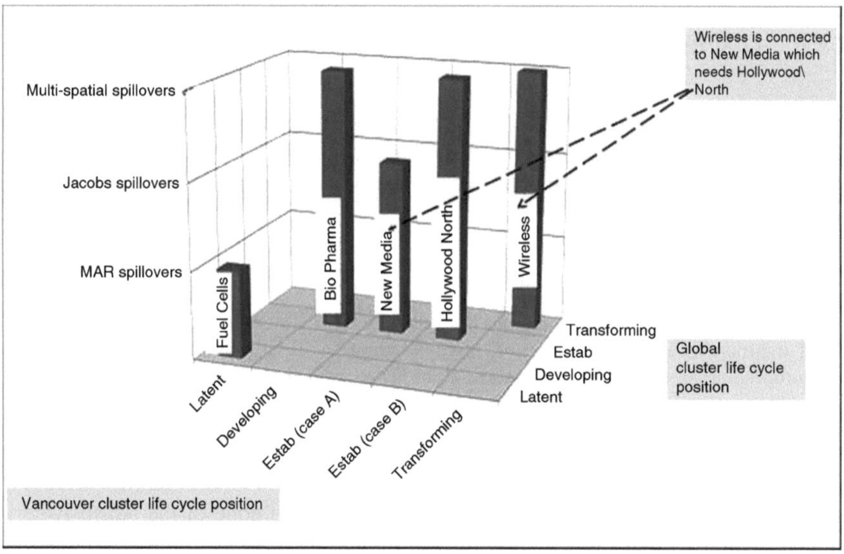

Fig. 17.6 Current lifecycle positions of Vancouver's clusters (Source: Wixted and Holbrook [forthcoming])

Based on Table 17.3 and Fig. 17.6 three phenomena should stand out. First, apart from the example of motion pictures the Vancouver cluster started early globally, in one case (fuel cells) leading the world to re-examine the technology. It would suggest that mostly these are not spin-out clusters but have their own independent trajectories. Second, apart from fuel cells all the clusters rely on a diversity of spillovers for inputs that at the very least are local but in many cases stretch across the North American continent both east and south. Though not presented in depth here the outputs of these clusters are marketed globally. Lastly, we can suggest that there something west coast about many of these activities which have a distinctly mid to late 20th Century origins in comparison to east coast US manufacturing or business services; which in Canada are based in Toronto and Calgary but not Vancouver.

The evidence can be summarised as follows.

- MAR–Jacobs. While there is some evidence for cross industry spillovers the strongest evidence is for multi-spatial spillovers in terms of inputs and for a global outlook in terms of sales.
- Agglomeration – the lifecycle evidence suggests that all the clusters are mostly of long standing in the Vancouver region and with the exception of the major motion picture activity all clusters have indigenous roots, suggesting that they are not spin-outs from more innovation intensive regions.

So why has Vancouver been able to be early or at least only slightly behind the frontier in these talent based activities?

17.6 Discussion: Vancouver on the Pacific

It is suggested in this chapter that a dominant factor in the emergence and continuance of Vancouver is its position on the west coast of North America, as well as being the most accessible Canadian airport to Asia and being in the same time zone as California.

In contrast to the current regional science philosophy, it was previously argued that transport systems were vital.

> The development of cities and regions has generally been associated with the development of transport. A good description of this association has been provided by overview of European urban history where the fortunes of places have been shaped primarily by their position in a transport system. (O'Connor and Scott 1992: 240.)

Prior to the arrival of the Boeing 707, Vancouver was where the trans-Canada rail line had placed it – literally and figuratively at the end of the line. It was simply an intermodal (rail/sea) transfer point and the northwest edge of the North American continent. At that time, Calgary was a bigger destination for air passengers in the early 1950s (Natural Resources Canada 1957). This changed with the introduction of the first mid-sized long haul aircraft (the 707s could safely fly over the

Rocky Mountains) and today the Boeing 777 and the Airbus 340 are specifically designed to fly the 'thin' routes and thus continue to change hub positioning. The change in aircraft has been assisted by changes in navigation and air route administration involving trans-polar routes which has reduced flying times between Europe and North America, and from the Asia Pacific region to both North America and Europe. A second technical shift has involved the refinement of engine performance, along with management of passenger and freight loads, to create 13–15 h point-to-point services. Extended range 747s, 777s and A340–500s have made it possible to fly directly between places like Chicago and Hong Kong, New York and Hong Kong and Los Angeles and Singapore. Taken together these changes have helped change the pattern of passenger traffic in North America.

Today, Vancouver is a major gateway city, being the fourth largest seaport in North America and more importantly for its human capital based activities it continues to develop as a major airline hub.

> The ocean rim character of the region's settlement provides another reason for the growth in air travel. In effect, the region is a series of nodes around the Pacific Ocean, with strong, outward looking commercial activities, and in many cases, very limited local hinterlands. Vancouver, Los Angeles, Sydney, Singapore, all share an unusual characteristic in that they have large concentrations of population, but with limited development of an inland hinterland (O'Connor and Scott 1992: 244).

Thus, aircraft technology and preexisting settlement patterns form important feedback loops that enhance or diminish a city's ability to attract traffic and talent.

> In the modern age there are co-evolutionary processes between aircraft technology, airline traffic management and city development, such that once established a major centre cities themselves become major draws of people which reinforces their status which then feeds back to drawing more people. It would seem that the character of the global geography of airports is shaped by the forces associated with global city development. That means the continuation in the significance of the global cities will maintain the dominant role of a few very busy airports in the global network (O'Connor 2003: 84).

Different innovation systems around the world have different human capital attributes and different technological advantages and their geographic positions differ markedly, but all three characteristics matter. Vancouver is almost unique as a local innovation system (LIS) in North America: it is not self-contained; it is dependent upon its transportation and communication links. The city and its LIS are a 'pivot point' between North America and Asia, unlike many of the other high-tech cluster areas in Canada and the US. Its major continental competition is in California, which has similar geographical attributes.

There are other strong LIS systems on the Pacific rim, 'city-states' such as Singapore, Hong Kong, Sydney, and some large Chinese port cities, but only a very few (Busan, Korea for example) are gateways to larger systems of innovation. A few of them are currently or have been manufacturing centres, others are entrepot trade and financial centres (Hong Kong and Singapore) and two in particular; Vancouver and Sydney – interestingly both at extremes of the Pacific system are based on human capital activity. For all of its history and current development trajectories there is clearly something unique about innovation systems on the

Pacific Rim? Is it likely that the behaviour of LIS on the Pacific Rim differ greatly from LIS in continental North America or Europe, or for that matter other diverse economies such as India or inland China? Is it possible, that because of the distances involved, the Pacific system of innovation can be thought of being as a chain of local systems of innovation, with innovation gateways where continental lines of communication intersect the Pacific Rim?

Polèse (2009) has commented on the development of a second band of economic activity along the Pacific coast of North America (as opposed to the primary concentration along the Atlantic Seaboard), and has compared it to the lack of similar development in the Mediterranean:

> The answer lies in the location – across the sea – of each continents principal trading partners. This is where we jump from a continental to a global perspective. The direction of international – intercontinental – trade influences the emergence of economic centres within each continent.... The West Coast of North America found itself facing dynamic trading partners on the other side of the Pacific, beginning with the emergence of Japan.. . . (Polèse 2009: 84, 85)

What then can policy makers derive from this evidence? It is clear that conventional innovation policy tools, such as support for R&D, are necessary, but not sufficient. The bio-pharma cluster in Vancouver certainly started through major bio-pharma R&D investments, but that by itself is not sufficient to explain its continued existence. Clearly conditions must also exist to attract and retain highly skilled workers (Florida's 'super-creatives').

In Vancouver these conditions exist: the natural setting is spectacular, but also the various levels of government have combined to provide infrastructure and amenities that above many in North America. The Economist Intelligence Unit regularly rates Vancouver as one of the most desirable places to live in the world. The infrastructure includes significant investment in mass transit, airports, hospitals, schools, and universities. But government policies have also favoured the development of significant cultural and recreational amenities: theatres, galleries, sports venues and green spaces. In the Canadian system of government many of these improvements fall to the city government. Both the City of Vancouver and Metro Vancouver (the regional government) have contributed directly to the development of the Vancouver LIS.

Vancouver's clusters increasing are benefiting from both the market demand and the cross-fertilisation possible from the presence of Hollywood North. Further the individual cluster stories have been shown to have benefited from industry associations providing some of the needed social capital for clusters that are fragmented and dominated by small enterprises to continue to seek and develop new opportunities (Petrusevich 2005; Reibling 2004).

But the evidence on the history of the clusters – replicating by means of entrepreneurial spin-offs from earlier firms suggests very specific policies for innovation. Experience appears to matter to a very significant degree. In such an environment that is always likely to lose larger businesses and research/ manufacturing endeavours to more central places encouraging a system of mentoring and experience needs significant policy focus. While it is difficult to generalise, the specifics of the Vancouver case indicate that clusters with low critical mass and distant from mega-regions appear to have particular dynamics. Firms seem to struggle to grow to any size and there is significant turning over of businesses and staff. Thus, the opportunities lie with speeding up the learning process for new graduates through fostering greater opportunities for experience.

17.7 Summary and Conclusion

This chapter is an attempt to both integrate literatures on national urban systems and innovation as well as the 'geographical turn' in economics with its emphasis on the particularity of place, which as Martin notes has its origins in Losch: 'If everything occurred at the same time there would be no development. If everything existed in the same place there could be no particularity. Only space makes possible the particular which then unfolds in time' (quoted in Martin 1999: 66). The former is needed as the latter (particularity) has increasingly squeezed out of economic geography and innovation studies a sense of asking questions on the larger distribution of technological and innovation clusters. However, the important literatures in bringing the macro and the particular together do not necessarily sit well together. The rapidly expanding research exploring MAR-Jacobs (industry specialisation versus diversity) relationships is valuable for differentiating sectoral and city characteristics but little of the work incorporates city systems. The recent new economic geography promoted by Krugman and others has some useful mathematical modelling of agglomeration and thus centripetal and centrifugal forces but empirical testing of the models seems limited and is quite problematic when it comes to innovation.

Innovation policy in Canada is driven by the linear nature of European settlement across the country. Most innovation policy initiatives focus on the concentration of manufacturing enterprises in southern Ontario, and to a lesser extent, southern Quebec. Vancouver is unique in Canada as Halifax, Vancouver's counterpart on the Atlantic coast, simply does not have the same order of magnitude of innovative or gateway activity (seaport or airport). This suggests that unique innovation polices need to be developed for Vancouver and its hinterland, in the context of its participation in the two dimensional array of local innovation systems around the Pacific Rim. What might they be?

What are Vancouver's competitive advantages in a Pacific context? Vancouver benefits from being part of Canada – a Westminster-style democracy with well established, and prudent legal and financial systems. It is seen as a haven for entrepreneurs around the Pacific, a place where they can store the fruits of their enterprise. Similarly it has the advantage of having an educational system that is based on English systems and which uses English as its language of instruction. Foreign students form an important component of both secondary and post-secondary education (and bring in significant revenues, since they pay premium fees). All of this suggests that innovation policy for Vancouver should focus on infrastructure, both physical (traditional as well as education and research related) and intangible. Governments, both local and national can and should invest to maintain Vancouver's growth, and to ensure that the gains made to date do not atrophy.

In an archipelago of cities mobility is harder (economic costs and loss of social networks), choices are fewer, and the local/ regional innovation system interface is different. In such an economy the key cities become more important than the industries in them. The quality of life and economic opportunities for people across a wide range of economic groups is crucial. The implications of this in Canada is that the country does not need regional policies – i.e. western Canada is not like central Canada but that a federation with the geographic spread and diversity such as Canada needs a radically different policy framework. Instead of the Canadian federal focus on the creation of 'new industries' it is time to start thinking about a regional cities development fund.

Acknowledgements The authors acknowledge the financial support of the Canadian Social Sciences and Humanities Research Council (SSHRC) for the Innovation Systems Research Network projects 1 (Innovation clusters) and 2 (Innovation and cities). They also gratefully acknowledge the comments of the reviewers.

References

Barnes, T. J. (2003). What's wrong with American regional science? A view from science studies. *Canadian Journal of Regional Science, 26*, 3–26.

BC Stats. (1994). *The Size of the Underground Economy*. September 1994 Issue: 94-09.

Beaudry, C., & Schiffauerova, A. (2009). Who's right, Marshall or Jacobs? The localization versus urbanization debate. *Research Policy, 38*(2), 318–337.

Beckstead, D., & Brown, M. (2004). *From Labrador city to Toronto: The industrial diversity of Canadian cities*. 1992–2002 Statistics Canada Catalogue no. 11-624-MIE — No. 003. Ottawa: Statistics Canada.

Boreham, P. R., Clegg, S. R., Emmison, J. M., Marks, G. N., & Western, J. S. (1989). Semi-peripheries or particular pathways: The case of Australia, New Zealand and Canada as class formations. *International Sociology, 4*(1), 67–90.

Bowden, J. (2000). Airline hubs in Southeast Asia: national economic development and nodal accessibility. *Journal of Transport Geography, 8*(1), 25–41.

Bunnell, T., & Coe, N. (2001). Spaces and scales of innovation. *Progress in Human Geography, 25*(4), 569–589.

Carlson, B. (2006). Internationalization of innovation systems: A survey of the literature. *Research Policy, 35*(1), 56–67.

Christaller, W. (1966). *Central places in Southern Germany* (C. W. Baskin, Trans.). Englewood Cliffs, NJ: Prentice-Hall.

Coe, N. (2001). A hybrid agglomeration? The development of a satellite-Marshallian industrial district in Vancouver's film industry. *Urban Studies, 38*(10), 1753–1775.

Cooke, P. (1992). Regional Innovation Systems: Competitive Regulation in the New Europe. *Geoforum, 23*(3), 365–382.

Cooke, P. (1998). Introduction: Origins of the concept. In H.-J. Braczyk, P. Cooke, & M. Heidenreich (Eds.), *Regional innovation systems* (pp. 2–25). Oxford: Oxford University Press.

Cooke, P., Heidenreich, M., & Braczyk, H.-J. (Eds.). (2004). *Regional Innovation Systems: The Role of Governance in a Globalized World.* (2nd edition). London and New York: Routledge.
DeBresson, C. (Ed.). (1996). *Economic interdependence and innovative activity: An input-output analysis.* Cheltenham: Edward Elgar.
Easton, S. (2004). Marijuana Growth in British Columbia. *Public Policy Sources*, No. 74. http://www.fraserinstitute.ca
Edquist, C. (Ed.). (1997). *Systems of innovation: Technologies, institutions and organisations.* London: Pinter.
Florida, R. L. (2002). *The rise of the creative class.* New York: Basic Books.
Freeman, C. (1995). The 'national system of innovation' in historical perspective. *Cambridge Journal of Economics, 19*(1), 5–24.
Frey, B. S., & Pommerehne, W. W. (1984). The hidden economy: State and prospects for measurement. *Review of Income and Wealth, 30*(1), 1–23.
Glaeser, E., Kallal, H., Scheinkman, J., & Schleifer, A. (1992). Growth in cities. *Journal of Political Economy, 100*, 1126–1152.
Harrison, C. (2007). *World internet connections: City-to-city.* Digital. http://www.chrisharrison.net/projects/InternetMap/. Accessed 27 September 2010.
Holbrook, J. A. D. (2006). Regional differences in innovation in knowledge-based industries. *International Journal of Technology, Knowledge and Society, 2*(4), 153–162.
Holbrook, J. A. D., & Wolfe, D. A. (Eds.). (2000). *Innovation, institutions and territory: Regional innovation systems.* Kingston: McGill-Queen's University Press.
Holbrook, J. A. D., & Wolfe, D. A. (2005). The innovation systems research network (ISRN): A Canadian experiment in knowledge management. *Science and Public Policy, 32*(2), 109–118.
Holbrook, J. A. D., Arthurs, D., & Cassidy, E. (2010). Understanding the Vancouver hydrogen and fuel cells cluster: A case study of public laboratories and private research. *European Planning Studies, 18*(2), 317–328.
Klepper, S. (2001). Employee start-ups in high-tech industries. *Industrial and Corporate Change, 10*(3), 639–674.
Lundvall, B.-Å. (Ed.). (1992). *National systems of innovation: Towards a theory of innovation and interactive learning.* London: Pinter.
Martin, R. (1999). The new 'geographical turn' in economics: Some critical reflections. *Cambridge Journal of Economics, 23*(1), 65–91.
Maskell, P., & Lorenzen, M. (2004). The cluster as market organization. *Urban Studies, 41*(5/6), 991–1009.
Natural Resources Canada (1957). *Atlas of Canada* (3rd ed.). http://atlas.nrcan.gc.ca/site/english/maps/archives/3rdedition/economic/transportationandcommunications/088. Accessed 23 February 2009.
Nelson, R. R. (Ed.). (1993). *National innovation systems: A comparative analysis.* New York: Oxford University Press.
Nordhaus, W. (2006). Geographically based economic data. http://gecon.yale.edu/research-papers. Open access file. Accessed August 2010.
O'Connor, K. (2003). Global air travel: Toward concentration or dispersal? *Journal of Transport Geography, 11*, 83–92.
O'Connor, K., & Scott, A. (1992). Airline services and metropolitan areas in the Asia-Pacific region 1970–1990. *Review of Urban and Regional Development Studies, 4*(2), 240–253.
OECD. (1999). *Boosting Innovation: The Clusters Approach.* Paris: OECD.
OECD. (2001). *Innovative Clusters: Drivers of National Innovation Systems.* Paris: OECD.
OECD. (2008). *The Internationalisation of Business: R&D Evidence, Impacts and Implications.* Paris: OECD.
Oinas, P., & Malecki, E. (2002). The evolution of technologies in time and space: From national and regional to spatial innovation systems. *International Regional Science Review, 25*(1), 102–131.

Petrusevich, M. R. (2005). *Social Capital Generators? A Case Study of Industry Associations within the Vancouver New Media Cluster*. Unpublished thesis, Master of Arts in the School of Communication, Simon Fraser University. http://www.sfu.ca/cprost/theses.html.

Phillips, P. (2002). Regional systems of innovation as a modern R&D entrepot: The case of the Saskatoon biotechnology cluster. In J. Chrisman et al. (Eds.), *Innovation, entrepreneurship family business and economic development: A western Canadian perspective* (pp. 31–58). Calgary: University of Calgary Press.

Polèse, M. (2009). *The wealth and poverty of nations*. Chicago: University of Chicago Press.

Reibling, S. L. R. (2004). *Measuring 'high tech' social capital in the biotechnology cluster located in Vancouver, British Columbia*. Master of Arts thesis in the School of Communication, Simon Fraser University. http://www.sfu.ca/cprost/theses.html.

Rodrigue, J.-P., Comtois, C., & Slack, B. (2009). *The geography of transport systems* (2nd ed.). New York: Routledge.

Ryan, C., Li, B., & Langford, C.H. (2011). Innovative workers in relation to the city: The case of a natural resource-based centre (Calgary). *City, Culture and Society, 2*, 45–54.

Sharif, N. (2006). Emergence and development of the national innovation systems approach. *Research Policy, 35*(5), 745–766.

Storper, M., & Christopherson, S. (1987). Flexible Specialization and Regional Industrial Agglomerations. *Annals of the Association of American Geographers, 77*(1), 104–117.

Wallerstein, I. (1976). Semi-peripheral countries and the contemporary world crisis. *Theory and Society, 3*(4), 461–483.

Wixted, B. (2009). *Innovation system frontiers: Cluster networks and global value*. Berlin: Springer.

Wixted, B., & Bliemel, M. (in preparation). Configurations of Global Factories: Product Modularity and Architectural Control.

Wixted, B., & Holbrook, J. A. D. (forthcoming). Living on the Edge: Knowledge Interdependencies of Human Capital Intensive Clusters in Vancouver. In D. Wolfe (Ed.), *Innovation and Creativity in Canadian City Regions*. Toronto: University of Toronto Press.

Wolfe, D. A., & Bramwell, A. (2008). Innovation, creativity and governance: Social dynamics of economic performance in city-regions. *Innovation: Management, Policy & Practice, 10*(2–3), 170–182.

Yeung, H. W.-C. (2005). Rethinking relational economic geography. *Transactions of the Institute of British Geographers, 30*(1), 37–51.

Chapter 18
User-Driven Innovation and Knowledge Integration in Elderly Care Services: A Community Integration Model

Koichi Ogasawara

Abstract This chapter shows key methodological factors in applying a user-driven co-creation approach to the domain of service innovations in elderly care, based on three field studies and action research. Through interpretation of the logical structure and service process mechanisms of elderly care, we focus on the relation between the values of elderly people and the knowledge and skills of experts to fulfil these values. The complexity of dimensions of values and categories of support functions is regulated by a moral game between the service user and experts during the process of service creation. Interpreting the user's status and values during service design by bringing together experts from different disciplines results in services that are easy for users to adopt and facilitates user accommodation of the process of service implementation. Recognising the values to be fulfilled, determining tasks and planning solutions, and fitting together process and method in service provision are conceptualised as 'designing'. This is not a matter of modulating parts and engineering processes, but of designing the most comfortable and creative solution that will add value for the user. This chapter proposes an active cluster net centred on user-driven co-creation as a community integration model that reflects innovative societal structuring in coordinated human service allocation.

18.1 Introduction: Co-creation in Elderly Care Service Innovation

'Co-creation' is a concept for radical business innovation theorised by Prahalad and Ramaswamy (2004). In the sense of gathering different types of knowledge for smoothing out the value-adding process, co-creation already emerged in the 1980s

K. Ogasawara (✉)
Tohoku Fukushi University, Sendai, Japan
e-mail: o-koichi@tfu-mail.tfu.ac.jp

H. Melkas and V. Harmaakorpi (eds.),
Practice-Based Innovation: Insights, Applications and Policy Implications,
DOI 10.1007/978-3-642-21723-4_18,

as an improved production method known as 'concurrent engineering'. In the late 1990s, it began extending into a new marketing approach and practice that contrasted with mass marketing or power marketing. This took place in the era of consumers' changing approaches to acceptance of transactions, of tending toward cross-media interfacing between providers and consumers thanks to the dynamic, web-related technological leap, and of shifts in consumer values regarding engagement and participation in creative developments. Along with 'co-creation' the burst in concepts such as 'crowd-sourcing', 'open innovation', 'user-driven', 'user-centric', 'value-based', and so on by the mid-2000s emerged as a strategic response, via a knowledge-driven value-creation methodology (Möller and Svahn 2006), to reducing risk of uncertainty in fluctuating markets and to creating new sources of profitability through concept designing and virtual collaboration with active consumers.

The co-creation model has evolved on the basis of a deep interpretation of 'user-centric' by Verganti (2009), which emphasises the key role of an interpreter, not simply being closer to the customers' and users' preferences in proposing value-added meanings during creative design. Value creation and disruptive innovation are perceived as being driven by observing, leaning, and proposing new messages and meanings through interplay between design-push and technology-push approaches. This interpretation provides the co-creation model with high feasibility in the domain of health and social services, especially elderly care, where behavioural phenomena and the care service interfacing with them are shaped by the users' cognitive uncertainty and by fluctuation in functionalities as well as by socio-relational circumstances.

For a dynamic innovation in health and social services, the fundamental question is how to break through a widely recognised barrier: the fact that the logic of commodities is not particularly applicable to service market processes. Health and social services are utilised continuously and comprehensively by the individual target, and services must support flexible adjustment of service contents to meet changing needs for support functionalities as determined by monitoring. This is an especially critical requirement in elderly care. Case solution processing is not really driven by approaches involving the notions of 'client' and 'consumption'. This is the reason why cross-disciplinary interpretation of values and mass commitments to designing services are necessary to maintaining the users' comfort zone in adoption of services and at the service provision interface.

This chapter attempts to apply the concept of user-driven co-creation in the service domain of elderly care, by establishing fundamental methodological elements and raising key factors for innovating in the field.

The chapter is based on results from three separately conducted studies. The first is a qualitative study on the integration of nursing and caring knowledge and skills in the service process of dementia/elderly care. Nine dementia care units in different institutional categories and one hospital specialising in dementia care were sampled. The background facts and evidence are available in NCSIRU-TFU (2010). In Japanese long-term care, there is a system of registered care professionals, *Kaigo-Fukushi-Shi*, in charge of mental and physical support service tasks; these professionals work together with registered nurses in care units and hospitals. The range of tasks of the *Kaigo-Fukushi-Shi* is almost equivalent to that of a practical nurse in other countries, while there is an increasing extension and depth of their knowledge and skills – a trend toward high professionalisation of care work in the context of dementia and malignant disease care.

This chapter also presents data on the skill formation of care workers; this data is based on a mass survey of task analysis and self-check assessments in terms of job performance ability. The survey was conducted in 2000–2002, at the same time as the kick-off of full implementation of the new Japanese long-term care system.

The designing of a model of comprehensive community service clusters took place during the period 2007–2010 under the auspices of an action research consortium led by Tohoku Fukushi University, in the city of Sendai in the mid-northeast region of Japan, as a part of the Japanese MEXT (Ministry of Education, Science and Technology) project 'Knowledge Cluster Initiative Global Cluster Linkage'. A follow-up action research project is now underway as a joint project between Tohoku Fukushi University, Laurea University of Applied Sciences (in Finland), and Videra Oy (Ltd.). One of the core questions of those original studies is how effective the improvements to care service quality and the service provision process are for users.

This chapter offers hypotheses for the core methodological elements of a user-driven approach in innovating elderly care service and service provision systems by articulating evidence from these qualitative studies (Ogasawara et al. 2011).

18.2 The Long-Term Care System and Need Surveying During Co-creation

Japanese findings from the implementation of the Long-Term Care for the Elderly Act (LTCA) are a good example of the necessity of user-driven co-creation in elderly care. LTCA was enacted in 1997, with implementation planned from the beginning of the fiscal year of 2000. The Act introduced a division between purchasing and providing to facilitate service consumption based on the principle of individual decisions and to enhance wider private sector involvement in the care provision system. This division between purchasing and providing was initially introduced in 1990, during the first stage of welfare legislation reform, in order to open up home care service delivery to the private sector. LTCA, as a major part of the second stage of reform, introduced an exclusive funding system for elderly care as a unique form of public insurance that is managed by municipalities. LTCA established a shift in approach to Japan's elderly care system, introducing a new type of public management that differed radically from Japan's post-war welfare service system.

While the LTCA has succeeded in increasing the volume of public spending on elderly care service and facilitating an expanding number of private-sector providers to participate in this quasi-service market, the size of the business and number of clients per provider has been continuously decreasing. Although the market as a whole is expanding, units are downsizing due to high costs. Compartmentalisation of services and de-skilling of manpower for the purposes of cost control has succeeded in putting an end to public debate on improving care service in a way that would offer more comprehensive and continuous service provision (Ogasawara 2007: 11–14). The target of a 2005 amendment to the Act was precisely to cope with this compartmentalisation and the downturn in service quality by introducing a system named 'community comprehensive care', which placed emphasis on the continuity of support service flow and seamless multi-expertise liaising. In terms of continuity, it validated a method of preventive care for groups who were at risk in terms of physical functionality and other groups, such as the slightly demented or those isolated from social interaction (Ogasawara 2007: 15–22).

The current approach of prevention is based on the medico-gymnastic assumption that socialisation and physical activation are effective methods of reducing the chance of risks being realized and reaching the critical zone of requiring care. Nevertheless, statistics show that less than 4% of the elderly population is actually assessed and classified in terms of risk group, in contrast to the Government's target of 8–12%. Less than 1% of the elderly population actually participates in preventive care programs, while the Government's target is 5%. When faced with this poor performance in institutionalised services, the Ministry of Health, Labour and Welfare points to workplace-related reasons, such as the heavy care planning loads of care managers, poor support facilities for care planning, inefficiency of diagnostic processes, the low capacity of preventive programs to meet the real needs of service users, and so on. (MHLW 2010: 12).

The low ratio of program participation should also be interpreted from the viewpoint of whether and to what extent the programs reflect the values that elderly people consider necessary to be met, or whether the effectiveness the programs afford is visible and acceptable to these users. The real reason for this low user participation is a kind of mismatch between the methodology itself and the real demands of the users. There is evidence that the ratio of participation is higher and continuity of involvement is steadier and longer in those cases where elderly people are involved in organising and managing the exercise programs, take on the role of supporting new members in the program, and are expected to assist in running or even leading exercise groups (Suzuki 2009).

There is other evidence of a positive correlation between enabling the participation of the demented elderly and decreasing their risk of being mentally uncertain and disoriented. Inviting them to participate in creation co-action with care workers has been assessed as being effective in decreasing the time spent on care and the workload volumes of service practitioners as well as decreasing the risk of accident (NCSIRU-TFU 2010: 136–146). The view that high utilisation of the remaining capacity of the elderly is the best way to enhance effective rehabilitation is widely shared by experts. In addition to enabling and vitalising the remaining functionalities of the user, inviting the user's involvement in continuous co-creative innovation processes on methods of care is the key to achieving higher effectiveness and efficiency. Co-creation through user participation is, thus, deemed a breakthrough approach to innovating a more user-friendly service process in the domain of care services.

18.3 Smartness of Life and Co-creation

Human health or a healthy life is conceptualised as a structure of elements embedded in daily life (Reed et al. 2004; Bond and Corner 2004; Bendelow 2009). Boylan classifies, from the perspective of global public health guidance, these elements into two levels of basic goods and three levels of secondary goods, according to the depth of embeddedness (Boylan 2008: 1–2). Basic goods are those necessary for human bodily action, and secondary goods are the possibility and ability to pursue, participate, utilise, and exert. His structure of embeddedness of goods is, thus, oriented to securing a subjective freedom of choice and capacity to participate in activities that form the basis of societal existence. Well-being certainly includes this notion of health, but from the social service perspective, one element is missing: the opportunity for commitment by somebody eligible to make decisions and choices to ensure that values, in the form of needs, are met through co-working. Health and healthy living are, therefore, enhanced through co-creation of values.

Life thus becomes an enabling and activating process for members of society, by preserving high-quality membership in a given societal circumstance and system of relations, regardless of age and mental or physical impairments. Four basic elements have been identified that enable and activate the lives of the elderly: fair access to opportunities despite economic status and social position; functionalities and participation; the cultural and moral circumstances of the surrounding community; and barrier-free communication. The lives of the elderly, on the other hand, are a kind of patchwork, in the sense that a series of self-determinations toward an unforeseen future creates a system for leaving success and satisfaction as well as regrets, grief, loss, and disappointments behind. Support service design for the elderly should be embedded as a failsafe.

User-driven service creation entails, therefore, observation and being observed, user-expert coupling, designing, co-working, and accepting. It is the process of creating a model of life through close contact with the reality of living.

18.4 Achieving Rationality Through a Moral Game

18.4.1 The Logic of Care Service

Care service is a typical personal human service in the sense that it is fundamentally a case-by-case and phase-by-phase service, due to the individuality of needs, the epidemiological nature of any evidence involved, and fluctuations in methodology. Care service is the process of coping with each individual case, so that satisfaction with and the utility of the service provided to the user can be artificially regulated.

Care service is also by nature an integrated service: it is mandatory that there be a variety of clinical activities provided by a multi-disciplinary care team, because of the complexity of elderly people's physio-mental conditions as well as individual life circumstances (this is the same in nursing; see McCray 2009: Chapts. 1–2). Therefore, the care service process is an organisational mechanism in which analysis of the causes and elements of problems, synthesis with care planning, monitoring and assessment of changes, and feedback to the planning process are harmoniously required throughout the process of determining eligibility, diagnosis, prognosis, and clinical treatment. This is a process of learning and feedback between service providers and users, in other words, a process of co-creation of matching the service to the case as well as possible through a concurrent service engineering approach.

This nature of the care service process suggests that care service is not based on the logic of commodities, where value in a certain specified dimension, like price, and a division between purchaser and provider are critical. Nor is the logic of care service well defined in the way ‘goods vs. service’ or ‘commodity vs. service’ are. Rather, the logic of care service operates consistently through the care process, even during the phase of determining the proper way of providing support goods and expertise to users in need of professional support to achieve their goals. Care service is a management process for supporting the user in his or her autonomous efforts to improve his or her health.

18.4.2 Moral Games

Because care is a practice that supports the user in his or her desire to meet his or her needs, the user and his or her family should implicitly or explicitly communicate the worth and convenience of expertise when receiving it. This is why needs analysis, service planning, sharing case information within a case conference, and monitoring are crucial for managing service processes rationally and for keeping risks to a minimum.

On the other hand, expertise must adhere to institutional rules regarding the goals and manner of service provision. Activities are also structured according to moral assumptions regarding their goal and the technical rationality of specialised care. What often happens during service planning and establishing the method of service provision is that they are shaped by the provider's assumptions, not the user's assumptions regarding his or her needs. The effectiveness of service outcome is often assessed according to pre-established criteria, and the utility for the user is easily interpreted according to an expert scale of improvement in physio-mental status, rather than being observed.

Therefore, care service is a moral game between user and experts, where service interfacing, processing, and outcome measurement are first and foremost elements of the politics of the properness and rationality of service engineering management. A moral game even exists between the user and his or her family and between different experts within a team. A sophisticated approach to arriving at best solution through this political game is ‘multi-dimensional evidence-based practice’ (MEBP), a way of acquiring various types of knowledge to be synthesised and of coordinating values (Petr 2009: Chap. 1). MEBP is a seven-step model for articulating multiple resources and knowledge from different perspectives on value criteria and actualising the best practices for the user, as well as for the various experts involved in the case.

Another approach, in addition to the service outcome satisfaction model, is networking based on *de-domesticity of target* and *spontaneity of expertise* (Ogasawara and Shimazu 2007: 133–135; Ogasawara 2009: 183–184). This approach is based on integration theory. De-domesticity of target means considering the goals of care service from a de-domestic perspective, as a way for stakeholders to share the value created by service processes and, with the exception of newly produced knowledge and innovated value, to understand to not privatise any fruits of a successful outcome to the individual participating experts. Spontaneity of expertise means autonomous lessons based on inviting cooperation from a neighbouring profession in order to achieve the goal of care in the best possible articulated way. This is a way of interpreting the logic of networking in the field of medical and care services, and is also a valuable approach for solving the politics of the moral game.

This is why a managerialist approach, including the concept of professionalisation (Clarke et al. 2000) is particularly attractive for interpreting the reality of care service process.

18.5 Re-interpreting Managerialisation and Managerialism

Managerialisation and managerialism are familiar concepts of manipulation in the New Public Management approach to public services. They are, therefore, nothing new in terms of the methodology of emphasising cost efficiency, input–output productivity, and compartmentalisation of organisational behaviour in public services. Managerialism is also a strategic concept for enhancing motivation of management and professionalism in public sector staff.

Through the process of managerialisation, the notion of ‘management’ or ‘manager’ has widely penetrated cost-control solutions, and is also used as a tool for bringing transparency to public service (Turpeinen and Haapasalo 2010). The notion is also applied in care service processing and effectiveness control at the very lowest levels, like ‘community care manager’ in the UK or ‘care manager’ in Japan. A Japanese ‘care manager’ has wide discretion in the elderly care service process. Without the ‘care manager’, care service planning and the clinical process do not proceed.

Managerialisation is thus necessarily combined with professionalisation of core expertise in a 'manager' who ensures the rationality of integrated provision of care. This means managing care is no longer neutral in terms of the basis and strategy of care and its estimated value. Rather, managing care service is the process of imagining and imaginary experimenting with how to get the best mix of knowledge and skills for the processes of service creation and practice.

Although standardisation of care methods and fundamental skills is advocated especially by those vocational associations representing qualified expertise, professionalisation in designing service plans and processes is clearly the core for getting high-quality management in care. A certain category of knowledge and skills needs to be standardised to make risks visible and to ensure the criteria of service activities, so that users are protected from inexperienced or de-skilled service practices. Standardisation is also necessary to clarify the purpose and objective of care from the perspectives of universal norms and moral rules. Standardising ethical requirements and the infrastructural base of care capabilities is always mandatory to ensure a recognised quality standard for users. Standardisation, on the other hand, tends to facilitate a modular service composition and compartmentalisation of care, which easily leads to provider-driven or expert-centric practice. Therefore, to enhance creation of user-driven care service, both elements of technical standardisation and human design have to be integrated into the practical process of service provision. A critical fact here is that it is always in the user's interest to drive this integration of technical rationality and solution design.

18.6 Values to Be Fulfilled and Skills

18.6.1 Dimensions of Values

The best mix of knowledge and skill adoption is based on the value-function matrix. In this matrix, the user's needs are manipulated as dimensions of values to be fulfilled, as suggested in Fig. 18.1. There are several different ways of classifying human needs or wants. Akio Kameoka proposed a five-tier model of satisfying service user wants (Kameoka 2006). In our qualitative research on the daily life of demented elderly people, we found that the realisation of the self as the highest level of satisfaction in Kameoka's model is not an independent variable apart from satisfying other lower-level wants, but rather the consciousness of being satisfied is the result of the other values being fulfilled (NCSIRU-TFU 2010: Chap. IV). Therefore, four relative dimensions of values are appropriate for classifying the goals of care to fulfilling the client's values through its functionalities, as suggested in Fig. 18.1.

The value of sustenance is absolutely necessary: receiving food, clothing, bathing facilities, sleeping space, protection from unbalanced nutrition, and so on – the fundamental matters of bed and bread. The value is communicated as a need for

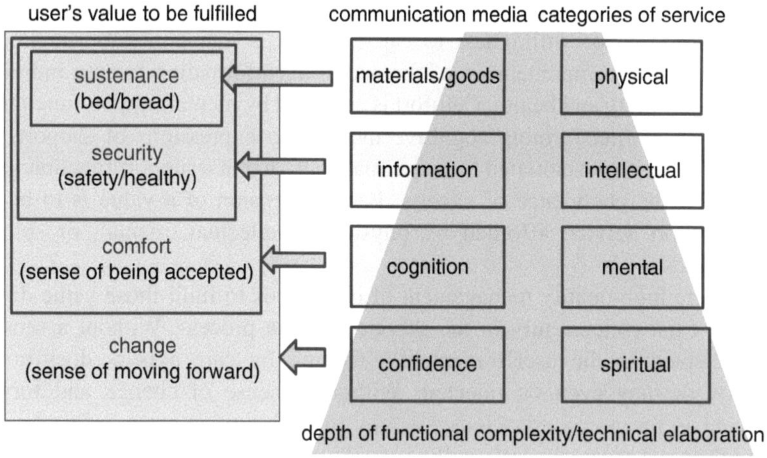

Fig. 18.1 Value-function matrix

goods and staff. The second tier of value is security: not to be harmed or exploited by somebody else, or security against violence and being neglected, or against misinformation or repeated behaviour that is disruptive to mental or physical health. Accurate information on timing, useful information that enhances behavioural change, rational knowledge on risks, and so on, are needs that should be provided in a transferable way here. The third value, a more implied one, is the comfort of being accepted, being touched, or being included, which is a cognitive and sensory sphere of human mentality, but certainly a critical value in and of itself. The fourth and scarcely observable value is change, which is a sense of moving forward or making improvements. This value is expressed as an emotional frame of confidence. Dignity, as the highest value of human life, is the apex of physical, mental, socio-cultural, and spiritual states in terms of those dimensions of values that are satisfied, so that dignity without sustenance, without security and comfort, and without a sense of change, can only be described as in vain.

18.6.2 Skills

There are communication media that correspond to the values through which the values are observed with expertise: materials, knowledge, cognition, and emotion. Service is the methodology and process of co-working and supporting someone's need to fulfil a value, and it is a characteristic of support service to allocate material communication media to those in need in a rational way.

Those communication media are transformed by skills into practical services. Sustenance is met by the support of a team of experts concerned with transforming material media into feasible service practices. Meals or drinking water are supported by the quality of care staff's skill at assessing the individual ability to, for instance, swallow, or willingness to eat. Security is maintained by intellectual interpretation and comprehension of the necessary information to ease individual worries about safety and health. Comfort is achieved by mental support functions in the expertise of transforming cognitive media into expressions of support. The sense of change is substantiated by a spiritual approach in a support function that is enhanced by the confidence of experts. Each dimension of a value is to be met by the support service afforded by physical, intellectual, mental, or spiritual functionalities.

Designing high-quality management of care needs to fulfil those value dimensions that exist concurrently in the service practice process. Without a sense of being accepted in the user's mentality, feeding by care experts does not go smoothly, or may even be rejected. Without a sense of change and forward movement in the user's consciousness, any provision of information on keeping healthy does not create a user's willingness to understand. The actual care service process has functional complexity, and the depth of complexity gradates from a one-dimensional service that fulfils a single value to four-dimensional services that simultaneously satisfy multiple values through elaborate technical rationality. Therefore, designing care service utilises sensory and emotional perceptions, or tacit knowledge, to formulate, and technical rationalisation to compose and roadmap. Design in this sense and of this structure inevitably takes the form of a concurrent creation process through co-working between service user and care manager and other experts, who mutually interpret relevance and technical feasibility.

In terms of skill development, incremental innovation is carried out day by day through step-by-step learning in order to decrease the degree of risk and increase user satisfaction. Deeper, disruptive innovation is enhanced through such service processes, as multidimensional skills need to be utilised to fulfil the user's complex values and a number of different competences combined to contribute to the care process. Even incremental innovation is based on the notion of 'client centric'.

18.7 Service Innovation

18.7.1 The Logic and Structure of Care Service

In the historical context, 'care service' has a specific logic and structure. Modern care service covers a wider range of practices, from preventive health promotion to terminal care, as well as incidental care. It also covers physical practice, mental health care, dementia care, spiritual and grief care, and even care for families, in the domains of home, community, facilities, and hospitals. For the best care, care professionals should acquire the practical knowledge and skills to be able to seamlessly collaborate with medical staff, rehabilitation experts, nurses, and

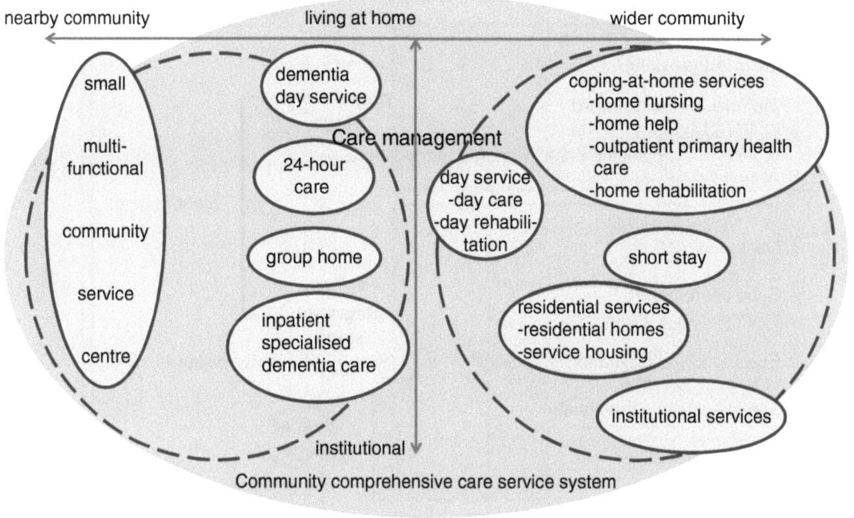

Fig. 18.2 Current elderly care/social services complex (community comprehensive care service net)

clinical psychologists. Therefore, there has been continuous pressure to upgrade the quality of competence of the people involved and expand functional flexibility.

There has also been a structural change in the care provision system toward more targeted activation, to community-based, integrated utilisation of resources and a coordinated service path. Figure 18.2 is a conceptual map of care services' functional allocation; it shows the network of various units of service in an integration framework. This net also features allocation of areas of liaison between authorities in the medical and co-medical fields, public health, and housing, and so on. The net is an easier-to-access and higher-usability complex of care services for the user, so that the needs and convenience of the user can be met by a nexus of services closer to the user's daily life and with a shorter wait for support. The net also enhances the capacity of care managers and care workers to expand the flexibility of their knowledge and skills to support the user's needs throughout his or her life.

Therefore, the logic and structure of 'care service' as a system should be seen as total management within the framework of a comprehensive community care network.

18.7.2 *Process and Mechanism Innovation*

The care service system is also structured as a complex of tasks. Each task is structured by a series of actions, each of which is based on elementary skills and knowledge. Taking the case of normal elderly care in an ordinary care facility, nine tasks, from identifying a case to successful conclusion of a case, form a service flow

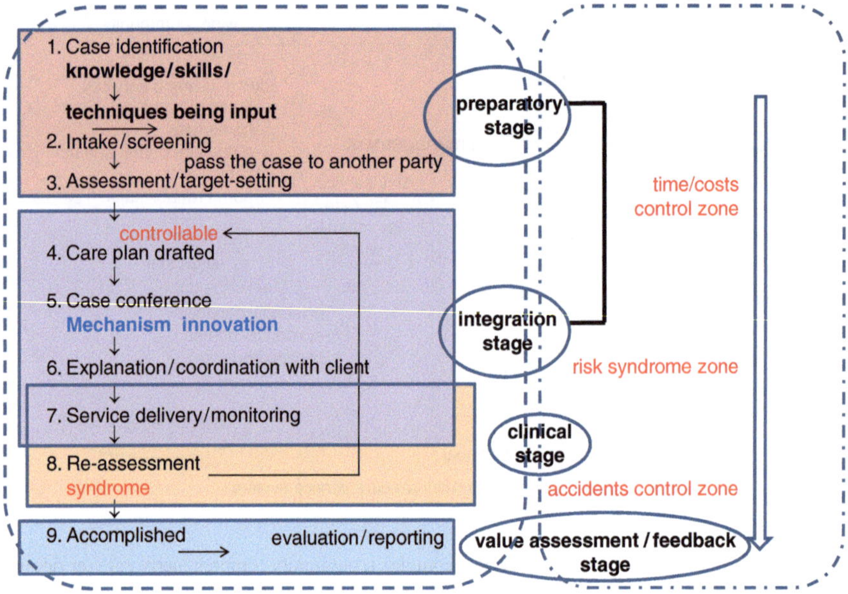

Fig. 18.3 Service process and mechanism innovation

process, as described in Fig. 18.3. Tasks 1–3 comprise the fact-finding and target-setting process, the preparatory stage prior to starting service delivery. At this stage, tasks are for the most part individually executed by the social worker, care manager, or public nurse who identify the case, conduct hearings, and observe the user's circumstances and strength, assessing the case for documenting the materials considered when drafting a service plan. The competence and experience of the experts involved are key factors for establishing the quality and effectiveness of this preparation.

Tasks 4–7, the care planning and service delivery process, are categorised as the integration stage. At this stage, a care team is built to share information, establish targets and a service roadmap, assign roles, and commence service and monitoring. Thus, the integration stage is for the most part collective work based on expert insight. Tasks 7 and 8 are service delivery and the reassessment process through which the contents and method of service are adjusted as needed in response to effectiveness, usability, fluctuations in the user's mental or physical status, and changes of circumstance. After one service span ends, reassessment is undertaken to the consider effectiveness of the past span and the goal of the next span. Tasks 7 and 8 are the core of whole service process, the stage of clinical teamwork. Task 9 is case conclusion through evaluation and reporting. This is the stage for clarifying what values have been fulfilled, and to feed the relevance of the service process and the values fulfilled back into improving care methodologies and designing processes for enhancing societal innovations.

During the preparatory stage, various experts itemise the many tasks to be undertaken: consultation and reporting; fact finding; observing the physical, mental, and spiritual liveliness of the user; evaluating the user's strength in everyday circumstances; assessment of the user's financial status; hearing the user's and family members' preferences regarding types of support; checking resource availability; planning hypothetical allocations in terms of time and staffing; and so on. From a service engineering point of view, these different tasks should relate in a path diagram fashion, in order to achieve quick, lower-cost adjustment to the user's needs. This stage is also critical to ensure high-quality flow and proper sharing of information by and among those experts involved in the integration stage, so the preparatory stage is the stage of feed-forward control of the integration process.

The integration stage starts bringing the diagram to fruition through the drafting of a care service plan. In the Japanese system, this drafting is done by a single care manager for each 90-day span of service provision. A care conference with the team of experts is held to share the goals of the case and to confirm the service plan. The feasibility of the plan is thoroughly examined during the conference. Thus, the plan drafting by the care manager and the subsequent team conference are procedures for collaboratively minimising foreseeable risks and hardships.

The integration stage also includes the care manager's explanation of the coordination procedure to the user and his or her family members. This is the time for the care manager to participate in the moral game. Having the discussion and motivation of the care team behind him or her, and understanding the technical rationality of the case, the care manager should gather the values of the user that need to be fulfilled and the moral justifications of the user that need to be supported. This complex game of verification by the care manager and his or her skill for gathering feedback and adjusting the team is the key to minimising risks and troubles in the process of care provision. Balance in consideration, adjustability according to merit, and rapidity in identifying the logic that will be accepted by the team members are the key factors to reducing the time and costs involved in kicking off service deliberation.

At the beginning of service delivery, room for adjustment still exists in terms of multi-expertise collaboration or a service's delivery process. Therefore, time and cost control are still part of the beginning of the clinical stage.

The clinical stage is the stage of the accident syndrome. Accidents may happen for various reasons: deficient risk identification communication gaps, lacks in preparation, bad feedback, or fluctuation in the user's conditions. Most accidents cannot be managed by simple feed-forward control. The ratio of accidents can be decreased when the care workers have a capacity for slight forward understanding. Understanding forward means observing a user's irregular moves or alterations in physio-mental condition. Quick adjustments to or replacement of service functions is, therefore, critical to reducing the accidents caused by shortage of skills or misleading information.

The value assessment and feedback stage is the stage for providing the drive for process innovation and for improving the service process in subsequent cases.

It is also the process of inputting renewed knowledge and even wisdom through which professionals can adjust to continuously evolving care methodologies.

Care service has this continuous stepping-up structure, where any innovation, either incremental or disruptive, is driven by efforts to create smarter functional cooperation, shorter allocation time, and lower costs, including risk basting. Therefore, smart engineering of roles and functionalities as well as efficient management of the cost-consuming process are the key methodological assets from a service management viewpoint.

The user-driven viewpoint is closely connected to this methodological asset in service management. User-driven innovation is based on the assumption that, the closer to the user's real needs we focus, the deeper we drive a value-creating innovation. As an innovation methodology, it contrasts with seed-driven innovation, which has resulted in a lot of wasted products or functions that are technically attractive, but are not successfully embedded in everyday life. User-driven innovation is a methodology of concurrent engineering by co-creating with users with specific needs, supporting them in meeting goals and fulfilling values or overcoming difficulties. In user-driven innovation, therefore, support and creative services offered by experts are always an indispensable part of transforming needs identified through close observation into technical feasibility through design. This improves accessibility, usability, and effectiveness by saving time as well as reducing risks and costs.

In applying the user-driven methodology to service innovations in elderly care, the methodology should enhance (1) quicker processing and closer communication in the process of diagram management at the preparatory stage, (2) deeper sharing of the target and smoother understanding of critical elements in the team care process at the integration stage, (3) just-in-need and on-time adjustments to problems cased in the clinical service delivery process, and (4) clearer feed-backing of the results of case evaluation to the service innovation and skill formation.

A user-driven methodology certainly enhances innovation of care service processing and process management, but it moreover stimulates innovation of the service mechanism itself by accumulating and integrating knowledge, by creating new skills, and by renewing the design techniques by which services are integrated into a complex. Innovation in any social mechanism, on even a relatively small scale such as elderly care service processes, extends its effects to other mechanisms. If a new care skill is innovated and innovations reached under a certain level of mass pressure to renew pedagogical contents, then education and training programs to develop the skills of professional care workers evolve. If an idea for designing care service processing is innovated, then the institutional system for allocating monetary resources and the method of evaluating the value and price of services should be reconsidered. The merits of user-driven innovation lie, thus, in two impacts: processing innovation in and of itself and mechanism innovation, with its extended influence on other societal innovations.

18.8 Capabilities and Skill Formation

18.8.1 The Importance of Fundamental Generic Capabilities

Service is a way of co-working with someone to achieve one's personal goals and well-being, and service should be innovated in a more scientific way, integrating user-driven methodology, to make the process as well as the mechanism of service provision more intelligent. Co-creation with the user is based on an *abduction approach*, which is neither a client-first induction approach nor a provider-first deduction approach. The professionalisation of support care provision and the managerialisation of the service process stem from designing the capacity to optimise this value-function linkage in a complex context. Therefore, the structure of the capacity in care expertise is also a critical factor in user-driven innovation.

There are generic capabilities in service work that are fundamental to human nature and familiar before the acquisition of taught knowledge, such as logical thinking, or trained acquisition, such as attitude and demand. These generic capabilities are instinct-oriented and sharpened through practice. They are the capability to relate with others and socialise oneself, the capability to comprehend and discipline one's self within a context of relations, the capability to determine the tasks being faced and to try to solve them, and the capability to co-work with others in the solving process. These are not capabilities that can be taught, but they can be activated through learning by self-practice. Supportive co-learning or outside guidance may be necessary, but this activation is still a process of self-acquisition.

As showed in Fig. 18.4, generic capabilities may be termed as indispensable capabilities for social and relational adaptation, which are the very basis for successful co-creation with the user and co-working within a team. Even if you have task determination and solution planning capability, you cannot co-create with

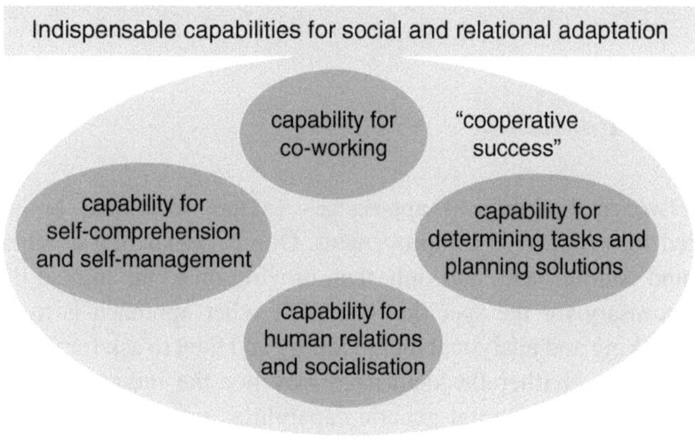

Fig. 18.4 The fundamental generic capabilities

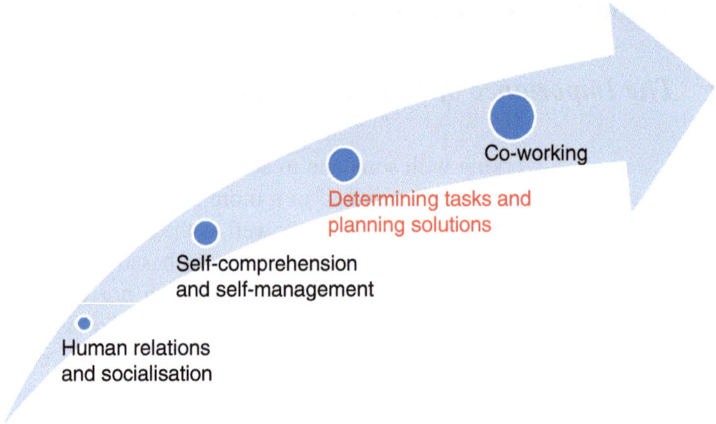

Fig. 18.5 Process of acquisition

a user if you lack relational and socialisation capability. If you have socialisation capability, but lack the capability for self-comprehension, you cannot co-work with others. These four capabilities are related to achieving cooperative success in the logical order described in Fig. 18.5.

To socialise with and discipline one's self for the purpose of developing continuous relations with elderly people, one must first take care of the precondition of close observation and determining the tasks that need solving. If the tasks are identified, planning of the solution process follows. In solution processing, co-creation with the service user and co-working with others has already begun. Professionalisation of care service expertise means bringing these capabilities from fundamental human nature to creating the first contact with the reality of elderly people's existence as such, and to try to truly understand one's role in working with them.

18.8.2 Skill Formation

There are two methodological approaches to the vocational knowledge and skills needed for productivity improvement. One is demanding a general attitude of service and commitment, and only then moving on to the logical thinking and creativity necessary for the specified task. The other approach is to first request theoretical thinking and analytical methodology and then to ask for the competency to achieve targets. Neither the deductive way nor the inductive way takes the indispensability of fundamental generic capabilities into account, leaving them a type of individual asset that everyone should already exhibit by the time of recruitment.

In health and social services, especially in care service, skill in person-to-person interfacing does not necessarily develop with length or type of experience. Different professionals who have performed the same job for the same length of time will exhibit wide differences in patterns of skill development (Ogasawara 2000; 2001 and 2002).

Figure 18.6 shows the skill formation variation among approximately 600 care managers with almost the same length of experience. The direct tasks are those care service planning and provision tasks that belong to the integration stage and clinical stage as shown in Fig. 18.3. Indirect tasks include consultation and advisory work as described in the preparatory stage in Fig. 18.3, documentation work, and administrative work. The figure suggests that, even at the same level of indirect work capability, there is a wide range of variation in skill levels in direct work, at least in the self assessment of each sampled care manager. Figure 18.7 shows skill

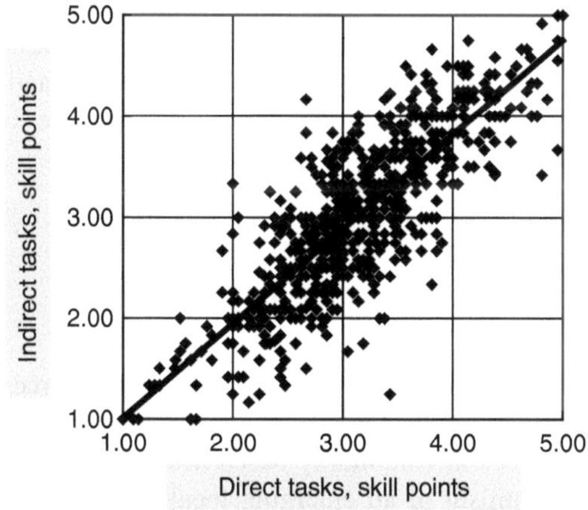

Fig. 18.6 Variation in care managers' skill formation

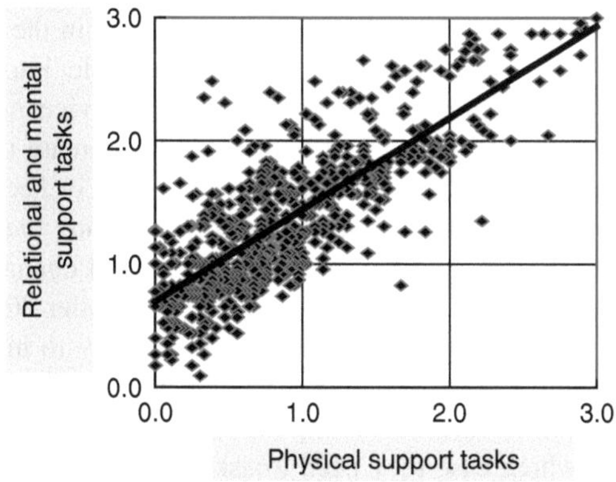

Fig. 18.7 Variation in home helpers' skill formation

formation variation among approximately 2,000 home helpers with different length of experience. Physical support skills relate to those support service activities – such as nutrition, moving, bathing, brushing teeth, activities, and so on – that are almost always part of the dimensions of sustenance and safety as depicted in Fig. 18.1. On the other hand, relational and mental support skills have to do with fulfilling values in the dimensions of comfort and change in Fig. 18.1, including technical advice offered to domestic care workers and regulation of the relationship between the user and his or her family members. The figure once again demonstrates rather widely scattered differentiation in skill formation.

Although there is a difference in background vocational qualifications for being a care manager, and Fig. 18.6 includes certain reflections of this difference, there is no difference in home helper background requirements. Instead, there is differentiation in the length of experience as a home helper, from beginners to veterans with careers of over 20 years. But the crucial fact here is that, despite career differences, there are still different patterns of skill formation at the individual level.

This evidence clearly indicates that fundamental basic capabilities are continuously trainable after recruitment and during work processes. Therefore, in care service work, co-creation and co-working are critical and indispensable as conditions for continuous improvement of a worker's generic capabilities and balanced skill formation.

18.8.3 Generic Capabilities and Service Nets

This phase of human capabilities is also the core driving force of networking to create service nets bringing together various types of expertise. Based on the case analysis of good practices in community-based health and social service nets, we have clarified the dynamism of an emerging local network of different types of expertise. As has already been mentioned, 'de-domesticity of target' and 'spontaneity of expertise' are the keys to successful net creation. Posing and sharing such targets de-domestically means that the target in and of itself is commonly acceptable, and not for merit-taking by individuals participating in the net, and that as a result the fruit is commonly valuable, and not dividable into each individual collaborator's pockets. Unlike current business nets, service nets in health and social care tend toward integration, not cooperation, to create new public value, like a decreasing number of elderly people staying in bed or keeping functionally impaired people from becoming socially isolated. Therefore, the target in itself is integrated in nature. It is almost the same logic of radical change as exists in the 'emerging value system-emerging new business nets' (Möller 2006: 989).

The other factor for creating service nets is spontaneity to invite experts from other fields to integrate efforts for the purpose of creating a comprehensive and continuous complex of services. Real experts have a real understanding of what they can do, and in which case they need to ask for others' help. A care network emerges based on the spontaneity of one expert eager to contact the other leading nodes and create a core network and then extend to the broadest range of community throughout the core network. The driving forces of this de-domesticity and spontaneity are close observation and commitment to the service user's real needs and values waiting to be fulfilled (Ogasawara 2009: 183–184). And we may conceptualise this kind of service net as a wisdom cluster for overcoming commonly faced difficulties and inducing societal innovation.

18.9 Creating a Model of a Comprehensive Community Service Cluster

In comprehensive community care, care service tends to be integrated with health and social services. As the support functionality for autonomous living, care service is also embedded in a community's critical path system. Care service also deepens its functional unity through social work, especially in the field of mental and psychiatric clinics, and also has an increasing role in dementia and terminal care. Care service is, thus, evolving as an articulated subsystem within the comprehensive community system of preventive, clinical, recovery, sustaining, and terminal support services.

The idea of 'community comprehensive' service has emerged, at least in the Japanese context, through the reformative initiative for creating an effective and efficient liaison between related services on the basis of small-scale geographical units, based on an individual junior high school area. The proposed is basically a provision-driven restructuring process. The action research this chapter was based on was conducted as alternative experiment to designing a prototype for a community comprehensive care service cluster closely based on users' and families' needs and values. The cluster is an R&D and innovation cycle mechanism that ensures user-driven innovation and co-creation by integrating knowledge resources within the community (Fig. 18.8).

The key functional factors of the cluster are, firstly, to set the needs-observation and – perception gateway as close to the users' daily life as allowed by the functional capacity of the community comprehensive care centre, so that the reality of the users' values and expectations are taken into the R&D process as smoothly and quickly as possible. The Community Comprehensive Care Centre (4Cs) is institutionally expected to perform an 'all-in-one' role within an individual junior high school area through the teamwork of an experienced public nurse, social worker, and supervisory care manager; through a long list of consultation and support programs funded by the municipality's long-term care budget; and through community networking. This commits the team to the daily lives of local inhabitants, from initial prevention to long-term care and from coping-at-home activities to coordinating intake into care facilities. Therefore, perception of needs ought to be timed pro-actively, not reactively.

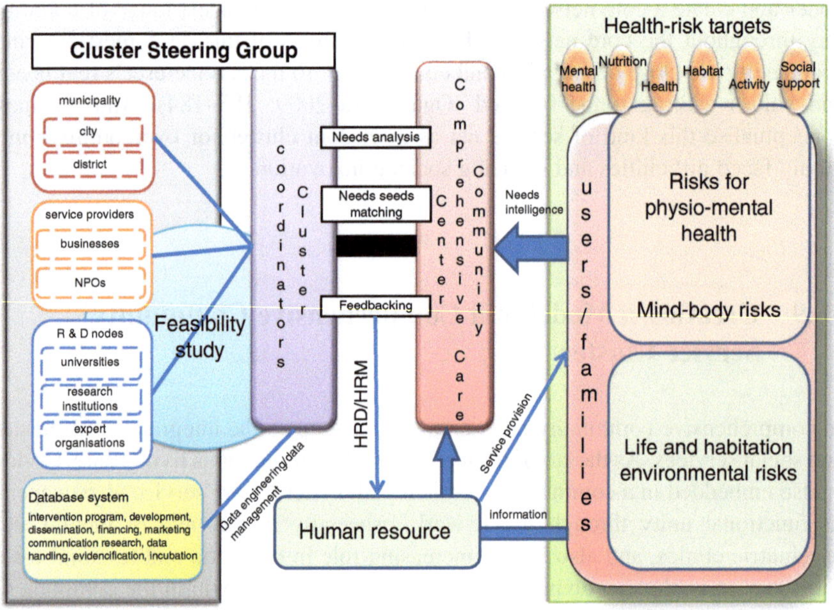

Fig. 18.8 Comprehensive and continuous innovation & incubation cluster (TFU Model)

Secondly, there is a need to benchmark the needs observed by the 4Cs and selectively transfer to the co-creation stage. The cluster should have a coordinating role, transforming those perceived needs into analytically received tasks for innovation. So the roles expected of the cluster coordinator are to benchmark and receive tasks from the 4Cs, to input tasks into an R&D consortium in a way that stimulates seeds and knowledge integration for an innovative solution, to coordinate the testing field for feasibility studies and databasing, to coordinate the process of transforming innovative solutions into service **appreciation** and skills, and to pass the fruits of innovation to 4Cs staff to implement in practice. The testing itself is a concurrent engineering process, where the 4Cs staff and users are already co-working.

Thirdly, it is critical to have a network foundation, a steering committee composed of leading experts in academia, business, the public sector, and the volunteer sector. This is not just a place of discussion, but rather an integrated, driving power for innovation based on de-domesticity of purpose and spontaneity or vulnerability of professionals. It is a kind of trusted core net extending its feet and roots into local wisdom.

Fourthly, broadly but clearly delimiting the domain of life-related risks is key to the cluster. Comprehensiveness does not mean taking every life-related problem faced by the user into consideration. The cluster focuses on enhancing a healthy life, not a successful life, so the concept of health concentrates on fundamental risk areas: mental health, nutrition, physical health, habitation, activity, and socialisation. As to risks, two comprehensive categories are established: physio-mental or mind-body risks and risks in living circumstances. The concept of health applied and categorisation of risks are based on the adoption of the four dimensions of values set in Fig. 18.1.

Finally, the action research to form the service cluster and feasibility study has been possible within the framework of the national 'Knowledge Cluster Initiative' project, for the purpose of enhancing integration of local knowledge and driving the forces of innovation, and this particular case is the only project assigned to community service clustering. Therefore, national funding support and exchange of know-how and ideas with other locals are mandatory. It is up to the city government to lead application of this research by collaborating with universities and local agencies, and also to coordinate the pre-conditions for the feasibility study by inviting the participation of administration-related service centres and agencies.

18.10 Summary and Conclusion: Reflections on the Policy Formation

The macro-micro distinction is a prevailing methodology in policy science. In the era of community comprehensive care, the mezzo level perspective is at the core of resource allocation planning and care system management. The de-centralisation agenda and community governance on the basis of social capital are the focus, as if those were the keystones to a new way of structuring society in the post-welfare state. Over the last three decades, there has been a long-lasting methodological transition from centralised redistribution systems to a community-based system of local feasibility. But there is also a widely shared feeling during the current global downturn and elderification of postmodern society that community- and municipality-based welfare systems easily suffer in terms of inequality and differentiation in the welfare people receive, from budgetary allocation for support services to increasingly large holes in existing institutional preparation.

So the task for social designers is to draw up a blueprint of a comprehensive community support system that consists at least of comprehensive human services composed of health, social, and care services in a consistent way; a comprehensive care service system offering support functions for life security, effective opportunities for preventive health and socialisation, and flexible services for living assistance, including human rights protection; and highly qualified housing services within the community.

User-driven innovation is also the way to create wisdom for designing a social system for smart living. Therefore, to embed in the local community a mechanism for observation-based perception of values, for a cluster-type consortium of solution processing, a feedback system to fulfil users' values should always exist together with the societal system design at the level of community and municipality, and should also exist together with new roles in central government and its agencies to back up and enhance societal innovation through co-working with municipalities and even individual projects. The so-called triple-helix model used in Finland for leading successful, united innovation is just one leading case, while analysis is underway to watch out for certain elements necessary for the next step of driving innovation (Sabel and Saxenian 2008; Teperi et al. 2009). The Japanese way of placing emphasis on community strength and a locally controlled social insurance system may be one way in a social context that has relied heavily on a publicly controlled service purchasing-providing system.

The critical thing here is not the types of integration at the three levels nor how roles are divided among the levels, but that the needs observed and received are not only needs from the reality of everyday life and a true expression of the values to be fulfilled though service process, but that thy are also always interpreted into policy needs that are categorical conceptualisations of the tasks for society to perform through allocating societal efforts and material resources based on a certain socially shared criteria of fairness. Therefore, needs are what service experts should take care of, and at the same time what the policy should take into consideration as a societal necessity.

References

Bendelow, G. (2009). *Health, emotion and the body*. Cambridge, UK: Polity Press.

Bond, J., & Corner, L. (2004). *Quality of life and older people*. Buckingham: Open University Press.

Boylan, M. (Ed.). (2008). *International public health policy and ethics*. New York: Springer.

Clarke, J., Gewirtz, S., & McLaughlin, E. (2000). *New managerialism, new welfare?* London: SAGE Publications.

Kameoka, A. (2006). MOT and service innovation. Resource document. http://www.research.ibm.com/trl/projects/sis03_2006/materials/20060310_ServInnovSymp_MOT_Kameoka.pdf. (In Japanese.) Accessed on 14 July 2011.

McCray, J. (2009). *Nursing and multi-professional practice*. London: SAGE Publications.

MHLW (2010). New managerialism. Tasks for the care programme. Ministry of Health, Labour and Welfare. www.mhlw.go.jp/shingi/2010/06/dl/s0615-6e_02.pdf. (In Japanese.) Accessed on 14 July 2011.

Möller, K. (2006). Role of competences in creating customer value: A value creation logic approach. *Industrial Marketing Management, 35*(8), 913–924.

Möller, K., & Svahn, S. (2006). Role of knowledge in value creation in business nets. *Journal of Management Studies, 43*(5), 985–1006.

NCSIRU-TFU (Nursing and Caring Skills Integration Research Unit at Tohoku Fukushi University) (2010). Integrating nursing and caring skills in the impaired elderly care. Final report presented to the Mizuho Foundation of Education and Welfare Research, March. (In Japanese.)

Ogasawara, K. (2000). Kaigo-Kanren-Bunya ni okeru Koyo-Noryoku Kaihatsu Shishinn no Sakutei ni Kansuru Kenkyu [Knowledge and skills development guidelines in elderly care]. The first year research report presented to Ministry of Health, Labour and Welfare. http://mhlw-grants.niph.go.jp/niph/index.html file list 00000029A0001.pdf. (In Japanese.)

Ogasawara, K. (2001). Kaigo-Kanren-Bunya ni okeru Koyo-Noryoku Kaihatsu Shishinn no Sakutei ni Kansuru Kenkyu [Knowledge and skills development guidelines in elderly care]. The second

year research report to Ministry of Health, Labour and Welfare on grants in aid. http://mhlw-grants.niph.go.jp/niph/index.html file list 200100017A0001-0018.pdf. (In Japanese.)
Ogasawara, K. (2002). Kaigo-Kanren-Bunya ni okeru Koyo-Noryoku Kaihatsu Shishinn no Sakutei ni Kansuru Kenkyu [Knowledge and skills development guidelines in elderly care]. The third/final year research report to Ministry of Health, Labour and Welfare on grants in aid. http://mhlw-grants.niph.go.jp/niph/index.html file list 200200070A0001-0015.pdf. (In Japanese).
Ogasawara, K. (2007). New way to cost and profitability of health and social services. In H. Erjanti (Ed.), *Refurbishing elderly care – strategy of cost efficiency in theory and practice*. Vantaa: Laurea Publications.
Ogasawara, K. (2009). Securing liveliness of the elderly. In H. Erjanti & K. Ogasawara (Eds.), *Refurbishing elderly care – The new stream and organizational transformation in Finland and Japan*. Vantaa: Laurea Publications.
Ogasawara, K., & Shimazu, N. (2007). *Chiiki Iryou-Kaigo no Nettowaku Kousou [Community network strategy of medical and care clinical services]*. Tokyo: Chikura-Shobo.
Ogasawara, K., Saito, H., Hagino, H. (2011). *Sekkyokuteki Kenkyo-Shakai Sa-bisu no Kousou –Finrando-Moderu tono Taiwa [Vision for positive health and social service dialogue with the Finnish model]*. Sendai: Fukushi-Kobo. (In Japanese.)
Petr, C. G. (2009). *Multidimensional evidence-based practice – synthesizing knowledge, research, and values*. London: Routledge.
Prahalad, C. K., & Ramaswamy, V. (2004). *The future of competition*. Boston: Harvard Business School Press.
Reed, J., Stanley, D., & Clarke, C. (2004). *Health, wellbeing and older people*. Cambridge, UK: Polity Press.
Sabel, C., & Saxenian, A. (2008). *A fugitive success – Finland's economic future*. Helsinki: Sitra.
Suzuki, R. (2009). Health promotion through community exercise programs. MA thesis at Graduate School of Comprehensive Welfare Studies. Sendai: Tohoku Fukushi University.
Teperi, J., Porter, M. E., Vuorenkoski, L., & Baron, J. F. (2009). *The Finnish health care system: A value-based perspective*. Helsinki: Sitra.
Turpeinen, T., & Haapasalo, K. (2010). Transparency of Finnish social and healthcare services: Maisema model – experiences and results. Presentation at TFU International Workshop: Integrated Preventive Health and Social Services – Sendai Service Cluster Model in International Perspective, 25–26 February 2010, Tohoku Fukushi University, Sendai, Japan.
Verganti, R. (2009). *Design driven innovation – Changing the rules of competition by radically innovating what things mean*. Boston: Harvard Business School Press.

Chapter 19
A Holistic Model of Innovation Network Management: Action Research in Elderly Health Care

Timo Järvensivu, Katri Nykänen, and Rika Rajala

Abstract Network management, in particular innovation network management, is a topic of increasing interest and scope. Research on innovation network management has offered various, but ultimately partial, theoretical and practical contributions. Trust and commitment have been identified as the basic elements of a functioning network, and we know that there are certain factors that foster or discourage their existence. Networks are different; each different network has its particular challenges. Indeed, practice-based innovations involve specific challenges for network management. Our aim is to look at innovation network management from a holistic perspective, bringing together the relevant but scattered viewpoints and contributions. We use action research to look at what managers can do to manage an innovation network. The resulting holistic model rises from one particular practice-based innovation context – elderly health care in Finland – but we argue that it is applicable in other contexts and innovation types as well.

19.1 Introduction

Network management is a topic of increasing interest and scope (Dhanaraj and Parkhe 2006; Ritter et al. 2004; Järvensivu and Möller 2009; Hibbert et al. 2008), also in the field of health care (Provan and Milward 1995; Provan et al. 2004; Nykänen et al. 2009). Orchestrating innovation networks is one particular field of this research (Dhanaraj and Parkhe 2006; Möller et al. 2005). Innovations are often thought to arise through processes that begin from basic research and are followed by applied research and development. However, quite often innovations have their origin in practice-based processes, defined by co-creating networks and intense knowledge exchange between users and developers that takes place at the point of

T. Järvensivu (✉) • K. Nykänen • R. Rajala
School of Economics, Aalto University, Helsinki, Finland
e-mail: timo.jarvensivu@aalto.fi

H. Melkas and V. Harmaakorpi (eds.),
Practice-Based Innovation: Insights, Applications and Policy Implications,
DOI 10.1007/978-3-642-21723-4_19, © Springer-Verlag Berlin Heidelberg 2012

actual usage (Dougherty 2004; Harmaakorpi and Mutanen 2008). This accentuates the importance of high-quality network management in the context of practice-based innovation.

Hibbert et al. (2008) recently offered a review of inter-organisational management research. They argue that the field has much to offer, but the contributions are scattered, and we lack a comprehensive management theory for inter-organisational networks. We need research that looks at inter-organisational management from a more holistic perspective, bringing together the relevant viewpoints. Our aim in this study is to follow this path.

Järvensivu and Möller (2009) recently introduced a metatheory of network management, with the aim to provide a framework for more comprehensive research in this field. They identified four contingent layers of network management research: socio-economic context, functions, tasks, and roles. While this metatheory helps us to locate our study in a wider metatheoretical framework – our study focuses mainly on the tasks of network management – the metatheory remains at a rather conceptual level. We aim at taking the discussion to a more pragmatic level.

We ask a simple question: What should a network manager do to improve and facilitate practice-based innovation? Our aim is therefore to formulate a holistic model of network management that managers can follow to improve the effectiveness of an innovation network. Although we search for a model of network management, we maintain the understanding that the phenomenon of networking is inherently complex and that networking situations are idiosyncratic (Hibbert et al. 2008). Our empirical focus is on elderly care in the Finnish context.

19.2 Innovation Network Management

Inter-organisational relationship management, including innovation network management that is of special interest, is a rich and complex field of research covering a wide range of theoretical disciplines and empirical contexts (Dhanaraj and Parkhe 2006; Ritter et al. 2004; Järvensivu and Möller 2009; Hibbert et al. 2008; Provan and Milward 1995; Provan et al. 2004). Although we aim for a holistic model of network management applicable to various innovation contexts, our empirical focus leans toward practice-based innovations. Practice-based innovations are defined by networking among multiple actors that engage in the exchange of development ideas and usage information and operate closer to the actual innovation usage context (Dougherty 2004; Harmaakorpi and Mutanen 2008).

19.2.1 Elements of Innovation Network Management

Innovation networks do not merely intend to make an invention or develop a technical or service formula or model; they seek actual changes in the real world. After all, an invention becomes an innovation only when it is adopted by users.

Innovation networks, including practice-based ones, are characterised by instability that arises from uncertainty. This uncertainty relates to the network members' future behaviour and the absence of an authority to ensure that network participants comply with the goals of the network. Cooperation in such networks is not automatic: a member's self-interest can lead to actions that are individually rational yet produce a collectively suboptimal outcome. (Dhanaraj and Parkhe 2006)

Innovation network management involves using appropriate governance mechanisms, developing inter-firm knowledge sharing routines, making appropriate relationship-specific investments, and changing the partnerships as they evolve while managing partner expectations (Dyer and Nobeoka 2000). In addition, innovation networks need informal leadership as well as sense-making of dispersed knowledge and emerging new ideas (Möller and Rajala 2007). Innovation network management can be characterised as an activity orchestrating a group of independent actors rather than tight coordination (Dhanaraj and Parkhe 2006).

Based on their review, Dhanaraj and Parkhe (2006) argue that there are three key tasks for managing an innovation network: ensuring knowledge mobility, managing innovation appropriability, and fostering network stability. The effectiveness of innovation networks is related to how well knowledge is mobilised in them (Dyer and Nobeoka 2000; Harmaakorpi and Melkas 2005; Dhanaraj and Parkhe 2006). Knowledge mobility, defined as the ease with which knowledge is shared, acquired, and deployed within the network, is important, because innovations cannot be created if the specialised knowledge required to produce the innovation is not shared.

Innovation appropriability is an environmental property that "governs an innovator's ability to capture the profits generated by an innovation" (Teece 1986: 610). It is important for the network to agree on a broad framework of appropriating along with the results of the innovation process, so that members do not try to capture potential profits opportunistically but instead see the innovation as the property of the whole network. Opportunistic behaviour is a sign of insufficient level of trust and can lead to lower commitment to joint activities, ultimately decreasing knowledge mobility and the potential for innovation (Dhanaraj and Parkhe 2006).

Fostering network stability is the third key task, since innovation networks are often loosely coupled. Without stability, there will be fewer possibilities to share knowledge and less likelihood of creating innovations (Dhanaraj and Parkhe 2006). Stability here does not refer to a static stability but rather to a dynamic stability where the goal is a positive input and output growth of the network (thus including stability in growth) while allowing for entry and exit of network members (thus also being dynamic).

Trust and commitment are key elements in fostering the functioning of innovation networks and their knowledge mobility, innovation appropriability, and stability (Harmaakorpi 2006). Finally, the success of an innovation network also depends on network membership and structure, that is, the size and diversity of the members and the density of the connections between them (Dhanaraj and Parkhe 2006; Doz et al. 2000; Lorenzoni and Baden-Fuller 1995).

19.2.2 Network Management in Practice

The perspectives presented above are helpful since they conceptualise key elements in innovation network management. However, they are not readily usable for managers in practice.

In a recent review of inter-organisational collaborative management, Hibbert et al. (2008) approach network management from a more pragmatic perspective, much in line with our research aim. They identify a range of theoretical viewpoints applied in inter-organisational management research: practice-oriented, micro-level research applying psychology, sociology, economics, political science/public administration; structurally-oriented macro-scale research applying economics, social network theories, political science and institutional theory; and process-oriented, intermediate scale, empirically grounded research with a focus on life-cycle, trust, and cooperative processes. These studies and perspectives have provided a range of insights that Hibbert et al. (2008) divide into six different categories (see Table 19.1).

Network management modelling can be divided into two basic categories: conceptualisation and prescription. Some theories obviously touch upon both, as prescriptions require at least some level of conceptualisation; even conceptual models are often followed by prescriptive ideas or advice for managers. The conceptual categories include life-cycle, stage, and phase models; analytical conceptualisations such as typologies, models, and diagnostics; and success and

Table 19.1 Categories of inter-organisational management studies (adapted from Hibbert et al. 2008)

Categories that help to conceptualise the nature of collaboration and identify management challenges		Categories that offer prescriptions or responses to management challenges	
Category	Examples	Category	Examples
Category I: Life-cycle, stages, and phases	Phases such as problem setting, selection, direction setting, getting engaged, learning to collaborate, structuring, stabilisation, dissolution	Category IV: Competencies, behaviours, and tasks	Network building capabilities; activities such as consensus building and problem solving
Category II: Analytical conceptualisations: typologies, models, and diagnostics	Network typologies with different categorising variables, such as hierarchical levels, and degree of risk or trust	Category V: Guidelines and process steps	Descriptions of best practices, contingencies of best practices, steps of effective networking
Category III: Success and failure factors	Lists of success factors promoting or inhibiting networking success, measurements using single or multiple criteria	Category VI: Tools and facilitation	Techniques for categories IV and V, such as project management techniques and group work facilitation

failure factors. The prescriptive categories outline the requirements for good network management, such as competencies, behaviours, and tasks; guidelines and process steps; and tools and techniques for facilitation.

The categorisation by Hibbert et al. (2008) is imprecise. The categories are not precisely defined but are overlapping and non-exhaustive. However, the categorisation is useful, because it provides an overview of the range of the research contributions in the field. It creates a helicopter perspective over the array of tasks that a network manager can face.

Hibbert et al. (2008) argue that the categorisation actually points toward a seventh category of network management: a summarising category that looks at the six partial categories as one network management framework that can be characterised as being 'holistic'. This holistic framework is neither fixed nor precise. Hibbert et al. (2008) explicitly state that this holistic approach "makes a fundamental assumption that collaboration is too complex and idiosyncratic for precise prescriptive remedies" (p. 405). It is useful precisely, because it provides the network manager with a general typology of research contributions that can be used as "handles for reflective practice" (p. 405).

19.2.3 Network Effectiveness

Effectiveness of a network is not easy to define; it is not our intention, nor would it be purposeful, to go into the details of it here. Our focus is rather on the *antecedents* of network effectiveness. We refer to Provan and Milward (2001) in the definition of effectiveness of a network since they have reviewed the concept in the context of public services, which is also the context of our empirical study. According to them, the effectiveness of a network is difficult to define and assess, because its measurement depends on the level of analysis, and it means different things to the network's various stakeholders. In the private sector, network effectiveness is easier to study through financial performance; for the public sector networks, the empirical context of our study, the measurement of effectiveness is more a complex issue. Public sector service delivery networks must be built and maintained at the organisational and network levels, but their overall network effectiveness will ultimately be judged by community-level stakeholders.

Analysed at the *community level*, actors such as funders, politicians, regulators, general public, and client advocacy groups may relate effectiveness to cost to community, building social capital, public perceptions that the problem is being solved, changes in the incidence of the problem, or aggregate indicators of client well-being. At the *network level* of analysis, networks' primary funders, administrative organisations, and members may also have varying effectiveness criteria, such as network membership growth, range of services provided, absence of service duplication, relationship strength (multiplexity, the number of ways member organisations are connected), creation and maintenance of a network administrative organisation (NAO), integration/coordination of services, cost of network maintenance, and member commitment to network goals. At the *level of individual member organisations* and participants, effectiveness may mean agency survival, enhanced legitimacy, resource acquisition, cost of services, service access, client outcomes, and minimum conflict for multiprogram agencies across multiple networks (Provan and Milward 2001).

The definition of effectiveness therefore depends on the level of analysis and the various potential perspectives of the network's stakeholders. In short, network effectiveness relates to the needs and goals of the network's stakeholders (its direct participants and its indirect beneficiaries) and is ultimately defined and enacted through collaboration among them. In the remainder of this chapter, we will look at the process of creating an effective network or, more particularly, how network effectiveness unravels through a process of fostering trust and commitment.

19.2.4 Trust and Commitment

Trust and commitment are in general the basic elements of a functioning network (Hunt and Morgan 1994b; Morgan and Hunt 1994). In particular, knowledge mobility, innovation appropriability, and stability in innovation networks also require their existence (Dhanaraj and Parkhe 2006; Harmaakorpi 2006).

All kinds of collaboration include some elements of trust between the collaborating actors (Ring and Van de Ven 1992), but actors become aware of trust and its need only when they become vulnerable during the course of cooperation or when they come across a problematic situation that contests the existence of trust (Mayer et al. 1995; Möllering et al. 2004). There is a relationship between trust and risk, the latter defined as the acceptance of uncertainty in cooperation (Luhmann 2000; Inkpen and Currall 2004; Mayer et al. 1995; Nooteboom 2007). Trusting actors are aware of the existing risk, but some level of risk must be taken to engage in social action; moreover, trust increases the tolerance of this risk.

The role of trust in cooperation is manifold. First, trust has direct benefits related to communication, conflict management, negotiation processes, satisfaction, and individual and unit level performance (McEvily et al. 2003). Moreover, trust induces positive interpretations of others' behaviour, resulting in improved cooperation (McEvily et al. 2003). Lack of trust can induce concealment and distortion of information; increase the likelihood of misunderstanding and misinterpretation, and result in lack of open discussion (Zand 1972).

Trust increases the likelihood of commitment in joint activities (Håkansson and Snehota 1995). Commitment can be seen as the willingness to learn about the other partner (Doz 1996); it makes actors more willing to invest their time, effort, and attention to collaboration, and increases their tolerance of risk (Inkpen and Currall 2004).

We look at innovation network management as a process that starts from the formation of a network and results in a solution being invented and disseminated. Trust and commitment have a similar processual nature; their evolution carries the network formation forward and enables the creation of the innovation. Both trust and commitment grow over time as actors learn to know each other and create shared values through the process of learning (Dwyer et al. 1987; Hunt and Morgan 1994a). At the beginning of a relationship, trust and collaborative goals create the climate for and shape interaction between the partners, while learning and trust co-evolve later in the relationship (Inkpen and Currall 2004; Laaksonen et al. 2008).

As trust between actors evolves over time, it shifts gradually from initial to evolved trust (Inkpen and Currall 2004). Initial trust exists before actors start cooperation. The shift from initial to evolved trust is a process of learning. In the case of high initial trust, the process starts with disclosing relevant information to other actors, moves on to acceptance of influence from these actors, and finally leads to the coexistence of trust and control (Möllering et al. 2004; Nooteboom 2007). However, as Nooteboom (2007) argues, trust goes beyond control, since more trust allows less control. This means that trust and control are substitutes to some extent. At the same time, one can claim that trust and control complement each other, since trust has its limits. There can be situations where trust alone is not sufficient, so that some level and type of control is needed.

19.2.5 Network Manager

The above discussion highlights key perspectives to innovation network management but leaves the agency of network management still open: *Who* is managing the network? It is not evident who will play the role of network manager. Doz et al. (2000) and Dhanaraj and Parkhe (2006) suggest that an innovation network may be led by an intermediate actor that initiates and develops the network. An actor may operate as an intermediate actor if it has prominence and power gained through individual attributes and a central position in the network structure and uses these to get dispersed resources and capabilities of network members together (Dhanaraj and Parkhe 2006). From the perspective of an intermediate actor, innovation network management can be defined as the set of deliberate, purposeful actions undertaken by the intermediate actor, as it seeks to create value (expand the pie) and extract value (gain a larger slice of the pie) from the network (Dhanaraj and Parkhe 2006).

However, we see this definition of network management as slightly problematic. It presumes that there exists an intermediate actor that is able to foster trust and commitment in a network. However, what if the level of trust is not sufficient to foster commitment building and networking, or what if no single actor has the capabilities to build the required level of trust? It may well be that increasing the level of trust and building commitment then falls upon the collective responsibility of network members, who together start building trust in an iterative process. Jackson and Stainsby (2000) suggest that if the number of members is sufficiently small, then it is easier for everyone to agree on managerial issues jointly, so that the managerial responsibility can become collectively shared.

In a public service network, the public sector may have the authority and interest to form networks and manage them on behalf of the beneficiaries (Jackson and Stainsby 2000) and thus take on the role of the intermediate actor. For instance, in Finland, the empirical context of our study, municipalities are responsible for organising basic health and social care services for their citizens; they retain this responsibility even if they outsource the production of the services. This gives the municipalities little option but to assume their role as an intermediate actor.

19.3 Methodology: Action Research in the Context of Elderly Health Care

Our goal is to understand the patterns of managerial work required to manage a network. We chose action research as our method, as it is well suited to address this type of a research goal (McNiff 1995; Drummond and Themessl-Huber 2007). Action research reveals the social structures, traditions, as well as power structures of the research community, all of which could not be revealed through plain interviews, since people either hesitate to reveal or are unaware of underlying practices and power relations (McNiff 1995). Social and power structures as well as traditions are all important when trying to understand network management in practice.

Action research is a social process; as such, it includes empowering the researched, reflecting on social issues, and reacting to challenges that threaten the change process (Gummesson 1991). Novelty, provocative new theories, innovative concepts, and relation to critical research are potential contributions of action research to traditional research (Gustavsen 2003). Action research is also ideologically close to network management and networks in general due to its emphasis on involving or engaging all relevant actors; both accentuate the role of customers in research or developed services.

Empirically, the focus of this study is on innovation networks in the field of health and social care, more specifically, elderly care. Health and social care includes many types of services (health care, social care, informal care), many types of experts (physicians, nurses, informal caretakers, managers), services at various hierarchical levels (primary, secondary and tertiary care), and actors from all sectors (public, private, non-profit, voluntary). This complexity makes the field a rich area for network management studies.

We studied the functioning of elderly care innovation networks in two cities in Finland. Elderly care services in Finland are increasingly under pressure to improve effectiveness as the population is aging rapidly, which made this a rich target for the study: there was a clear need and ambition for developing innovations. However, the research context was also challenging for our network-oriented study, since the Finnish health and social care sector is fairly hierarchical by nature. This is due to the fact that the responsibility for organising basic health and social care rests on the municipalities. Most of the services are produced by the municipalities, although the services may be produced also by the private market (which is being done increasingly, but most of the basic services still are produced publicly). This hierarchical nature of the empirical context imposed some restrictions on the functioning and management of the studied networks, as we shall describe.

In collaboration with the two cities involved, starting in early 2008 and continuing until spring 2011, we organised more than 100 workshops in eight different networks. The number of participants in a single workshop ranged from about 10–100, representing mostly the public sector but also NGOs, volunteers, customers, and, in some cases, the private sector. The aim of the workshops was to develop and diffuse solutions to various problems related to both service quality and functioning of networks. Since theory is an integral part of action research (Gummesson 1991; White 2004; Turnbull 2002), we brought our theoretical ideas on network management into the workshops and later reflected on what we have learned in terms of theory.

Elderly care innovations can be typified using various perspectives, such as: solutions targeted at formal and/or informal care; solutions that are based on markets, networks and/or hierarchies; solutions at different levels such as institutional, intermediate, and/or micro level; and technological and/or service related solutions (Djellal and Gallouj 2006). Our eight case networks touched upon all of these types. Two networks in our study focused on improving home care services and selecting supporting technologies, and one dealt with informal care services. One of the networks conceptualised and initiated regional service and development networks for elderly care services, comprising of public, private and non-profit actors. One network aimed at creating a city-level strategy for the future of elderly care services and for improving hierarchy-network relations. One of the networks targeted service delivery after an acute care episode, and one looked at new structures and services that should help the elderly stay at home as long as possible. One network was in charge of developing a one-stop information service for the elderly and their relatives.

The first four networks started in early autumn 2008, and the last four started by the beginning of 2009. One of the networks finished its operations after 1 year, when an innovation project with similar goals was quite suddenly started at a higher level of hierarchy, ultimately consuming the stakeholders' interest to continue this network. External project funding for the network building will cease by the end of April 2011, but we expect at least five of the seven remaining networks to continue even without this external funding.

We collected and analysed various forms of data from the workshops and other events (such as phone calls as well as formal and informal meetings) related to the networking processes. The data include more than 1,500 pages of text, including researchers' field notes and diaries, presentations, meeting memos, various plans and reports written during the project, e-mails, and transcribed focus group interviews. In order to analyse the vast amount of data, we constructed process diagrams to illustrate the development of the networks. In the diagrams, we highlighted all of the workshops and other key events in terms of what their contents and key results were, if everything went as expected or if there was need for improvisation, what group work methods were used and how they worked, what kind of feedback was given by the participants, and our own feelings and conclusions as researchers and facilitators.

From the diagrams, which visualised all of the key data, we then identified the following patterns. First, we searched for process-like patterns such as planning, action and assessment steps and their iterations (Khanlou and Peter 2005). We then identified patterns in various network-related key elements, such as managerial tasks, actor roles, and levels of trust and commitment. We did this by iterating and redrawing the diagrams if or when new insights occurred. Finally, we collected the key insights and put them together. Based on this analytical process, we constructed our network management model, which is described in the following section.

19.4 Results: A Four Stage Model of Network Management

The main result of our empirical study is a holistic model of innovation network management, depicted in Fig. 19.1 and Table 19.2. This model outlines the managerial activities required to improve the functioning of a network. It is an ideal representation of an optimal progress of network management – a simplification with an attempt to illustrate the potential feedback relations between the activities, outcomes, and network situations.

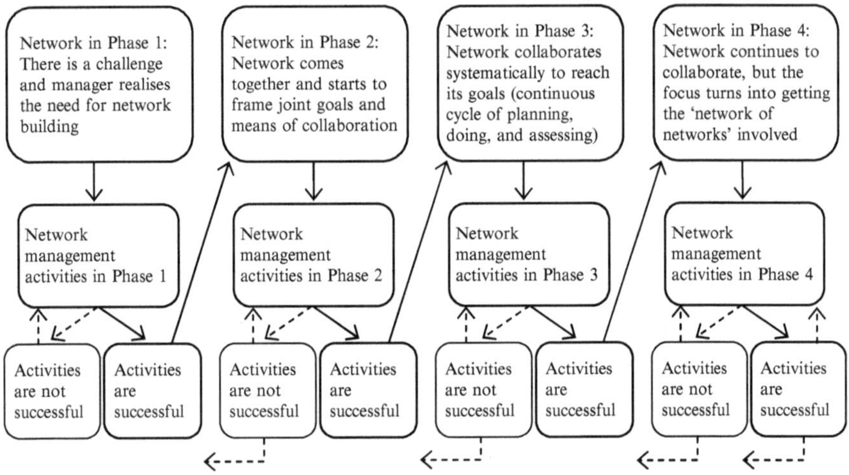

Fig. 19.1 General model of network development through network management activities

19.4.1 The Holistic Model of Innovation Network Management

This model incorporates the six categories of network management introduced by Hibbert et al. (2008). First, it conceptualises network management into a process with four phases (Category I: Life-cycle, stages, and phases). The model includes elements of conceptualisation and analysis, such as assessing the network's situation and structure (Category II: Analytical conceptualisations) and identifying success factors (Category III: Success and failure factors). Moreover, the model prescribes competencies, behaviours, tasks, guidelines, steps, and tools and techniques to seize opportunities and tackle challenges (Categories IV, V and VI).

The phase model is based on the process of building a network, helping it work and produce results, and then spreading its results. It rests on a situation–activities–outcomes typology (see Fig. 19.1). The model is iterative: each phase has its own iterative situation–activities–outcomes sequence, and there are also back-and-forth loops between the different phases.

It is important to note that the model downplays the complexity and dynamics of networking. In reality, the network that is being managed is permeable. The network gains and loses members, is embedded in a 'network of networks', and may include sub-networks within it. In short, a network is, in many ways, in constant flux. The network manager can assess the network at a certain situation and use the insight gained from the model to adapt to opportunities and challenges proactively or reactively (Hibbert et al. 2008). We propose that the network as a whole, rather than any single authority, should jointly assess the network's progress and decide on a respective follow-up.

As such, the number of the phases in the model is not important. In reality, one can witness the occurrence and reoccurrence of the phases as well as back-and-forth loops, so that a network may in reality seem to have more than the four basic phases identified in our model. Nevertheless, it is conceptually useful to carve out the situations that accentuate the need for differing managerial activities. The model, as we have conceptualised it, has four phases, with Phase 3 divided into two sub-phases. First, there is a challenge – a need for an innovation – and a realisation that a network is required in solving the challenge. Next, the network is brought together; goals and networking means are agreed upon; and collaboration is coordinated and facilitated. During these phases, the network participants reflect upon the process, learn from their experiences, use the gained insight to solve the original challenge, and finally engage in diffusing the solution(s).

The model is created with a network in mind. As the network is being managed, boundaries (of membership, of mutual understanding etc.) are inevitably created. Even though boundaries are created, the network never exists in isolation but is embedded in a 'network of networks'. In our empirical research, this network of networks is the overall health and social care service mix of the elderly. If at any point the network faces substantial changes in the network or its boundaries (e.g., if network membership changes significantly or an external shock changes the

Table 19.2 A holistic model of innovation network management

	Phase 1: Identification of challenge that requires a network, organising first network meetings	Phase 2: Agreeing on network-level goals and means of collaboration	Phase 3a: Systematic planning and doing, according to agreed goals and means	Phase 3b: Systematic assessment of collaboration and the developed solution(s)	Phase 4: Diffusing the solution(s)
Activities	– The person that originally identifies the challenge is responsible for taking the first steps	– Coordinating the first network meeting(s)	– Coordinating and facilitating recurrent collaboration forums for systematic planning and doing	– Coordinating and facilitating recurrent assessment forums	– Bringing the solution to relevant diffusion forums/structures within the 'network of networks'
	– Identifying the knowledge and knowhow required to solve the challenge	– Motivating the importance of (a) the challenge, and (b) the need for network building	– Enabling the creation of trust and commitment: empowerment, openness, etc.	– Agreeing on the means of assessment	– Discussing with the 'network of networks' to gain a better understanding of the situation
	– Identifying actors that have the required knowledge and knowhow	– Allowing the network members to learn from/about each other	– Coordinating and facilitating dialogue and communication within the network	– Collecting and assessing information together	– Fostering trust and commitment within the 'network of networks'
	– Organising first network meeting(s) and inviting and motivating key actors to join meeting(s)	– Facilitating open and honest discussion	– Responding to member turnover: familiarising new members into the network	– Re-assessment: Is network meeting its goals? Is network functioning as planned? Does the network have required resources? Is there a need to change goals/means/membership? Is there still a need for this network?	– 'Selling' the solution, or motivating the need for adopting the solution

	– Discussing with the key actors to gain a better understanding of the situation	– Agreeing on the goals, means, and structure of the network	– Communicating with the 'network of networks' to gain a better understanding of its needs and requirements	– Celebrating achievements and learning from mistakes	– Evaluating the success of innovation diffusion
	– Initial planning of network goal and means	– Re-evaluating required knowledge and knowhow and key actors		– Agreeing on corrective measures together	– If needed, building or improving the forums/structure of innovation diffusion
	– Choosing (a) network manager(s) to take on next steps	– Inviting and motivating new network members			– The solution may keep on evolving as it diffuses; the innovation process may continue
		– Network starts to take joint responsibility for managing the network			
Success criteria	– Key actors are involved and motivated to participate in discussions	– Trust begins to evolve and support commitment	– Trust and commitment continue to evolve; members are empowered	– Members are motivated and empowered to execute the assessment plan	– Network members agree on and commit to a diffusion plan
	– Key actors are willing to trust that network building is a good way to solve this type of challenges, and this challenge in particular	– Required knowledge and knowhow, i.e. actors that have these, are brought together	– Members know each others' expertise, goals, and needs	– Network collects and processes assessment information together	– The 'network of networks' is empowered in the diffusion process

(continued)

Table 19.2 (continued)

	Phase 1: Identification of challenge that requires a network, organising first network meetings	Phase 2: Agreeing on network-level goals and means of collaboration	Phase 3a: Systematic planning and doing, according to agreed goals and means	Phase 3b: Systematic assessment of collaboration and the developed solution(s)	Phase 4: Diffusing the solution(s)
	– There exists an initial vision of required knowledge and knowhow to solve the challenge	– Network members know each other and start to understand each others' needs	– Agreed goals (in terms of both collaboration progress and outcomes), means and structure of the network are followed	– Progress of the collaboration is assessed: evolution of trust and commitment	– The solution(s) created by the network is (are) spreading in the 'network of networks'
		– The network agrees on goals, means and structure of the network	– Network structure is stabilised; member turnover is taken into account	– Outcomes of the network are assessed: has the network produced valuable solution(s) to the original challenge(s)	
		– The agreed goals concern both the collaboration progress (how the network functions) and the outcomes (e.g. amount and quality of expected solutions as well as their diffusion rate)	– Network communicates effectively with the 'network of networks'		

network's goals), a new network is created from the viewpoint of the model, and one may need to start over from the beginning or from an earlier phase.

19.4.2 Phase 1: Identification of a Challenge That Requires a Network; Organising First Network Meetings

Phase 1 focuses on the starting situation in which someone has identified a challenge lying ahead, realises that an innovation is needed to solve the challenge, and knows that a network is needed to create such an innovation. In this situation, as one network member described in an interview, "You have a task, problem, an issue to solve, then you should think what you need and who you need to solve the challenge." Identifying the key knowledge and knowhow needed in the network and discussing with key actors related to the challenge at hand are among the first key steps.

Managerial responsibility may, or may not, rest on a 'manager' during these early steps. In principle, these managerial steps can be carried out by anyone who identifies a challenge and starts to discuss it with some key experts. However, at some point during the first phase, we suggest that it is beneficial to agree more explicitly on the managerial responsibilities. This may mean that, instead of the whole network, a single actor or some actors together will take a more explicit role as network manager. The Finnish health and social care sector is largely hierarchically organised due to the dominant role of the municipalities. This means that the existence of (hierarchical) control and authority cannot and should not be disregarded. In this type of a context, at least the early phases of network building may be easier if network management is performed by, or at least initiated by, top managers that have a high enough hierarchical rank (see also Jackson and Stainsby 2000).

Typical mistakes in Phase 1 are failing to identify the key actors that have the knowledge required to comprehend the challenge ahead and relying only on one's own or on one organisation's knowledge of the situation. It is also important to note the strings and obligations that potential network members may have that can affect their possibilities to join a network.

One may need to organise one or several meetings or small-scale workshops with relevant key actors to assess the situation and determine the optimum way of creating the membership list and going forward with the network formation. Organisational leaders can sometimes be too quick to fill in the list of names without properly analysing the dependencies involved, as we witnessed in one of our case networks: "When we collected names for the [network's] list, the list was mostly done by [three of the top managers]. We should have been more careful and used a 'snowball' sampling by relying on people from many hierarchical levels. It was actually odd that some of the middle managers did not know that their bosses enrolled their staff into the network, and then the middle managers started to question the naming procedure... We should have also understood that the potential members themselves are the best people for knowing who should be members and who shouldn't."

A key success factor is that by the end of Phase 1 the actors involved so far trust that network building is the right way to address this type of a challenge and the identified challenge in particular. Before moving on to Phase 2, the key actors involved should also agree on an initial vision of the goals and means of network building as well as on the required knowledge and knowhow that the network needs to bring together. This is necessary for the initial network member to be able to involve the right actors in the network. Without initial goals and means, the network is unable to identify the knowhow and resources needed for the creation of the innovation.

19.4.3 Phase 2: Agreeing on Network-Level Goals and Means of Collaboration

Phase 2 focuses on the first network discussions to frame the goals, scope, structure, and means of the network. A key point in such discussions is to familiarise members with each other and with each other's needs. As one of our informants nicely pointed out, "At first, everyone was driving their own unit's issues before we found this common viewpoint. In the workshop, we started to think about our current situation and the challenges we had, and we seemed to have the same idea that something needs to be done. From this, we got the first ideas [for development]. It took a few more workshops to find these joint goals that we have now."

The goals and means of the network as well as network membership or at least the initial members of the network are first planned in Phase 1 but should be openly re-assessed in Phase 2 to guarantee everyone's commitment to the plans, especially since the network will likely incorporate new actors in this phase. In addition to the concrete needs or goals regarding outcomes (problems or diseases covered, population served, services rendered, etc.), the network should also discuss and agree on the means of cooperation, that is, *how* the network should operate in order to reach the more concrete outcomes. It is likely to be easier to negotiate on the goals and means with a network having high initial trust, since this means that less time needs to be dedicated to learning and trust creation at this point than in the case of low initial trust. We witnessed cases where a successful network was built on both low and high initial trust. High initial trust does not always mean that joint goal setting will be successful. Sometimes starting from a conflicting situation forces the network to discuss the purpose of the network and get to know each other well right at the beginning, which, according to our experience, diminishes the likelihood of later conflicts during networking.

Productive discussions during the second phase may be difficult if (1) the invited members have low initial trust toward each other; (2) there is not enough trust in networking as a suitable method to solve the challenges that lie ahead; (3) or the invited members do not regard the challenge as important enough to attract their attention. An interviewee noted: "Everyone in a network needs to feel that they gain and win. You have to work on the goals and means enough so that they are clear to everyone. There's sure to be a lot of collaboration that people feel that they just have to do, so that they just can't see the point of that collaboration for themselves or their organisation. But part of this is also about insecurity, about the feeling that 'I wish that I could just relax at work by myself at my own organisation.'"

The first meeting(s) during the second phase will require not only careful coordination but also facilitation and motivation, so that all the relevant people are invited and committed to participate in the discussions. It is important that the network is open to new members throughout the networking process. This is especially important in this phase, where the goals and means of the network are still being discussed. If the network of networks is informed about the network early on, it is more likely that they will have a more positive outlook on the networking process, which in turn will have a positive effect on the dissemination of the innovation later on. If they hear about the networking opportunity after the fact, they may feel quite negative about the possibilities of the network to solve the challenges that lie ahead.

The hierarchical nature of the Finnish social and health care sector challenges the trust creation and goal setting process of the second phase. For instance, if a network has been assigned a broad overall goal, or it has many issues on the agenda, its members may feel that someone from the outside should choose a narrower focus for them. This sentiment was expressed quite often in our case networks: "The focus has been a bit lost every time. There has been a huge number of themes to cover. Everyone is playing a different tune. As if we would have gathered a symphony orchestra and different musicians together without a conductor who would draw everyone together to say, 'Let's play this'. This kind of a network will never get anything accomplished. There has to be that someone who will take control of the process, a conductor to hold it together."

The network may need guidance or help from outside for choosing its goals and means. We acknowledge that, in some situations, hierarchical linkages within an organisation can be so strong that some level of authority and control need to be used to ensure successful networking. However, we would be careful to jump to the conclusion that an outside manager should be in charge of choosing the goals and means for a network. How much should a manager decide on behalf of the network? Should the manager instead facilitate a process in which the network itself draws the conclusions? For the manager, it quite often seems to be easier to just decide, but this tends to lead to decreased commitment. We suggest that it is more beneficial in the long run to support the network in finding its own conclusions. This will take more time but should lead to stronger commitment.

By the end of Phase 2, the best case scenario is that the network members know each other's needs and competencies, and the goals and means of the network have been agreed upon in such a way that responsibility to fulfil the agreements rests collectively upon each and every network member. Success depends on having the right people involved, as one of our informants summarised: "We had here just the right, key people present, who can take matters in their hands and can change the way things are if they want to. In this way, this is a very successful network." Success can be assessed also by the breadth, depth, and relevance of knowledge that has been brought together to solve the challenge lying ahead. The network should now be empowered to move on operatively.

19.4.4 Phase 3: Systematic Planning, Doing, and Assessing

The third phase (Phases 3a and 3b) builds on the goals and means agreed upon during Phase 2 through a systematic process of planning, doing, assessing, and re-assessing. This is indeed practice-based innovation in action: performing the tasks of knowledge creation, mobilisation and appropriation. Innovations are created iteratively in an ongoing multi-actor collaboration process. The beginning of the networking process is spent on discussing and agreeing on goals, means, and motivations. Now, the network can finally concentrate on going ahead more concretely while remembering trust and commitment building: "In the first workshops, people were asking where the concrete results are. It was slow to go forward in such a big group. First, issues were quite abstract—people do not know each other and come from different units. It is admirable how the people got over the uncertainty [of the beginning]—no one leaves and says that this is not going to work. But after a year of work, I have understood that networking is a process—it is not daunting anymore."

Key challenges in Phase 3 are to coordinate and facilitate recurrent networking forums, respond to member turnover (familiarising new members, motivating participation), support productive discussion instead of unproductive debate, and in general to communicate well. It is easy *not* to coordinate and communicate well, as this researcher's diary note exemplifies: "We communicated about the workshops poorly... We should have constantly kept in mind how to get [the network] forward and who to keep informed of and/or involved in the process."

It is important to note that planning, doing, and assessing need to be adapted to a network mode of operation instead of using them in a hierarchical sense. In other words, the network participants need to be empowered to do all of these tasks themselves, together. The network should be on top of the situation in all important questions and decisions, although actors from outside the immediate network (such as consultants and experts, key decision-makers, or a steering group) can, and often should, be used as a help. The typical hierarchical mindset is that a manager should decide on the questions of *what* and *with which resources,* and only the *how*-question should be delegated to the lower ranks of the organisation. However, the networking mindset is different. The network should be in charge of deciding not only on the *how*-question but also on *what*, *when*, and *with which resources*. The outside experts' and managers' job then is to enable the network to find answers to these questions.

Much effort should thus be placed on 'enabling': empowering activities, not forcing action; asking open questions, not closing answers; facilitating open dialogue, not encouraging closed debate; and opening the time and space for network members to get to know each other. This type of enabling is important so that trust and commitment have room to evolve. One of our informants described such feelings: "In the workshop, we had enough time to focus on the issue and were allowed to throw in ideas freely. I got the feeling that I could really make a change, be part of development work, participate, and bring my own expert viewpoint into the development." Success can then be assessed as the degree to which members are empowered and the way in which trust and commitment evolve.

The enabling mindset also concerns the fact that network members should preferably learn to facilitate and enable the network by themselves and not become facilitated and enabled by outsiders. It is quite easy for a network used to the hierarchical mode of operating to seek help from outside. In such a situation, it is also easy for an outsider to slip into the facilitator's role, as we noted in our research diary: "We received positive feedback that the workshop was facilitated well so that the discussion was concrete and productive. However, the problem was that the facilitation was done by us researchers so that we did not empower the members to facilitate the discussion themselves, and there is a danger that they will become more and more passive in the process... The paradox is that things may proceed quicker at the level of concrete outcomes when someone capable facilitates, but if the network's level is not intentionally developed, the members' own network management capabilities do not evolve."

Knowledge mobility requires that members should learn to know each other in depth (each other's expertise, goals, needs, etc.), which in turn requires rich dialogues that occur frequently enough. Network membership needs to stabilise at this point at the latest; without stability, members cannot get to know each other well enough, and knowledge does not flow. However, stability does not mean that the network should become static. Rather, networks are dynamic by nature, and stability should be understood here as dynamic stability (see also Dhanaraj and Parkhe 2006).

In meetings and workshops, various dialogue facilitation techniques are helpful in improving knowledge mobility. For instance, it may be a good idea to try out a new group work method instead of relying on an old way of arranging a meeting. These techniques can help people to free themselves from their habitual means of operating, as we note in a researcher diary: "This 'learning café' method seemed to work. The people were discussing lively, and [the top managers that were present] also seemed to like the method and its results. The method brought with it lively discussions and new ideas. It also empowered the people to discuss instead of working under pressure from outside control or guidance."

Commitment building requires enabling as well: managers must ensure that members have resources and time to join network meetings, and the network needs to take into account and respect members' other commitments. Meetings should be organised frequently enough, so that the ideas are kept fresh in mind:

"Especially in the beginning of collaboration, there should not be too long intervals between meetings, or otherwise the point gets lost, like 'where were we again?'".

Commitment and trust are helped not only by enabling collaboration but also by agreeing on assessment criteria together and conducting the assessment together, as well as discussing results openly. The assessment must concern both the networking process and the results of the cooperation: *how well* the network operates and *what* solutions the network has produced.

Since joint assessment is important but often forgotten, we separated the doing and assessment parts of Phase 3 into separate sub-phases (3a and 3b); in practice, however, they are intertwined and simultaneous. We have observed that systematic joint assessment, and joint carrying out of the processes and decisions that follow, is often an undervalued effort. This diary note exemplifies the case in question: "Is there anything left of the results of the SWOT analysis that we did at the large scale workshops last autumn? The analysis resulted in development goals, to which we promised to get back afterwards. Have we got back? Has this group of people been producing yet again more and more development goals [instead of solving the goals from the SWOT analysis]? Somehow it feels that [this network] is inventing again new development targets (= the 'what' is being asked), but the network is not solving old goals or at least we are not following up on them together. So joint follow-up is at least missing (= the 'how' questioning is missing)."

The innovation network does not, and cannot, operate in isolation but is embedded in a larger network of networks. Therefore, Phase 3 also must involve communication with actors from outside the particular innovation network. The network of networks naturally involves some actors that are linked to the innovation process but do not participate in the networking process actively; others are only very loosely or indirectly linked to the network. During Phase 3, the network may communicate with this network of networks, but the communication is related mainly to gathering user insight and ideas for development. As soon as the network assesses its solutions as being ready for launch, Phase 4 – or the *diffusion* of the solutions – is ready to begin.

19.4.5 Phase 4: Diffusing the Solution(s)

Phase 4 is about diffusing or spreading out the solution(s) invented by the network into the wider network of networks. The success of the network is ultimately determined by whether the network of networks adopts the solution(s). In fact, an invention truly becomes an innovation only after it has been adopted into wider use.

There are several possible challenges in this phase. The network of networks may not always share the network's understanding of the importance of the challenge in question, and thus does not buy the solution either. Another network may have existed to solve the same challenge(s), making the network's solution obsolete. The network of networks may also have changed during the creation process – even so much that the solution is outdated already when it is finished.

To solve challenges like these, open discussions in an atmosphere of trust with the network of networks are needed in order to gain the wider audience's commitment to adopting the solution. This requires the network to participate in or organise wider forums for discussion and to create other support structures for the dissemination.

The solution must be 'sold' to the audience; this is most likely to succeed through informing and involving the network of networks in the process from the beginning. As part of the innovation diffusion, the network should also keep in mind that there may also be a need to diffuse some networking competence in the network of networks. As a particular example, innovation adoption is dependent on the openness of the network of networks toward new ideas. The network should engage in building a more open innovation culture if needed: "We should take it [networking] into the structures and development processes, so that this is the way we do things around here. That it is part of our mission or way of operating. Networking should be included in our job descriptions—this is part of the job that we invest time for networking."

Success in Phase 4 can be measured by the degree to which the original network members agree and are committed to the means of diffusion; how well the network of networks is empowered in the process; and, ultimately, by the adoption rate of the innovation.

19.5 Summary and Conclusion

The key contribution of this study is the holistic network management model depicted in Fig. 19.1 and Table 19.2. The model builds on trust and commitment as the key elements of a functioning network, and introduces four phases required to systematically foster trust and commitment. We have shown how earlier theoretical contributions can be put into practice within one holistic framework. We believe that our model has both theoretical and practical contributions.

The first theoretical insight is that the practice of network management can now be understood as a holistic 'whole'. Our model brings together such complex concepts as trust, commitment, and management tasks, but does not reduce any of these into a particular and thus inevitably partial variable of network success, as previous research has tended to do. It is not any single element that counts but the comprehensiveness and adaptability of the whole. The second theoretical insight is that once we look at the reality through a holistic lens, we are better able to put some of the more particular theoretical conceptualisations (such as detailed phase models, success factors, managerial steps, etc.) into a context.

A third key finding, theoretically and practically relevant, is that the network manager's mindset is one of an enabler rather than a decision-maker. Hierarchical managers are usually good at decision-making, such as deciding on what should be done, how it should be done, and who should do it, but they may struggle with an enabler's mindset. Decisions coming from a manager rather than from the network itself can gradually decrease the network's commitment to innovation. The more decisions are made *for* the network, the more dependent the network becomes of decisions made for it. A manager who desires to foster trust and commitment in a network must enable the network to become its own decision-maker.

Much in line with the work by Dhanaraj and Parkhe (2006), our model highlights the importance of fostering knowledge mobility and (dynamic) network stability. Trust-supporting activities will open opportunities for knowledge sharing and collective learning during each of the four network building phases. In our model, network membership gradually stabilises so that by the third phase, there should be enough network stability to support the evolution of trust and commitment, as well as the sharing of in-depth knowledge. Our empirical study did not highlight innovation appropriability, although Dhanaraj and Parkhe (2006) identified it as one of the three key elements of innovation network success. This may be due to the fact that the actors in our study mainly represented the public sector, which is not motivated to acquire benefits for itself but rather to produce societal value.

A more practical finding from our study relates to the nature of innovation network management within a rather hierarchical context, such as the Finnish health care sector where municipalities play a dominant role in organising and often producing basic services. In such a context, the existence of hierarchical control and authority should not, and cannot, be disregarded. Authority and control may even be important in reducing ambiguity and uncertainty during the early phases of networking (Jackson and Stainsby 2000; Mandell 2001).

The model depicted in this chapter awaits further validation through practice-based application. This should be done by keeping in mind that the reality of innovation networks is full of complexities and dynamics that the model cannot fully portray or predict. However, as Hibbert et al. (2008) describe, a holistic model such as this is not intended to be precise for all situations but to offer "handles for reflective practice" for a manager facing them. The model is a simplification and powerful precisely because it *is* a simplification.

Acknowledgements This research was funded by the Academy of Finland and the European Social Fund. We would like to thank the editors for their insightful and detailed comments on an earlier version of this chapter.

References

Dhanaraj, C., & Parkhe, A. (2006). Orchestrating innovation networks. *Academy of Management Review, 31*(3), 659–669.

Djellal, F., & Gallouj, F. (2006). Innovation in care services for the elderly. *Service Industries Journal, 26*(3), 303–327.

Dougherty, D. (2004). Organizing practices in services: Capturing practice-based knowledge for innovation. *Strategic Organization, 2*(1), 35–64.

Doz, Y. (1996). The evolution of cooperation in strategic alliances: Initial conditions or learning processes? *Strategic Management Journal, 17*(1), 55–83.

Doz, Y., Olk, P., & Ring, P. (2000). Formation processes of RandD consortia: Which path to take? Where does it lead? *Strategic Management Journal, 21*(3), 239–266.
Drummond, J., & Themessl-Huber, M. (2007). The cyclical process of action research: The contribution of Gilles Deleuze. *Action Research, 5*(4), 339–357.
Dwyer, R. F., Schurr, P. H., & Oh, S. (1987). Developing buyer–seller relationships. *Journal of Marketing, 51*(2), 11–27.
Dyer, J., & Nobeoka, K. (2000). Creating and managing a high-performance knowledge-sharing network: The Toyota case. *Strategic Management Journal, 21*(3), 345–367.
Gummesson, E. (1991). *Qualitative methods in management research*. Newbury Park, CA: Sage Publications, Inc.
Gustavsen, B. (2003). New forms of knowledge production and the role of action research. *Action Research, 1*(2), 153–164.
Håkansson, H., & Snehota, I. (1995). *Developing relationships in business networks*. London: International Thomson Business Press.
Harmaakorpi, V. (2006). The regional development platform method as a tool for regional innovation policy. *European Planning Studies, 14*(8), 1085–1104.
Harmaakorpi, V., & Melkas, H. (2005). Knowledge management in regional innovation networks: The case of Lahti, Finland. *European Planning Studies, 13*(5), 641–659.
Harmaakorpi, V., & Mutanen, A. (2008). Knowledge production in networked practice-based innovation processes – Interrogative model as a methodological approach. *Interdisciplinary Journal of Information, Knowledge, and Management, 3*, 87–101.
Hibbert, P., Huxham, C., & Ring, P. S. (2008). Managing inter-organizational relations. In S. Cropper, M. Ebers, C. Huxham, & P. S. Ring (Eds.), *The handbook of inter-organizational relations* (pp. 391–416). Oxford: Oxford University Press.
Hunt, S. D., & Morgan, R. M. (1994a). Organizational commitment: One of many commitments or key mediating construct? *Academy of Management Journal, 37*(6), 1568–1587.
Hunt, S. D., & Morgan, R. M. (1994b). Relationship marketing in the era of network competition. *Marketing Management, 3*(1), 18–28.
Inkpen, A. C., & Currall, S. C. (2004). The coevolution of trust, control, and learning in joint ventures. *Organization Science, 15*(5), 586–599.
Jackson, P. M., & Stainsby, L. (2000). Managing public sector network organizations. *Public Money & Management, 20*(1), 11–16.
Järvensivu, T., & Möller, K. (2009). Metatheory of network management: A contingency approach. *Industrial Marketing Management, 38*(6), 654–661.
Khanlou, N., & Peter, E. (2005). Participatory action research: Considerations for ethical review. *Social Science & Medicine, 60*(10), 2333–2340.
Laaksonen, T., Pajunen, K., & Kulmala, H. (2008). Co-evolution of trust and dependence in customer–supplier relationships. *Industrial Marketing Management, 37*(8), 910–920.
Lorenzoni, G., & Baden-Fuller, C. (1995). Creating a strategic center to manage a web of partners. *California Management Review, 37*(3), 146–163.
Luhmann, N. (2000). Familiarity, confidence, trust: Problems and alternatives. In D. Gambetta (Ed.), *Trust: making and breaking cooperative relations* (electronic ed., pp. 94–107). Oxford: Department of Sociology, University of Oxford.
Mandell, M. P. (2001). Collaboration through network structures for community building efforts. *National Civic Review, 90*(3), 279–288.
Mayer, R., Davis, J., & Schoorman, D. (1995). An integrative model of organizational trust. *Academy of Management Review, 20*(3), 709–734.
McEvily, B., Perrone, V., & Zaheer, A. (2003). Trust as an organizational principle. *Organization Science, 14*(1), 91–103.
McNiff, J. (1995). *Action research: Principles and practice*. London: Routledge.
Morgan, R. M., & Hunt, S. D. (1994). The commitment-trust theory of relationship marketing. *Journal of Marketing, 58*(3), 20–38.

Möller, K., & Rajala, A. (2007). Rise of strategic nets – New modes of value creation. *Industrial Marketing Management, 36*(7), 895–908.
Möller, K., Rajala, A., & Svahn, S. (2005). Strategic business nets – Their type and management. *Journal of Business Research, 58*(9), 1274–1284.
Möllering, G., Bachmann, R., & Lee, S. (2004). Understanding organizational trust – Foundations, constellations, and issues of operationalisation. *Journal of Managerial Psychology, 18*(6), 556–570.
Nooteboom, B. (2007). Social capital, institutions and trust. *Review of Social Capital, LXV*(1), 29–53.
Nykänen, K., Järvensivu, T., & Möller, K. (2009). *Towards a more social perspective of network management: An action research study on trust and commitment.* Electronic conference proceedings of the 25th IMP Conference. www.impgroup.org/uploads/papers/7356.pdf. Accessed 10 January 2011.
Provan, K. G., & Milward, H. B. (1995). A preliminary theory of interorganizational network effectiveness: A comparative study of four community mental health systems. *Administrative Science Quarterly, 40*(1), 1–33.
Provan, K. G., & Milward, H. B. (2001). Do networks really work? A framework for evaluating public-sector organizational networks. *Public Administration Review, 61*(4), 414–423.
Provan, K. G., Isett, K. R., & Milward, H. B. (2004). Cooperation and compromise: A network response to conflicting institutional pressures in community mental health. *Nonprofit and Voluntary Sector Quarterly, 33*(3), 489–514.
Ring, P., & Van de Ven, A. (1992). Structuring cooperative relationships between organizations. *Strategic Management Journal, 13*(7), 483–498.
Ritter, T., Wilkinson, I. F., & Johnston, W. J. (2004). Managing in complex business networks. *Industrial Marketing Management, 33*(3), 175–183.
Teece, D. (1986). Profiting from technological innovation: Implications for integration, collaboration, licensing and public policy. *Research Policy, 15*(6), 285–305.
Turnbull, S. (2002). Social construction research and theory building. *Advances in Developing Human Resources, 3*(4), 317–334.
White, A. (2004). Lewin's action research model as a tool for theory building: A case study from South Africa. *Action Research, 2*(2), 127–144.
Zand, D. (1972). Trust and managerial problem solving. *Administrative Science Quarterly, 17*(2), 229–239.

Chapter 20
Practice-Based Innovations at 'Sendan No Oka': Motivation Management and Empowerment Management

Hiroo Hagino

Abstract Sendan No Oka, Tohoku Fukushi University's affiliated nursing home, has a strong reputation in Japan for its high quality health care and social services, and its excellent business performance. But at its establishment, these were only distant goals. The staff at Sendan No Oka have together tackled many difficult challenges and instituted a wide variety of useful innovations. Management focuses on motivation and empowerment for both clients and staff have been the two pillars of Sendan No Oka's innovation ecosystem. With motivation management and empowerment management based on expertise and seamless professional collaboration, incremental, practice-based innovation has led to the achievement of high-level health care and social services at reasonable cost. Based on these achievements, Sendan No Oka is now stepping on to attempt radical innovation in client-oriented, comprehensive community care.

20.1 Introduction

Extension of average life expectancy resulting from advances in medical and social care has led to rapidly aging populations in developed countries. As a result, health care and social services for the elderly have for some years been a particular dynamic business sector in these countries (Djellal and Gallouj 2006). Not only in western countries, but also in developed Asian-Pacific countries, the population is aging very rapidly. For example, by 2030 the aging index (the number of people aged 65 and over per 100 youths under age 15) is expected to exceed 100 in all developed countries, and pass 200 in several European countries and Japan (Gavrilov and Heuveline 2003). This means that there will be a huge need for elder care services in the near future. In addition to quantity, the quality of elder

H. Hagino (✉)
Tohoku Fukushi University, Sendai, Japan
e-mail: hiroo@tfu-mail.tfu.ac.jp

H. Melkas and V. Harmaakorpi (eds.),
Practice-Based Innovation: Insights, Applications and Policy Implications,
DOI 10.1007/978-3-642-21723-4_20,

care also must be improved. The new elderly – with high education, high income, and a strong sense of their rights – will request new kinds of care. On the contrary, economical projections for developed countries are not optimistic. We must respond to greater demands for elder care with decreasing fiscal resources.

Policy makers must innovate in establishing systems and services. Service providers at the front line of elder care also must innovate to provide these services more efficiently and effectively. Developing higher quality care services responding to various individual needs amidst budget cuts will obviously require practice-based innovation in health care and social services. This chapter's aim is to contribute to this process.

20.2 Theoretical Background

In considering this aim, we have to understand that innovation management for providing services differs from management for production and delivery of physical products. While the practical distinction between providing services and physical products is not always clear (Cook et al. 1999), recent announcements by companies, such as IBM, indicate that services have become a core element in business models and represent an increasing share of revenues. Indeed, services have become the engine of future growth for firms and organisations in the twenty-first century (Vargo and Lusch 2004; Matthing et al. 2004). However, innovation and product development have traditionally been studied in the context of manufacturing industries, and existing theories of innovation have often been developed on this basis (Alam and Perry 2002; Sundbo 1997). Therefore, we see an increasing interest in service sector innovation.

In this chapter I examine the nature of innovation. The 'nature of innovation' refers to how radical is the change that innovation evokes (e.g., Anderson and Tushman 1990). The basic classification of innovation as *incremental* or *radical* can be supplemented by considering the extent, or scope, of innovation. It has been argued that there is a tendency in service industries toward small, scattered changes instead of large-scale innovation; meaning that innovations in service firms are less radical, rapidly implemented, and copied than in manufacturing (Cowell 1988; Scarbrough and Lannon 1989; Voss et al. 1992). Furthermore, Sundbo (1997) claims that these small incremental innovations may be difficult to differentiate from normal organisational learning processes. Thus it is sometimes claimed that service firms do not innovate, or that they innovate in ad hoc, haphazard fashion (Kelly and Storey 2000; Dolfsma 2004).

Anderson and Tushman interestingly claim that the 'innovation cycle' is reversed in service innovations (cf. Anderson and Tushman 1990). In technology, radical innovation is typically followed by a period of ferment and the eventual development of a dominant design, where after incremental innovations appear which enhance the existing dominant design. In services, however, incremental innovations leading to efficiency improvements accumulate, whereupon more radical process innovations occur (Barras 1986; Dolfsma 2004).

Certainly we can find radical innovations in the service sector in the context of developing new services. There are innovations based on existing value creation systems, including current technology, know-how, and architecture. Such innovations are more or less incremental. However, such innovations can be radical when they involve a new element or a new combination of old elements (cf. Schumpeter 1934). These kinds of radical innovations may also change the whole value creation system (Möller et al. 2005). In many cases these kinds of radical service innovations have been fostered by open, innovative thinking. Network management among value creating entities enhances innovation ecosystems, where incremental innovations develop to become radical innovations.

As a case in point I introduce Sendan No Oka (later appearing as *Oka*) of Sendai, Japan, focusing on its practice-based innovations. Oka is a nursing home with a strong reputation for dealing with multiple, difficult challenges. As Anderson and Tushman (1990) have observed, service innovations at Oka started with incremental steps, which can be difficult to distinguish from normal organisational learning processes. However, Oka executives deliberately chose to start in a very primitive way, making incremental basic innovations toward their ultimate goal of radical innovation in health care and social service; and with adequate management, they have achieved their aims.

While it may be more difficult to export service concepts than products to others societies and cultures, I believe the practice-based innovations at Sendan No Oka have universal implications, for elder care innovation, and different fields, as well.

20.3 Tohoku Fukushi University Group

Tohoku Fukushi University (later as 'TFU') is a Japanese private university. Japan's biggest traditional Buddhist group, 'Soto-Syu' established TFU in 1875 as a school for Buddhist monks and has managed TFU since then. In 1995, TFU established its subsidiary, Social Welfare Association Tohoku Fukushi Corporation. In 2000, it also founded a subsidiary medical arm (Medical Association Tohoku Fukushi Corporation). These two arms are different corporations, and this difference was important in the innovations at Sendan No Oka discussed below.

The aim in establishing these affiliated organisations was to put into practice the health care and social service concepts and practices coming out of research and education at TFU. At that time Long Term Care Insurance (Japanese social insurance for long-term care mainly for the elderly, later as LTCI) had not been established. And social services were only available to a small number of disadvantaged individuals, with quality far from the standard of service for which TFU students were prepared in their undergraduate and graduate courses. Many students were disappointed at the reality of social service after starting their careers in this field.

Fig. 20.1 Sendan No Yakata, a nursing home of TFU

In 1996, TFU established its first affiliated nursing home, Sendan No Mori. Later it founded more nursing homes, including Sendan No Yakata (Fig. 20.1), a Finnish style nursing home that is part of the Sendai-Finland Well-being Center. These facilities exemplify TFU's policy of: 'Amalgam of Theory and Practice'. TFU tries to provide health care and social services based on its academic research and development work. In the course of this real-world practice, many challenges have been faced, which in turn have served as the seeds of research and innovation at TFU. Such a virtuous spiral is the driving force of innovation at TFU, and has informed the education of many social workers, certified care workers, clinical psychologists, nurses, occupational therapists, clinical artists, and physiotherapists at TFU.

Sendan No Oka (*'oka'* meaning 'hill' in Japanese) is one of these affiliated institutions. It provides elder care mainly within the Japanese LTCI framework.

20.4 Long-Term Care Insurance ('Kaigo Hoken')

Before introducing Oka, I will briefly explain Japanese Long-Term Care Insurance (Fig. 20.2). LTCI is an important income source for Oka, and frames Oka's business and innovation.

In response to social changes (changing status of women, trend toward nuclear families, etc.) and increased demand for health care and social services for the elderly, LTCI was introduced in 2000. With LTCI, health care and social services for the elderly were partly integrated and socialised. Previously, these services for the elderly were provided separately. The former was mainly covered with health insurance and the latter was available only to those with very low income and no

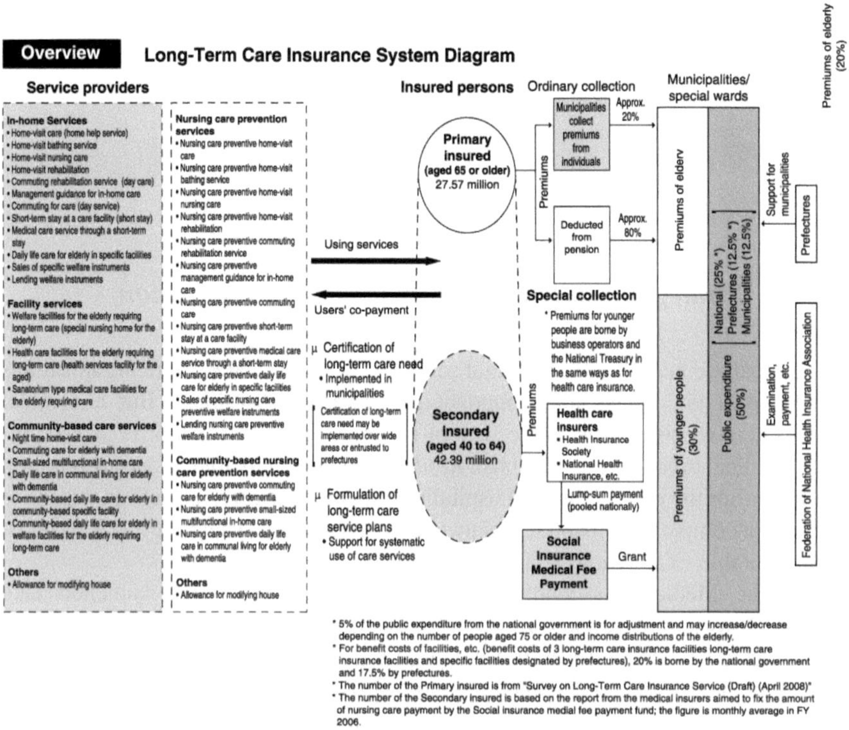

Fig. 20.2 Overview of Japanese long-term care insurance system (Source: Japanese Ministry of Health, Labour and Welfare 2010; http://www.mhlw.go.jp/english/wp/wp-hw3/dl/10-01.pdf)

family to rely on. This situation resulted in hospitalisation for many who could have been helped with less-costly social services. Through LTCI, social services became available to many who formerly would have depended on more costly health care.

LTCI provides three kinds of residential (institutional) care. *'Tokuyou'* ('Welfare facilities for elderly requiring long-term care') are nursing homes for those who can no longer stay at their own homes, but do not need health care. These often become the final place of residence for the elderly. *'Ryoyogata'* ('Sanatorium type medical facilities for the elderly') are for elderly clients requiring special health care. Clients include elderly who are not critically ill, but still need significant health care, and those who suffer from a chronic disease and cannot stay at home. The Japanese Ministry of Health, Labour and Welfare is trying to abolish these kinds of facilities that also increase 'social hospitalisation' (patients are hospitalised for a long time for nonmedical reasons, such as family circumstances and a shortage of nursing care facilities) and health insurance expenditures.

The other type of institutional care facility is the *'Rouken'* ('Health care facility for the elderly requiring long-term care'). Oka is this kind of a nursing home, offering health care and rehabilitation. The *Rouken* functions as an intermediate facility between hospital and client residence. After passing the acute stage in hospital, some elderly persons are provided with health care, rehabilitation and social services at *Rouken*, as a step to returning to their homes. But in reality, with a shortage of *Tokuyou*, *Rouken* often serve as *Tokuyou* in many areas.

20.5 Sendan No Oka: Practice-Based Innovation

20.5.1 Strategy from Incremental to Radical Innovation

In the Japanese health care and social services system, social welfare organisations benefit from tax breaks and abundant governmental subsidies. But a medical organisation receives little in the way of governmental subsidies or breaks. Typically, medical organisations establish *Rouken* with hospitals so that one corporation can look after the elderly during hospitalisation and later care, as well. In Japan we call such a complex of Hospital, *Tokuyou* and *Rouken* as a 'triple set', which can be very profitable (Yamamoto and Sugita 2008).

But the Medical Association Tohoku Fukushi Corporation launched Oka (Fig. 20.3) as an independent *Rouken* because TFU had no associated hospital at that time (TFU later founded Sendan Hospital in 2008). In rough terms, if a social welfare organisation builds a nursing home with a cost of $2 million (USD), it can receive a subsidy of about $1.8 million, with remaining debt of around $0.2 million. But in the case of a medical organisation, the subsidy would be $0.2 million, with remaining debt of $1.8 million. So from its establishment, Oka was in a critical financial situation, with huge debt (Doi 2006).

Fig. 20.3 Sendan No Oka

Oka was designed to operate with 50 staff members for 100 beds, but the financial situation did not allow this number to be hired, and Oka launched with the minimum staff required by Japanese regulations (33 staff members). Oka also was unable to hire senior staff with great experience and skill because of costs. Oka thus started with a small number of newly graduated staff, but this difficult situation became the seed of various later innovations at Oka (Doi et al. 2004a).

At the initial stage the quality of care at Oka was far from satisfactory. Innovative solutions were needed to overcome various challenges emerging in the course of daily care. The executives of Oka discussed the strategy to solve these challenges, and Oka started with the exact opposite policy with radical innovations (Doi 2006).

The care philosophy of Oka is in one sense very conservative. It puts more emphasis on providing five basic health and social services (support for waking up, eating, excretory control, keeping clean, and activity) plus rehabilitation service safely and steadily. The loss of various functions (handicaps, etc.) can lead to decreased quality of life (later as QOL) for the elderly. But by providing various supports and managing the living environment, Oka attempts to decrease the impact of handicaps on client QOL. But at the initial stage, this philosophy was difficult to implement to any significant degree (Doi 2006).

Another TFU-associated nursing home had already shown the way introducing radical innovations (Sendan No Mori – 2004). Some of these (small-scale, multi-functional in-home care, etc.), in fact, were later officially instituted in Japanese LTCI.

The few experienced staff members at Oka were also highly motivated by their dreams for offering health care and social services matching their own ideals. They hoped to implement advanced, radical innovations in client-oriented holistic care or community-based care. But these dreams seemed largely unrealistic given Oka's initial financial situation. Oka executives were also cautious about attempting new approaches that, without adequate support, might introduce risks to clients' physical or mental health. The priority was providing stable, safe, and reliable service. Oka executives groped for ways to maintain basic quality of daily care, while also implementing innovative holistic care to improve QOL. Oka executives decided to start with very simple, incremental innovations toward their eventual goals. Even though their first attempts did not appear, at a glance, innovative, their method of ensuring a stable infrastructure while slowly moving toward their eventual goals was itself, innovative. As value creation net theory predicts, Oka's incremental innovation eventually resulted in radical innovation.

20.5.2 'Motivation Management' for Sendan No Oka Staff

At Oka, they put more importance on 'how' to achieve practical aims rather than focusing on process-oriented documentation, systems, and so on. As a tool in this process, Oka utilised two management strategies: 'motivation management' and 'empowerment management'.

First, for motivation management, Oka executives began with information sharing (Doi 2004). They bravely opened all aspects of the operation, including the financial situation, to staff members, discussing the situation thoroughly and explaining why they were understaffed and their tasks were so hard. This open sharing and understanding motivated the staff. Feeling like stakeholders, the staff dedicated themselves in improving service from the bottom up.

To improve care quality and compensate for the lack of manpower, Oka introduced 'multi-disciplinary collaboration' (Doi 2004). Japanese LTCI requires an overall minimum of one staff member for every 3.1 clients. It is very difficult to provide a satisfactory level of care with this minimum number of staff, but Oka had to manage at this staffing level. Taking into account holidays and personal days off, the actual daytime average was one staff member to 8.5 clients, and one staff member for 25 clients at night. To handle this number of clients with inexperienced staff, all professionals (nurses, occupational therapists, dieticians, etc.) shared in the tasks of care workers and collaborated across disciplines (multi-disciplinary collaboration). Executives took the lead in this collaboration. The director (a medical doctor) willingly helped with meal support (Fig. 20.4). The vice director (an occupational therapist) and other executives often worked the unpopular night shift.

Figure 20.4 shows the director's poor service skills. But his courage in revealing his weaknesses motivated all the staff and gave them confidence, especially the care workers. The director's personal character also facilitated frank exchange of information and the building of trusting relationships. The sympathetic attitude of management and the willingness of professional staff to volunteer and help across

Fig. 20.4 Meal support by the director (a medical doctor)

boundaries of duty and position further enhanced the sense of collegiality as joint stakeholders and the development of warm, informal relationships between all staff and executives.

This collaboration had other effects, as well. For instance, a dietician helped with meal support. Beyond helping to cover staffing needs, food service experience yielded understanding that led to important improvements in how she did her own job. Since care workers outnumber other staff, involving them as colleagues and co-decision-makers unleashed a powerful, enthusiastic source of many incremental practice-based innovations which led to more radical innovations over time.

20.5.3 'Empowerment Management' for Sendan No Oka Staff

As mentioned above, the initial quality of care at Oka was far from satisfactory. Even though highly motivated, if staff has no means to realise their ambitions, it comes to nothing, or, even worse, increases stress. Empowerment is thus a key second pillar of innovation.

Empowerment started with the structuring of unit care. At both ends of each floor, the Oka building has bedrooms for 20–25 individuals surrounding a common unit area (Fig. 20.5). Ideally each unit would serve a smaller number of clients. All the trained staff members believe that each person has intrinsic value and should not be classified by function or ability level. In society, various people with various abilities live together. In a nursing home also, a mix of people would ideally live together. But minimal staffing forced Oka to set aside this philosophy and classify

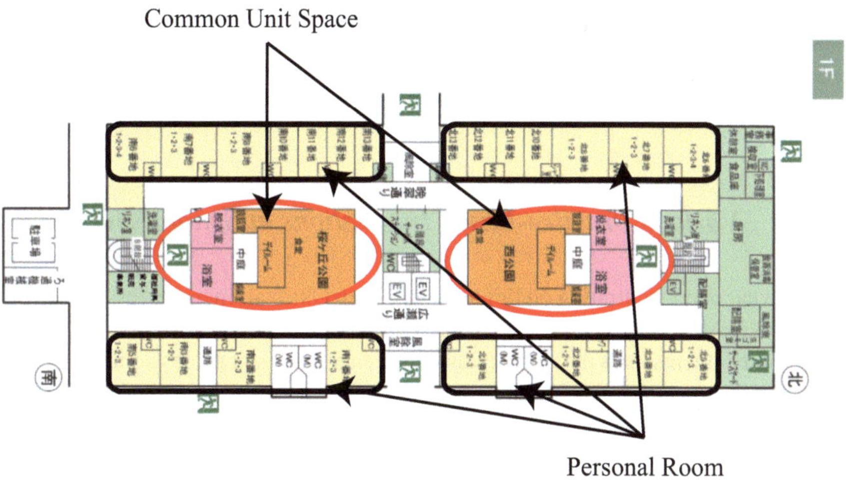

Fig. 20.5 Floor map of Sendan No Oka

clients based on their remaining abilities and type of support needs. This was like a disruptive innovation at Oka.

The Oka units were classified into six units – one day rehabilitation service unit and five residence units (Oomori 2004):

- A unit for ambulatory clients with dementia
- A unit for dementia patients who can no longer walk
- An intensive rehabilitation unit for helping clients return to their homes
- A unit clients requiring health care
- A unit for bedridden clients

This is obviously a provider-oriented classification that is far from the ideal of social service or client-centred innovative care. The lack of client involvement in service design would seem to hinder co-creation of innovation (Gadrey et al. 1995; de Brentani 2001; den Hertog 2000).

But this classification was essential to providing safe health care and social services for 100 clients with a small, inexperienced staff. By classifying clients based on support needs, the staff in each unit could focus on specific skills and knowledge. At the same time, Oka's executives persuaded their staff to give up radical approaches to care (e.g., holistic care) and instead concentrate on the three fundamental support areas (eating, excretion, and bathing). Limiting the care menu and focusing on clients with common needs allowed inexperienced staff to master their main tasks more quickly – undistracted by the multiple demands of more holistic care with a wider variety of clients.

Avoiding reshuffling of personnel from unit to unit not only helped speed the learning curve for specific tasks and responsibilities, but facilitated the quick development of intimate human relationships in each unit. This was also an example of empowerment management at Oka, in which the staff were organised and supported in ways that facilitated the quick development of expertise and sense of competency which could later blossom in innovative developments.

It is believed at Oka that high motivation and empowerment based on expertise are the keys to an innovation ecosystem, and they have tried to develop these through concrete methods with multi-disciplinary collaboration, as explained above.

20.5.4 Various Practice-Based Innovations at Sendan No Oka

After 1 year of concentrating on limited tasks, each Oka staff member, especially the care workers, had matured dramatically. Confident in their own developing abilities, care workers themselves moved to reduce the general tasks of professionals, so these specialists could put as much energy as possible into improving the quality of life of clients through their specialties.

This cooperative participation of all staff based on their established and developing expertise provided Oka's infrastructure for innovation. Motivation management and empowerment management yielded incremental innovation through such

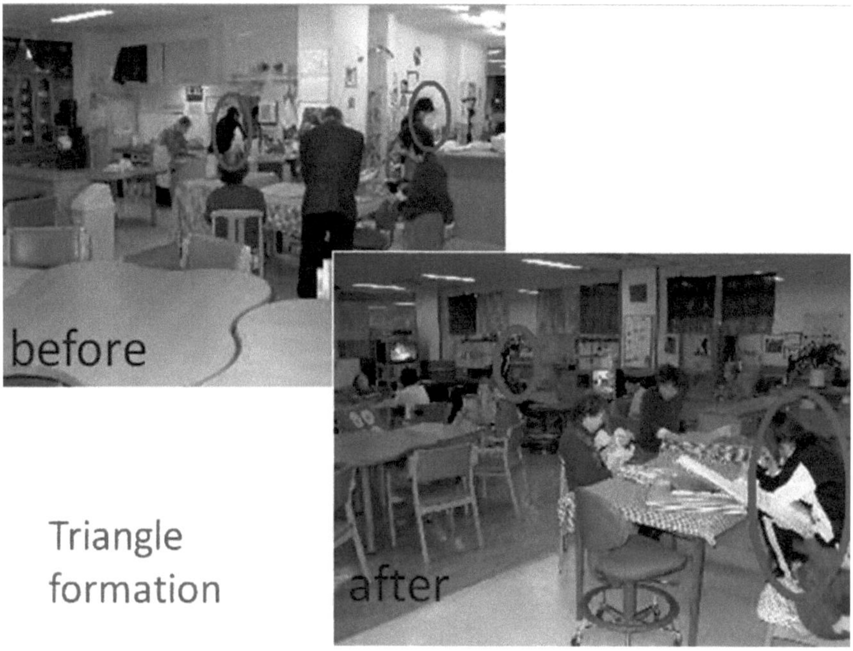

Fig. 20.6 Triangle formation

concrete initiatives as information sharing, seamless collaboration of professionals, limiting task scope, development of skills and knowledge, and so on. Oka intentionally took the 'long way' to its goal, starting with a service unit concept which seemed anti-innovation, but which actually provided a solid basis for the seeds of later innovation.

Following this initial stage, Oka undertook its first reshuffling of personnel – taking into account each staff member's ability, specialty, experience, character, and human relations. Through this shuffling, new fusion of specialties and experience yielded new seeds for innovation. As a result, many small ideas based on the bringing together of different skills and knowledge were generated in each care unit. The frank, open atmosphere among the Oka staff, formed through information sharing and collaboration, helped these small innovative ideas to develop.

For example, Fig. 20.6 shows a care unit. Daily documentation of each client's schedule is an important task in any nursing home. Previously the staff did their scheduling with their backs to clients, so they were not able to see what was happening behind them. There were risks of accident, and with minimal staffing, problems were inevitable. One day a care worker suggested that the staff members conduct this record-keeping while facing the clients and positioning themselves in a triangle formation to eliminate 'dead' (unobserved) positions on the floor. Like a zone press in football, this allowed a small number of individuals to cover a large

Fig. 20.7 Shoe shelf with handrail

space. Using this system, a few staff members could watch many clients, and the number of accidents decreased dramatically.

Figure 20.7 shows the shoe shelf at Oka. In the Japanese culture, we take off our shoes when entering residential buildings. But the act of removing shoes is difficult for many elderly clients and creates a high risk of falling. An occupational therapist at Oka noticed this and installed handrails at appropriate positions in the shoe area, thus decreasing accidents. Likewise, oral hygienists – at that time rarely employed in Japanese nursing homes – played an important role in improving oral hygiene, decreasing aspiration pneumonia, and motivating clients to enjoy meals and conversation. So many of these small innovations at Oka result from an expert eye being applied to concrete, daily care issues in a context of high motivation and empowerment among the staff members. While unfortunately, no precise records have been kept, these few anecdotes are just a small indication of the many helpful ideas implemented as a result of the motivated staff empowered to apply their expertise.

20.5.5 Sendan No Oka Radical Trial Innovation

With its innovation infrastructure and many incremental innovations in place, Oka embarked on its first radical steps, following the anticipated sequence of Barras (1986) and Dolfsma (2004), who suggested that more radical process innovation occurs after incremental process innovations. Here I will introduce three radical innovation trials in health and social services at Sendan No Oka:

- Excursions with bedridden clients conducted by inexperienced staff – successfully accomplished;
- Units going 'diaperless' – making a dramatic change for clients and positively contributing to Oka's financial situation; care workers were not only able to cut costs, but created new business and profit, which is radically innovative;
- Trial for community-based care, which had been Oka's goal since its establishment; this trial is definitely radical because it is beyond the LTCI framework.

Starting in a very incremental way, Oka has reached these radical innovations.

20.5.5.1 Trial in Health Care Needy Unit

The staff at Oka began with high hopes of implementing radically innovative, client-oriented health care and social services. Difficult conditions made this impossible in early days, but as the staff developed expertise and implemented incremental innovations they began exploring ideas for more radical innovations. Oka executives, considering finances, staffing needs, expertise, safety, and regulations, responded affirmatively to a number of staff proposals.

The first trial of client-oriented, radically innovative care was conducted in a unit for clients with serious health care needs. The staff in this unit were very enthusiastic about client-oriented care, but almost all the clients were bedridden and required special health care for conditions such as percutaneous endoscopic gastrostomy. Staff numbers and skills were not initially sufficient for an innovative trial. Figure 20.8 shows a first year scene (no. 1) in this unit. Many clients simply spent the day dozing in the common unit. They had little hope or pleasure in their daily lives.

After a year of learning and development, the staff felt ready to try an innovative change. They felt that cooking together with clients in the unit, even those with serious health issues, could contribute to improved human relations (Fig. 20.8; no. 2). The clients showed great interest in this activity and participated as actively as their conditions allowed – chopping ingredients, offering advice, etc. The smiles of clients gave great energy to the staff, and they were motivated to further develop this and other joint activities.

As these efforts increased, the clients also felt greater motivation. They set a goal with the staff of leaving the facility for an excursion; often considered impossible for clients who are bedridden and in need of special health care. But both the staff and clients made efforts to realise this common goal. After a year of planning, they made this excursion a reality (Fig. 20.8; no. 3). High motivation of both the staff and clients combined with growing expertise, multi-disciplinary collaboration and incremental innovations made it possible.

Fig. 20.8 First (1), second (2) and third (3) year of the trial

20.5.5.2 Going Diaperless and Exploring New Business Opportunities

The next case is an innovative attempt to go diaperless. Of course, diaper use is sometimes necessary for the elderly, but it can seriously harm one's dignity. In this case, no formal planning was involved. It simply evolved out of the focus of Oka staff on personal dignity. Many authors postulate that successful services can and must be systematically planned to avoid *ad hoc* processes (Ramaswamy 1996; Cooper and Edgett 1999; de Brentani 2001; Froehle et al. 2000; Fitzsimmons and Fitzsimmons 2000). But Oka rejects this notion and emphasises staff initiative from a client-oriented viewpoint. Care units can try whatever they believe will be effective and efficient, as long as there is adequate discussion and testing of new products or procedures.

This trial started with one woman. When she came to Oka, she was at Care Level 4 on the Japanese LTCI scale, which means almost bedridden. She needed diapers and showed no motivation to improve her quality of life. But after she had come to Oka, the staff developed a close relationship with her and learned her chief complaint. Using diapers was an unendurable shame and insult to her dignity. So the staff decided to try to help.

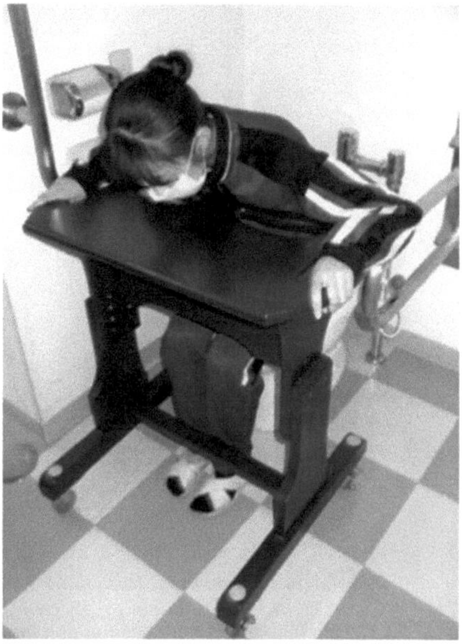

Fig. 20.9 Handmade equipment for toilet use

Care workers took the central role. They began carefully monitoring her daily actions, a key to high quality care. They kept precise record of her meals, tea, water, and other intake, and the number and amount of her excretions. Eventually this allowed them to accurately predict her excretory patterns and be sure she had the necessary stimulation and access to toilet facilities. This included collaboration with an occupational therapist who designed toilet aids affording her a stabler position while using the toilet (Fig. 20.9). Urine pads helped with occasionally unpredictable incontinence, and the urine pads and underpants finally replaced the dreaded diapers.

This raised the client's sense of dignity and increased her motivation for daily life. She made great efforts at rehabilitation to regain the ability to wash her underwear herself, and she succeeded. She also started rehabilitation to walk again and recovered enough to walk independently. She began washing all her clothes and drying them in her room herself (Fig. 20.10). This success began with a client-centred focus on human dignity and depended on multi-disciplinary collaboration. It was also a testament to the Oka staff of the great improvements possible with high client motivation – another demonstration of the importance of motivation and empowerment management.

The 'diaperless' innovation has also had a tangible financial effect at Oka. Figure 20.11 shows the monthly cost of diapers. On average nursing homes in Japan spend a million Japanese yen (about 9,000 Euros) per month per 100 beds. The red numbers in the blue boxes show the number of diaper styles used. Initially, Oka used only ten styles of diaper. One and a half years later, in early 2002, the average cost of diapers rose to a million yen, the national average. Then Oka introduced ten more kinds of diapers and started its diaperless initiative (excretory

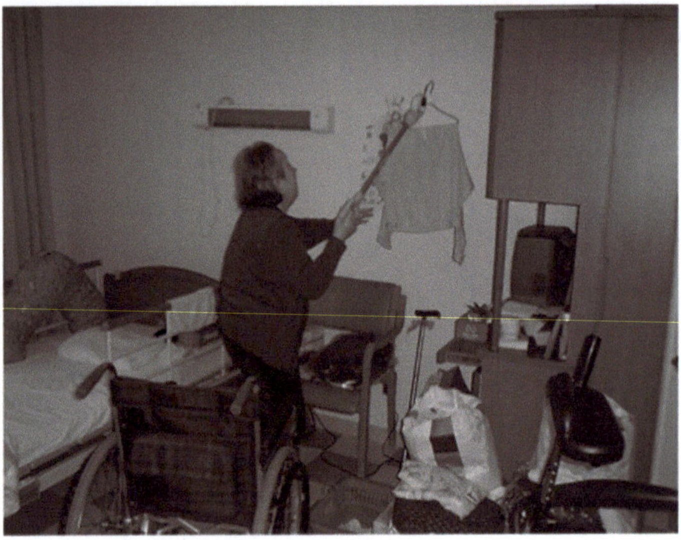

Fig. 20.10 The client started washing and drying her clothes in her room

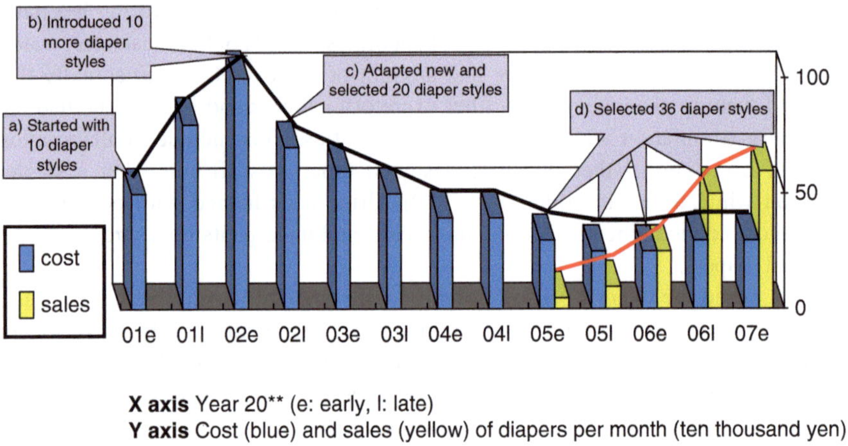

Fig. 20.11 Change in average monthly diaper expenses and sales at Sendan No Oka

control with pads, diaper fitting, and adequate opportunities for toilet use). This resulted in a drastic reduction in diaper expenses and helped improve the financial situation at Oka.

Beyond savings, a new retail business of delivery diapers and urine pads with personal fitting was created. Clients and families using Oka's daytime rehabilitation services and other clients, families who heard of Oka's innovations, asked Oka for fitting and sale of their diapers and urine pads. The yellow bars in Fig. 20.11 represent diaper sales. Normally, delivery costs would cut into income, but as the

Oka staff must visit clients' homes for health care and social services, no additional cost is incurred. This business has been very beneficial for clients and families, and at the same time become a highly profitable business for Oka.

20.5.5.3 Trial for Community Comprehensive Rehabilitation

Based on these and other practice-based innovations, the financial condition of Oka improved. Oka was eventually able to increase the number of staff and has begun even more client-oriented health and social services. The staff could spend more time interacting with each client, and with enhanced knowledge and relationships with clients, were able to implement more changes and more motivation management-based rehabilitation. Many clients fear rehabilitation, because it was originally severe exercise for injured warriors, not enjoyable activity. But as Oka focused on motivation management, many clients changed their attitudes toward rehabilitation and saw it as useful and enjoyable. Social workers, care staff and occupational therapists cooperate in motivating clients and providing safe, enjoyable and effective rehabilitation experiences. At Oka, the staff try to embed rehabilitation in clients' daily life. Ongoing rehabilitation woven into daily life and supported by client motivation and staff expertise can be much more effective than special, short-term, training (Oomori 2007). Bettencourt et al. (2002) claim that multi-party collaboration and client-provider co-creation of new care services can yield innovative approaches to the improvement of elder care, and this seems to be the case at Oka.

Increasing staff not only elevates the quality of Oka service, but can create monetary benefit. For example, Oka has many occupational therapists as staff members. LTCI reforms in 2005 have, in most cases, rendered the welfare equipment sales and rental business unprofitable. But Oka offering these services with effective adjustment and advice from knowledgeable occupational therapists has developed a relationship of high trust with clients and families and still makes a high profit in the business. Its investment in staff has paid off in terms of service and business, thus strengthening the whole organisation and supporting more innovation. Oka's innovation ecosystem itself may seem neither radical nor systematic, but it is certainly creating impressive radical innovation.

Based on the skills and knowledge of its maturing, growing staff, Oka has been expanding innovation beyond the frame of LTCI. Oka Plus is one of these initiatives. Most of Oka's revenue comes through the provisions of LTCI. The staff understand that client needs sometimes extend beyond LTCI, but it is financially difficult to provide major, long-term services inapplicable to LTCI. Oka Plus (Fig. 20.12) was established partly to provide such community-based service. In line with the European shift from institutional care towards open care services (Djellal and Gallouj 2006) there are strong trends toward community-based care in Japan, as well. The main three businesses of Oka Plus are preventive care services covered under LTCI, committed preventive care services funded by the local government, and other services which are not provided by either of those, but are helpful to the elderly in the community. Oka Plus is providing mainly physical

Fig. 20.12 Sendan No Oka Plus

rehabilitation and accident prevention services, but it also provides other, related services through constructing a place (*ba* in Japanese) for the elderly in their community. Oka Plus is one method for realising Oka's goal of client-oriented, holistic, community-based comprehensive health care and social services.

20.6 Mr. T and Comprehension of Oka's Practice-Based Innovations

In conclusion, I will offer one more, remarkable example of the success of Oka's motivation and empowerment management employing staff expertise and multi-disciplinary collaboration.

A client here called 'Mr. T' moved to Oka when he was 93 years old. His physical situation at that time was very severe. He had a left thighbone fracture, frequent strokes, a brain tumour, angina pectoris and prostatic hypertrophy. On entry, his activities of daily living (ADL) log looked like this:

- Eating – on bed with face down position
- Excretion – diaper (almost incontinent)
- Bathing – totally dependent on lift equipment
- Clothes change – totally dependent

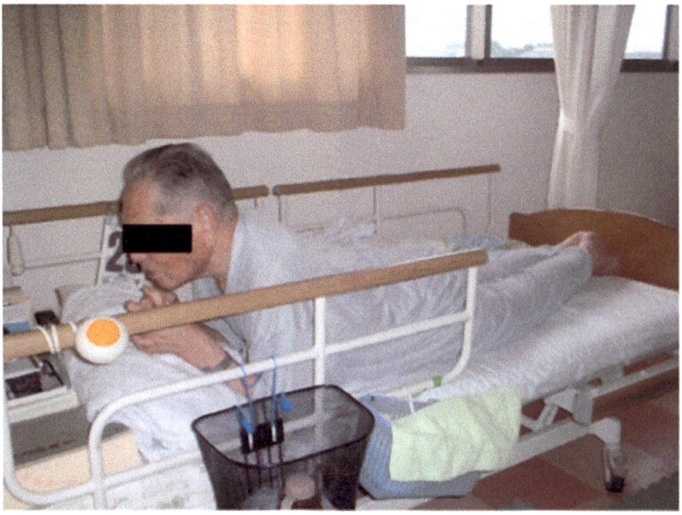

Fig. 20.13 Mr. T was always in bed with his face down

Mr. T seemed at a loss and appeared to have little interest in living. He was disgusted with rehabilitation and protested against it. His leg had been wounded in World War II and he could not bend his legs. That made it impossible for him to stand up independently and made him dependent in most aspects of daily life. He was always alone and in bed, even in the daytime (Fig. 20.13). He had no motivation or energy. He was desperate and emotionally dependent. His attitude was a heavy burden for his family and made it impossible for them to care for him in his son's home.

The Oka staff started by observing his daily life and working to build a human connection with him in hopes of building his motivation. Once, he whispered, without seeming to notice it himself, "If only I could stand up by myself." To help him realise this hope, care workers and an occupational therapist analysed his physical abilities and range of motion, and set up a pole in an appropriate place in his room. Using this pole (Fig. 20.14), he could stand up by himself.

Mr. T faced a not-uncommon situation. As a result of his handicap, he lost much more than physical function. He lost his hope, his motivation, and even his life with his family. But the Oka staff believed that they could intervene to prevent lost physical function from creating a second huge loss in quality of life. The Oka staff used their expertise to help him regain some hope, which resulted in growing trust and motivation.

Next Mr. T expressed his hope to sit in a chair and eat meals together with others, but his rigid leg was a big challenge. An occupational therapist modified a chair by hand to create a custom-made chair for Mr. T. With this chair he could escape from his bed (Fig. 20.15). The bed was no longer his home. He could eat meals together and communicate with other clients. His motivation increased dramatically. A very simple service created a profound effect. Motivation is the main engine for

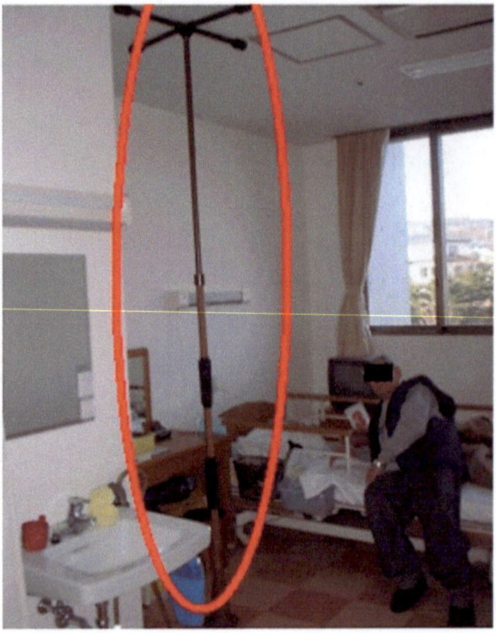

Fig. 20.14 Pole for standing up independently

Fig. 20.15 Mr. T could sit with the others in the unit

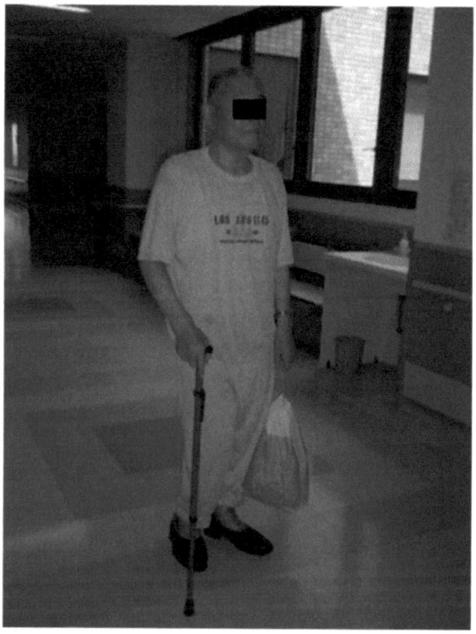

Fig. 20.16 Independent walking

Table 20.1 Mr. T's ADL transition in 28 months

	At entering	28 months	At leaving
ADL	Dependent		Almost independent
Eating	On bed with face down position		Independent on chair
Excretion	Diaper (almost incontinent)		Independent
Bathing	Totally dependent on lift equipment		Independent, normal bath
Clothes change	Totally dependent		Independent

improved quality of life. Each small success increased his motivation. He, together with the Oka staff, continued setting new goals, one after another. He began using the toilet independently and even started voluntary walking rehabilitation.

After 28 months at Oka, Mr. T finally could walk independently. His positive attitude also changed his relationship with his family. His son willingly welcomed his re-born, positive-thinking father back to live in his home. Mr. T independently walked out of Oka on his way to rejoining his son's family (Fig. 20.16). Mr. T's ADL transition is summarised in Table 20.1.

20.7 Conclusion and Discussion

Mr. T's case is remarkable and not all clients recover like him. But his case is one example of Oka's innovative approach to health care and social services, which yield a high level of care at manageable cost. This case, like Oka's overall development, consisted of incremental steps that eventually yielded remarkable results.

Sendan No Oka's innovations started in a difficult situation, as almost emergency measures. Initial operations were far from the innovative, client-oriented, high quality service and ground-breaking community care service that eventually developed. A casual observer might have seen the incremental changes as normal, provider-oriented, organisational learning processes. But Sendan No Oka steadily accumulated small innovations, and established an innovation ecosystem based on motivation management and empowerment management. This may be Oka's most radical innovation.

Referring again to Mr. T, why did his motivation improve? In ordinary Japanese nursing homes, staff are so busy that they can afford little time to communicate with clients, or maximise their expertise. Motivated, empowered staff can, in turn, provide the support necessary to motivate and empower clients, thus yielding great fruit. But it is very difficult for single fruit to grow in a vacuum. An innovation ecosystem is needed for a host of advances to be practically realised. Oka had humble beginnings and grew only incrementally, but this was a key in developing staff capable of motivating and empowering themselves and applying multi-disciplinary collaboration and ever-growing expertise for their clients' benefit. Mr. T's 'miracle' parallels Oka's own story of great achievement by way of incremental steps.

Of course, there are some aspects of the Oka story which may be culturally dependent, like multi-disciplinary collaboration. But Oka's real-time approach to solving immediate challenges and the development of an innovation ecosystem have universal implications for facilitating innovation in other areas and fields.

References

Alam, I., & Perry, C. (2002). A customer-oriented new service development process. *Journal of Services Marketing, 16*(6), 515–534.

Anderson, P., & Tushman, M. L. (1990). Technological discontinuities and dominant designs: A cyclical model of technological change. *Administrative Science Quarterly, 35*, 604–633.

Barras, R. (1986). Towards a theory of innovation in services. *Research Policy, 15*, 161–173.

Bettencourt, L. A., Ostrom, A. L., Brown, S. W., & Roundtree, I. (2002). Client co-production in knowledge-intensive business services. *California Management Review, 44*(4), 100–128.

Cook, P., Goh, H., & Chung, H. (1999). Service typologies: A state of the art survey. *Product and Operation Management, 8*, 318–338.

Cooper, G., & Edgett, J. (1999). Critical success factors for new financial services. *Marketing Management, 5*(3), 26–37.

Cowell, D. (1988). New service development. *Journal of Marketing Management, 13*(3), 296–312.

de Brentani, U. (2001). Innovative versus incremental new business services: Different keys for achieving success. *Journal of Product Innovation Management, 18*, 169–187.
den Hertog, P. (2000). Knowledge-intensive business services as co-producers of innovation. *International Journal of Innovation Management, 4*(4), 491–528.
Djellal, F., & Gallouj, F. (2006). Innovation in care services for the elderly. *Service Industries Journal, 26*(3), 303–327.
Doi, K. (2004). Unit care and team approach. *With, 21* (August). http://www.tfu.ac.jp/tushin/with/bk1.html. Accessed 21 April 2011. Sendai: TFU corresponding school.
Doi, K. (2006). *Practice of job education in care rontier – Case of Sendan No Oka*. Research Report of Academic Frontier Promotion, Project B group in 2005. Unpublished report. Sendai: Tohoku Fukushi University, Kansei Fukushi Research Center.
Dolfsma, W. (2004). The process of new service development – Issues of formalization and appropriability. *International Journal of Innovation Management, 8*(September), 319–337.
Fitzsimmons, A., & Fitzsimmons, J. (2000). *New service development: Creating memorable experiences*. Thousand Oaks, CA: Sage Publications.
Froehle, M., Chase, B., & Voss, A. (2000). Antecedents of new service development effectiveness: An exploratory examination of strategic operations choices. *Journal of Service Research, 3*(1), 3–17.
Gadrey, J., Gallouj, F., & Weinstein, O. (1995). New modes of innovation: How services benefit industry. *International Journal of Service Industry Management, 6*(3), 4–16.
Gavrilov, L. A., & Heuveline, P. (2003). Aging of population. In P. Demeny & G. McNicoll (Eds.), *The encyclopedia of population* (Vol. 1, pp. 32–37). New York: Macmillan Reference USA.
Japanese Ministry of Health, Labour and Welfare. (2010). *Annual health, labour and welfare report 2008–2009*. Tokyo: Ministry of Health, Labour and Welfare. http://www.mhlw.go.jp/english/wp/wp-hw3/index.html. Accessed 21 April 2011.
Kelly, D., & Storey, C. (2000). New service development: Initiation strategies. *International Journal of Industry Management, 11*(1), 45–62.
Matthing, J., Sandén, B., & Edvardsson, B. (2004). New service development: Learning from and with customers. *International Journal of Service Industry Management, 15*(5), 479–498.
Möller, K., Rajala, A., & Svahn, S. (2005). Strategic business nets – their type and management. *Journal of Business Research, 58*, 1274–1284.
Oomori, T. (2004). Unit care and facility management 1–2. *With, 19–20* (May–June). http://www.tfu.ac.jp/tushin/with/bk1.html. Accessed 21 April 2011. Sendai: TFU corresponding school.
Oomori, T., & Matsuura, A. (2007). Rehabilitation in care unit. *With, 42* (March). http://www.tfu.ac.jp/tushin/with/bk1.html. Accessed 21 April 2011. Sendai: TFU corresponding school.
Ramaswamy, R. (1996). *Design and management of service processes*. Reading, MA: Addison-Wesley.
Scarbrough, H., & Lannon, R. (1989). The management of innovation in the financial services sector: A case study. *Journal of Marketing Management, 5*(1), 51–62.
Schumpeter, J. A. (1934). *The theory of economic development*. Cambridge, MA: Harvard University Press.
Social Welfare Association Tohoku Fukushi Corporation Sendan no Mori. (2004). *Documents of Sendan no Mori practices*. Unpublished reports. Sendai: Sendan no Mori.
Sundbo, J. (1997). Management of innovation in services. *Service Industries Journal, 17*(3), 432–455.
Vargo, S. L., & Lusch, R. F. (2004). Evolving new dominant logic for marketing. *Journal of Marketing, 68*(1), 1–17.
Voss, C., Johnston, R., Silvestro, L., Fitzgerald, L., & Brignall, T. (1992). Measurement of innovation and design performance in services. *Design Management Journal, 3*(Winter), 40–46.
Yamamoto, K., & Sugita, T. (2008). Emerging oligopolistic hospitals and social welfare corporations in Japan. *The Quarterly of Social Security Research, 43*(4), 343–353. National Institute of Population and Social Security Research.

[illegible] (2008) [illegible] Different [illegible] *Journal of Product Innovation Management* [illegible]: 409–427
den Hertog, P. 2000. Knowledge-intensive business services as co-producers of innovation. *International Journal of Innovation Management* 4 (4): 491–528
Djellal, F. & Gallouj, F. (2001). Innovation in care services for the elderly. *Service Industries Journal* 21 (4): 50[illegible]–81[illegible]
[illegible]

[illegible]

Chapter 21
Innovation Capability and Its Measurement in Finnish SMEs

Minna Saunila, Juhani Ukko, and Hannu Rantanen

Abstract The importance of the development of an organisation's innovation capability for its success is highlighted in current literature. This study provides a comprehensive description of the status quo of Finnish SMEs regarding the different dimensions of innovation capability, expertise, work wellbeing, and the measurement of issues related to innovation capability. The study also clarifies whether the results of the studied issues differ depending on the size of the organisation, the industry and the organisational position. The most important insight of the study is that the perceptions of executives and employees differ significantly with regard to all the studied aspects. The study suggests that both academics and practitioners should focus on the development of new methods and practices for enhancing innovation capability, and especially the measurement of issues related to it.

21.1 Introduction

The valuation of innovations has increased in small and medium-sized enterprises (SMEs). Today, innovations are mainly created in practical contexts, where many different sources of information are exploited in solution-centred processes (Harmaakorpi and Melkas 2008). Many innovation systems are about to change from a Science, Technology and Innovation (STI) mode towards a Doing, Using and Interacting (DUI) mode. STI is a mode of innovation that focuses on codified knowledge and science-based learning. DUI is a mode of innovation where the focus is on tacit knowledge, organisational learning and user needs (Berg Jensen et al. 2007). The DUI mode requires a new kind of innovation culture for an organisation, implying abilities of producing innovations by everyday work.

M. Saunila (✉) • J. Ukko • H. Rantanen
Lahti School of Innovation, Lappeenranta University of Technology, Lahti, Finland
e-mail: minna.saunila@lut.fi

H. Melkas and V. Harmaakorpi (eds.),
Practice-Based Innovation: Insights, Applications and Policy Implications,
DOI 10.1007/978-3-642-21723-4_21,

Innovation is viewed as an evolutionary process within an organisation to adopt any change pertaining to a device, system, process, policy, or service that is new to the organisation (Calantone et al. 2002). Thus, innovation can be regarded as an organisational capability because it is an act that deploys resources with a new ability to create value (Yang et al. 2009). Developing this innovation capability is important, as innovation plays a key role in the survival and growth of organisations (Francis and Bessant 2005). In order to manage innovation capability, it has to be measured. The importance of measurement is especially true for innovations where there is need to bring clarity to a fundamentally creative process (Skarzynski and Gibson 2008).

The purpose of this study is to provide a comprehensive description of the status quo of Finnish SMEs regarding the different dimensions of innovation capability, expertise, work wellbeing and the measurement of issues related to innovation capability. The study also clarifies whether the results of the studied issues differ depending on the size of the organisation, the industry and the organisational position. The study shows that the most significant differences can be found between the executives and the employees. The study also contributes to issues around innovation capabilities that academics and practitioners should focus on in the future.

21.2 Innovation Capability

The successful operation of organisations in almost all industries is becoming highly dependent on their ability to produce innovations (Tidd et al. 2005). The current literature presents many classifications for innovation (see, for example, Wan et al. 2005), and some of them are presented below. There are two key aspects of innovation: the degree of innovation and the scope of innovation. The degree of innovation is divided into radical and incremental innovation. The scope of innovation capability consists of technical innovation and administrative innovation (Lin et al. 2010). Innovation is a dynamic capability (i.e., a capability which allows the organisation to integrate, build, and reconfigure internal and external competences to address rapidly changing environments; Teece et al. 1997) with multiple dimensions (Sáenz et al. 2009; Lin et al. 2010). Therefore, the concept of innovation capability is intangible by nature. It cannot be measured directly but by evaluating dimensions closely related to innovation capability.

21.2.1 Dimensions of Innovation Capability

Several earlier studies have presented dimensions related to innovation capability. According to Lawson and Samson (2001), the elements making up an innovation capability are vision and strategy, harnessing the competence base, organisational intelligence, creativity and idea management, organisational structure and systems, culture and climate, and management of technology. Skarzynski and Gibson (2008) divide innovation capability into four parts: leadership and organisation, people and skills, process and tools, and culture and values. The first part stresses that the leaders and organisation share a common vision of innovation. The second part includes a disciplined approach to building innovation capabilities across the organisation. The third part includes, for example, supporting tools to enable an idea generation pipeline and portfolio management. Innovation capability also requires a collaborative, open culture and incentives that reward challenging current actions, which forms the fourth part. Tura et al. (2008) define innovative capability through three subcategories: openness/creativity, knowledge/expertise, and operationalisation capability. The first subcategory comprises the capabilities needed to exceed the existing solutions and search for new possibilities. The second subcategory covers the capabilities to acquire the knowledge needed to build innovation. The third subcategory describes the capabilities to find and introduce applications, so that the organisation exploits the achievable knowledge base. Tidd et al. (2005) list the following components of an innovative organisation: shared vision, leadership and the will to innovate the appropriate structure, key individuals, effective team working, continuing and stretching individual development, extensive communication, high involvement in innovation, external focus, creative climate, and learning organisation. These different approaches and findings have been combined to a conceptual framework of dimensions of innovation capability developed by Paalanen et al. (2009) and further refined by the authors. In this study, innovation capability is divided into five dimensions. The dimensions are presented in Fig. 21.1 and discussed in detail below.

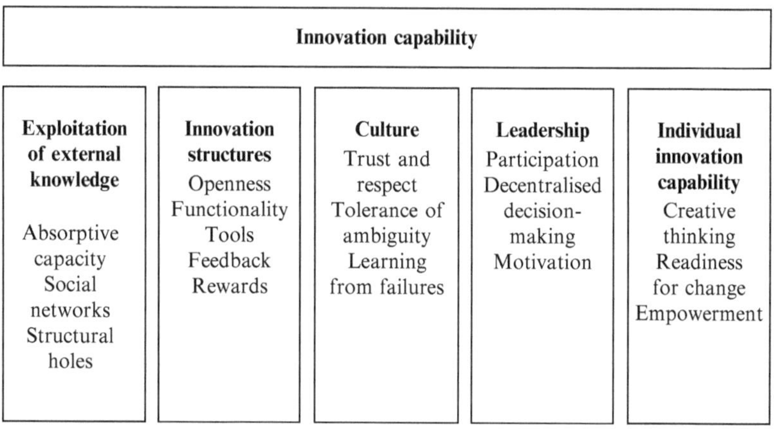

Fig. 21.1 Dimensions of innovation capability in the study (adapted on the basis of Paalanen et al. 2009)

21.2.1.1 Exploitation of External Knowledge

Absorptive capacity means the organisation's capacity to acquire and exploit new knowledge (Cohen and Levinthal 1990). According to Granovetter (2005), innovations are created in networks in social interaction of actors. Interaction with suppliers, customers, public assistance agencies, industry associations, foundations and the like can provide missing inputs into the learning process, which the firm itself cannot provide (Romijn and Albaladejo 2002; Lawson and Samson 2001). According to Panayides (2006), an organisation oriented towards a client relationship can enhance its innovation capability as it will become more creative in its methods of operation, will seek new ways for doing things and trying out new ideas, and will be the first to market with new products and services. People across organisations have more alternative ways of thinking. These structural holes give them more options to select between new ideas (Burt 2004).

21.2.1.2 Innovation Structures

There is a common agreement on the necessity of flexible structures for innovation (Dobni 2008; Nahapiet and Ghoshal 1998; Subramanian and Nilakanta 1996). According to Subramanian and Nilakanta (1996), decentralised and informal organisational structures facilitate innovations. They also propose that the flexibility and openness of structures help to encourage new idea generation. In the dynamic environment, the organisational structure will need to be more open than precisely defined, more emotionally-inclusive than rationally-inclusive, more interactive than integrative, more temporal, more flexible, and trust- and informality-based (Wang and Ahmed 2003). Communication channels facilitate knowledge sharing by exchanging experiences. It requires tight social dealings to transfer tacit knowledge (Nahapiet and Ghoshal 1998). Reward systems are powerful motivators and foster creative behaviour (Lawson and Samson 2001). Bringing innovation to every workstation requires practical tools, processes, and mechanisms that the employees can use day by day to turn innovation into the organisation's capability (Skarzynski and Gibson 2008).

21.2.1.3 Organisational Culture

Every organisation has its own culture, also known as style, character, or way of action. For organisations, culture is like personality for individuals. It is hard to change, but it can be developed. Innovative organisations tolerate ambiguity, but they do not take unnecessary risks (Lawson and Samson 2001). Organisations need to be tolerant of the mistakes that will occur and allow for recovery and learning from failures (Wan et al. 2005; Lawson and Samson 2001). Innovation is more likely in a situation where people attribute high levels of integrity, competence, reliability, loyalty and openness to others, and view others as equals. Creating this environment requires the employees to understand their roles, and then develop their creative and independent sides further (Dobni 2008). Knowledge transfer requires tight social dealings. Social capital is situated in relationships between individuals (Nahapiet and Ghoshal 1998). Organisations with high innovation capability often have a learning-by-doing effect, which makes it hard for competitors to imitate and buy this know-how (Cavusgil et al. 2003). For employees to be motivated to innovate, there must be a culture that both supports and rewards innovation (Wan et al. 2005). Mutual trust and respect create an atmosphere that encourages individuals to try new ideas without fear of failure and its consequences (Lampikoski and Emden 1999; Wan et al. 2005).

21.2.1.4 Leadership

Building a successful, independent capability for innovation is basically a leadership challenge. Building an innovation infrastructure means breaking the boundaries that usually separate people, ideas and resources, and creating cross-boundary conversation and collaboration (Skarzynski and Gibson 2008). Today, leadership is seen as a skill to direct the employees' energy towards the right direction instead of giving orders and instructions (Lampikoski and Emden 1999). The management must strike a balance that allows employees to act on good ideas (Dobni 2008). A participative and consultative management style is crucial for innovation (Wan et al. 2005; Yukl 1998). The responsibility for innovation needs to be widened to the people who need it to complete their tasks and spread throughout the organisation's businesses and functions. Decentralisation allows quick action and flexibility (Wan et al. 2005). The managers' active investment in increasing the employees' possibilities to participate in the development of the organisation facilitates innovation capability (Lampikoski and Emden 1999). Participative leadership boosts the employees' trust, commitment and appreciation of the managers. They are also more motivated to do their tasks (Yukl 1998).

21.2.1.5 Individual Innovation Capability

According to Skarzynski and Gibson (2008), innovation is a skill that can be taught. For example, empowerment affects the execution of innovation, as it drives employees to go above and beyond what is normally expected of them (Yang and Choi 2009; Dobni 2008). Yang and Choi (2009) present four dimensions of empowerment: autonomy, responsibility, information, and creativity. Empowerment reflects the fact that the individual can affect the outcome of his work. So, as business realities change, the employee's behaviour and actions need to be adjusted accordingly. The very essence of innovation is to get the employees to think creatively, become adventurous, and take managed risks (Amabile 1997; Dobni 2008). Creative thinking includes the following points: the individual has new perspectives on problems, is willing to take risks, and has tolerance for ambiguity (Amabile 1997).

21.2.2 *Innovation Capability, Expertise and Work Wellbeing*

Expertise is the senses, abilities and wills (=motivation) of individuals. Senses and abilities are not enough if there is no motivation to direct the expertise to action. Motivation is also connected to an individual's wellbeing and his/her experience as a member of the work community (Viitala 2005). The expertise of individuals, systems and ways of action, culture and atmosphere, and networks, contacts and partners form the expertise of a work community (Otala 2003). Physical and mental wellbeing are requisites of an individual's innovativeness. According to Salojärvi (2006), work wellbeing is formed by expertise, health, work community, and work environment.

Expertise, work wellbeing and the nature of the work community enable innovations (Salojärvi 2006). The factors that affect the development of the expertise of individuals to the capability of the organisation are very similar to the dimensions of innovation capability presented above: interaction, structures, patterns, systems, and leadership (Viitala 2005). If this resource crumbles, the same happens to innovation capability and thus the competitiveness of the organisation. Organisations have to pay attention to the work wellbeing and expertise of the employees, so that sufficient creativity and innovativeness can be guaranteed. Creativity and innovativeness are, after all, the only significant ways to survive in the global economy (Salojärvi 2006).

21.3 Measurement of Innovation Capability

It is obvious that improving the innovation capability is the key to acquiring sustainable competitive advantage for an organisation (Shan and Zhang 2009). According to Skarzynski and Gibson (2008), innovation measures can help managers in two ways:

- First, to make informed decisions based on objective data;
- Second, to help align goals and daily endeavours with the near- and long-term innovation agenda.

Thus, it is significant to evaluate the organisation's innovation capability accurately, and to find the key factors influencing the organisation's innovation capability (Shan and Zhang 2009).

Intangibles are hard to concretise, and that is why they are also hard to measure and manage (Bontis 2001; Kujansivu et al. 2007; Marr 2007). However, everything is measurable. The real question is whether it is worth measuring, if the accuracy is not good enough and the measurement costs are higher than the attained advantages. When measuring intangible things, indirect measures are usually used. Indirect measurement is used when something cannot be measured directly. If the phenomenon itself cannot be measured, then something closely linked to the phenomenon has to be measured. Indirect measures can be divided into objective and subjective measures.

21.3.1 Objective Measurement of Innovation Capability

Generally, it seems that the current innovation capability measures can be divided into input measures and output measures. Input measures evaluate how the innovation activities have been arranged and how resources are allocated to them. Input measures include the funds used in R&D activities, new product development, and education (Skarzynski and Gibson 2008; Tura et al. 2008). Input measurement is problematic, because it tells how much is devoted, not if anything has been accomplished. Input measures also underestimate smaller innovation activities. Smaller organisations do not have opportunities to invest in R&D. That is why input measures do not reflect the actual innovation capability (Romijn and Albaladejo 2002).

Output measures evaluate the effects of innovation capability. It is hard to express all kinds of innovations quantitatively, and output measures usually measure the results of successful innovations (Tura et al. 2008). Output measures mainly include the organisation's patents and licenses. The problem of output measures is that they are only suitable for certain types of innovations and organisations. They are not suitable for small or service organisations (Romijn and Albaladejo 2002). Output measures do not measure the economic value of all kinds of innovations, either (Tura et al. 2008). All in all, the current type of measurement is especially difficult for practice-based innovation, resulting from the intangibility of the phenomenon.

There are some common characteristics in the current objective innovation capability measures. They are focused on industrial and technology innovations, and service innovations have no proper measures (Skarzynski and Gibson 2008; Tura et al. 2008). The current measures do not recognise, either, that organisations are of different sizes, and they operate in very different business areas (Carayannis and Provance 2008) or in the public sector. The best ones of the current measures are the ones which pay attention to both the inputs and outputs of innovation (Romijn and Albaladejo 2002). The input and output measures actually evaluate the innovation performance, and therefore also subjective measurement is needed to capture the overall innovation capability.

21.3.2 Subjective Measurement of Innovation Capability

Innovation capability has traditionally been measured via questionnaires or other subjective assessment models. There are various models for the measurement of innovation capability or factors closely related to innovation capability.

- Panayides (2006) has studied the antecedents and consequences of innovation capability of logistics service providers. He defines five antecedents of innovation capability: trust, bonding, communication, shared value, and empathy, including 18 items. The items are measured on a seven-point Likert scale.
- Dobni (2008) has created a procedure for the measurement of innovation culture (divided into innovation intention, innovation infrastructure, innovation influence, and innovation implementation) in organisations, which includes some of the dimensions of innovation capability presented above. The procedure includes 86 items applied with a seven-point Likert scale.
- Prajogo and Ahmed (2006) have studied human-based practices (namely leadership, people management, knowledge management, and creativity management) that provide innovation stimulus, and technology-based practices (namely technology management and R&D management), underpinning the capacity to innovate as part of their study of the relationship between innovation stimulus, innovation capacity, and innovation performance. The whole instrument includes eight scales consisting of 34 items and utilising a five-point Likert scale.
- Calantone (2002) have studied the learning orientation, composed of commitment to learning, shared vision, open-mindedness, intra-organisational knowledge sharing, and innovation capability. The construct includes four dimensions with 17 items of learning orientation measured by a seven-point Likert scale.
- Alegre and Chiva (2008) present an instrument concerning an organisation's learning capability. The instrument includes five dimensions (experimentation, risk taking, interaction with the external environment, dialogue, and participative decision making), measured by 14 items. The measurement scale is applied using a seven-point Likert scale.
- Bettencourt (2004) presents a questionnaire about change-oriented organisational citizenship behaviours. All constructs are measured with seven-point scales. The construct includes seven dimensions: core transformational leadership behaviour, contingent reward leadership behaviour, leader-member exchange quality, organisational commitment, learning goal orientation, performance goal orientation, and change-oriented organisational citizenship behaviours, measured by 40 items.

The previous studies do not take into account the measurement perspective, and do not capture the concept of innovation capability comprehensively. It is not enough to measure just one or two particular aspects of an organisation's innovation capability. This is especially true for practice-based innovation.

21.4 Research Methodology

This chapter presents a quantitative survey study conducted in Finland in spring 2010. The first objective of the study is to provide a comprehensive description of the status quo of Finnish SMEs concerning the different dimensions of innovation capability, expertise, work wellbeing and the measurement of issues related to innovation capability. The second objective is to clarify whether the results differ depending on the size of the organisation, the industry and the organisational position.

The survey data of the study were gathered from Finnish SMEs with a web-based questionnaire. SMEs with less than ten employees were excluded from the sample. In Finland, there are 8,214 SMEs employing 11–249 persons and having revenue less than 50 Meuro. A sample of 2,400 SMEs was randomly selected. A representative of both management and employees received an invitation to participate in this study. Thus, 4,800 questionnaires were sent. A total of 4,050 questionnaires reached the respondents, as 750 e-mail addresses were invalid.

The survey was conducted in four waves. One week after the first mailing of the survey, reminder questionnaires were sent out. This process resulted in a total of 311 responses, which equals the response rate of 7.68%. The background information of the respondents is presented in Table 21.1.

To create the research questions, a literature review was conducted. The data for eight variables were collected by the questionnaire: innovation capability (divided into exploitation of external knowledge, innovation structures, organisational culture, leadership, and individual innovation capability), expertise, work wellbeing and measurement. Each of these was operationalised through 3–9 research questions (originally in Finnish). The questionnaire was pre-tested by a small group of academics. This resulted in minor changes to the presentation of the questionnaire.

Each of the questions was measured by a five-point Likert scale, ranging from 1 (strongly disagree) to 5 (strongly agree). The data were analysed first with SPSS

Table 21.1 Background information of the respondents

		n	%
Revenue (Meuro)	2–5	141	45.3
	5–20	135	43.4
	20–50	35	11.3
No. of employees	10–49	224	72.0
	50–249	87	28.0
Industry	Industrial	145	46.6
	Service	159	51.1
	No response	7	2.3
Organisational position	Executive	222	71.4
	White-collar worker	68	21.9
	Blue-collar worker	12	3.9
	No response	9	2.9

software. The means and standard deviations for all the questions of eight variables were formed first. Second, comparison of the means was conducted by analysis of variance (ANOVA). The differences between the size of the organisations, different industries and different organisational positions were studied.

21.5 Description of the Data

In this section, a description of the entire data is presented by means and standard deviations. As regards the results of different dimensions of innovation capability presented in Table 21.2, it can be stated that the issues concerning the exploitation of external knowledge (means 3.75–4.13), culture (means 3.72–3.97), and leadership (means 3.6–4.02) seem to be on a satisfactory level. In the dimension of

Table 21.2 Means and standard deviations of the research questions

Research question	Mean	Std. dev.
Exploitation of external knowledge		
U1 My work community encourages gaining knowledge through external contacts	4.00	1.042
U2 We have developed our ways of action by comparing our operations to other organisations	3.75	0.985
U3 We develop our actions together with our stakeholders (customers etc.)	4.13	0.864
Innovation structures		
I1 Co-operation works well in our organisation	3.84	0.852
I2 We have a clear way of processing and developing ideas	3.32	1.032
I3 The employees get feedback for their ideas	3.82	0.828
I4 Our reward system encourages ideation	2.86	1.062
Culture		
K1 Our organisation actively seeks new ways of action	3.72	1.040
K2 Our organisation has the courage to try new ways of action	3.83	0.941
K3 When experimenting with new ways of action, mistakes are allowed	3.86	0.860
K4 The employees have the courage to disagree	3.97	0.820
Leadership		
J1 The managers encourage initiatives	3.72	0.912
J2 The managers give positive feedback	3.60	0.844
J3 The managers pass employees' ideas to the upper levels of the organisation	3.67	0.864
J4 The managers participate in ideation and development	4.02	0.781
Individual creativity and innovation capability		
L1 The employees are willing to participate in development	3.70	0.830
L2 It is easy for the employees to adopt new ways of action	3.12	0.959
L3 The employees know how to be critical towards current ways of action when needed	3.97	0.734

innovation structures, the mean for the question "we have a clear way of processing and developing ideas" was only 3.32, and for the question "our reward system encourages ideation" even lower, 2.86. This indicates that the innovation process is not clear in these Finnish SMEs, and they have failed in rewarding good ideas. With regard to the dimension of individual creativity and innovation capability, the question "it is easy for the employees to adopt new ways of action" had the lowest mean, 3.12.

The issues concerning the expertise presented in Table 21.3 seem to be on a satisfactory level in the Finnish SMEs, with the exception of the question "there are practices for transferring tacit knowledge". This question had the mean 3.12. Also the issues of work wellbeing seem to be on a satisfactory level. The only question in this dimension that was not considered as good was "the number of working tasks is suitable" with the mean 3.29. When focusing on the big picture of the measurement

Table 21.3 Means and standard deviations of the research questions (continued)

Research question	Mean	Std. dev.
Expertise (know-how)		
O1 All employees have a possibility for education	3.72	1.072
O2 We have instructions and responsible persons for work orientation	3.66	1.061
O3 The employees are encouraged to be multi-skilled	4.03	0.740
O4 Voluntary learning and development of expertise are supported in our organisation	3.78	0.881
O5 There are practices for transferring tacit knowledge	3.12	1.055
O6 In our organisation, learning is an investment, not an expense	3.80	0.924
Work wellbeing		
T1 The employees prosper in our organisation	4.00	0.687
T2 The employees are treated equally	3.87	0.915
T3 The employees are appreciated for their work	3.80	0.706
T4 The number of working tasks is suitable	3.29	1.015
T5 The quality, demands and responsibility of tasks are suitable	3.73	0.807
T6 There is an opportunity for flexible working and working hours in our organisation	3.89	1.053
Measurement		
M1 Our organisation has measures for evaluating development	2.89	1.088
M2 Exploitation of external (customers, competitors etc.) knowledge is evaluated or measured in our organisation	3.02	1.066
M3 The functionality of internal processes is evaluated or measured in our organisation	3.44	1.068
M4 The development of ways of action is evaluated or measured in our organisation	3.18	1.032
M5 Leadership practices are evaluated or measured in our organisation	2.85	1.078
M6 The employees' ideas are evaluated or measured in our organisation	2.94	1.062
M7 The employees' work wellbeing is evaluated or measured in our organisation	3.58	1.101
M8 The employees' expertise is evaluated or measured in our organisation	3.48	0.958
M9 The measurement information is used for developing the actions and operations of our organisation	3.37	1.047

of the issues related to innovation capability, it can be stated that the questions of this dimension reached much lower means (2.85–3.58) than the other dimensions. The means for the questions "leadership practices are evaluated or measured in our organisation", "our organisation has measures for evaluating development" and "the employees' ideas are evaluated or measured in our organisation" were less than three, which indicates that they were perceived as unfavourable.

As a summary of this section, it can be stated that the different dimensions of innovation capability appear to be on a satisfactory level. The most essential challenges for the future seem to be the linkage between rewarding and ideation, clarification of the innovation process, and the adaptability of employees concerning new ways of action. Regarding the expertise and work wellbeing, the most essential challenges seem to be the development of the practices for transferring tacit knowledge and the handling of the number of working tasks. The biggest challenge for the future appears to be the measurement and evaluation of the issues related to innovation capability, innovation processes, and innovations. It appears that the organisations do not have appropriate methods for the measurement and evaluation of these issues.

21.6 Findings

In this section, it is investigated whether differences can be found between organisations of different sizes, between industry- or service-oriented organisations, and between executives and employees. The differences have been studied by comparison of means, where the analysis of variance has been utilised.

First, the differences have been studied according to the amount of revenues. The revenue of the first group is 2–5 million euros, that of the second group is 5–20 million euros, and the revenue of the third group is 20–50 million euros. The

Table 21.4 Comparison of the means depending on revenue

Research question	Mean			F-value
	2–5 Meuro	5–20 Meuro	20–50 Meuro	
Exploitation of external knowledge				
U2 We have developed our ways of action by comparing our operations to other organisations	3.60	3.85	3.97	3.061*
Expertise (know-how)				
O2 We have instructions and responsible persons for work orientation	3.53	3.70	4.03	3.036*
O5 There are practices for transferring tacit knowledge	3.29	2.91	3.09	4.343*
Measurement				
M3 The functionality of internal processes is evaluated or measured in our organisation	3.28	3.49	3.91	4.826**

*0.01 $< p \leq$ 0.05; **0.001 $< p \leq$ 0.01; ***$p \leq$ 0.001

significant differences concerning all the questions are illustrated in Table 21.4. As Table 21.4 indicates, a significant difference was found only in four questions. In the dimension of exploiting external knowledge, a significant difference was found only in the question "we have developed our ways of action by comparing our operations to other organisations". It seems that the larger companies are more familiar with or have more resources for benchmarking than the smaller ones. Regarding the dimension of expertise, significant differences were found in the questions of "we have instructions and responsible persons for work orientation" and "there are practices for transferring tacit knowledge". Tukey's post-hoc test reveals that a significant difference concerning the first question was between the smallest and the largest companies, and concerning the latter question between the smallest and middle-sized companies. This indicates that the guidance for work has been done more correctly in large companies than in the smaller ones. Furthermore, it seems that tacit knowledge is easier to transfer in small companies than in large ones. In small companies the employees are more familiar with each other, and they usually operate in the same environment, which may explain this result. As a whole, it can be stated that the revenue-based size of the company does not seem to have affected the perceptions of the respondents on innovation capability, expertise, work wellbeing, or measurement.

Next, the differences were studied between companies that have 10–49 employees and companies with 50–249 employees. Contrary of previous studies (e.g., Çakar and Ertürk 2010), the results suggest that the number of employees does not have a significant effect on innovation capability, employee expertise, work wellbeing, or measurement of issues related to innovation capability. As Table 21.5 illustrates, significant differences were found only in three questions. The companies employing over 50 people seem to have a possibility to offer more flexible work and working time in comparison to smaller companies. The larger companies are also more capable to evaluate leadership practices and work wellbeing than the small companies. However, the evaluation of leadership practices is on the whole at a rather poor level. Generally, it can be stated that the size of the company, including both the amount of revenues and the number of

Table 21.5 Comparison of the means depending on the number of employees

Research question	Mean		F-value
	10–49 empl.	50–249 empl.	
Work wellbeing			
T6 There is an opportunity for flexible working and working hours in our organisation	3.78	4.16	7.589**
Measurement			
M5 Leadership practices are evaluated or measured in our organisation	2.77	3.05	3.938*
M7 Employees' work wellbeing is evaluated or measured in our organisation	3.46	3.90	9.343**

*$0.01 < p \leq 0.05$; **$0.001 < p \leq 0.01$; ***$p \leq 0.001$

Table 21.6 Comparison of the means depending on the industry

Research question	Mean		F-value
	Industry	Service	
Individual creativity and innovation capability			
L3 The employees know how to be critical towards current ways of action when needed	3.87	4.08	5.633*
Expertise (know-how)			
O1 All employees have a possibility for education	3.48	3.96	15.261***
O6 In our organisation, learning is an investment, not an expense	3.65	3.94	7.293**
Measurement			
M8 The employees' expertise is evaluated or measured in our organisation	3.60	3.36	4.432*

*$0.01 < p \leq 0.05$; **$0.001 < p \leq 0.01$; ***$p \leq 0.001$

employees, do not have a remarkable effect on the perceptions of the respondents on innovation capability, expertise, work wellbeing, or measurement.

In Table 21.6, the significant differences between industry- and service-oriented companies are presented. In total, four significant differences were found. Concerning individual creativity, it seems that the employees have better abilities to be critical of inappropriate practices in service companies than in industrial companies. In contrast to industrial companies, the service companies seem to have better possibilities for education, and learning is seen to be an investment rather than an expense. However, the industrial companies are more capable of evaluating the expertise of their employees. Traditionally, the tasks and types of expertise are more specified in the industrial sector, and are thus easier to evaluate, for example, by a capability matrix. Generally, it can be stated that the orientation of the company on industry or service has not had a comprehensive effect on the perceptions of the respondents on innovation capability, expertise, work wellbeing, or measurement.

Next, the differences between executives and employees – including both white-collar and blue-collar workers – were studied. As Table 21.7 shows, significant differences were found in all the questions concerning the dimensions of innovation structures, culture, and leadership. A significant difference was also found in the dimension of exploitation of external knowledge in the question "we have developed our ways of action by comparing our operations to other organisations". In contrast, significant differences were not found in any of the questions concerning the dimension of individual creativity and innovation capability. In all the significant differences in Table 21.7, the perception of the executives was more positive than the perception of the employees. The results indicate that there is a lot of work to do around innovation capability in the companies. Especially the areas of innovation structures, culture, and leadership need to be enhanced to fill this gap and to keep the perception of the employees at least on a satisfactory level.

When focusing on the dimensions of expertise, work wellbeing and measurement presented in Table 21.8, significant differences can be found in all the

Table 21.7 Comparison of the means depending on the position of the respondent

Research question	Mean		F-value
	Exec.	Empl.	
Exploitation of external knowledge			
U2 We have developed our ways of action by comparing our operations to other organisations	3.84	3.48	7.989**
Innovation structures			
I1 Co-operation works well in our organisation	3.94	3.56	10.319***
I2 We have a clear way of processing and developing ideas	3.42	3.00	9.067**
I3 The employees get feedback for their ideas	3.96	3.43	24.170***
I4 Our reward system encourages ideation	2.95	2.58	6.919**
Culture			
K1 Our organisation actively finds new ways of action	3.86	3.30	16.832***
K2 Our organisation has the courage to try new ways of action	3.97	3.42	21.170***
K3 When experimenting with new ways of action, mistakes are allowed	4.00	3.45	24.033***
K4 The employees have the courage to disagree	4.09	3.61	20.561***
Leadership			
J1 The managers encourage initiatives	3.84	3.39	14.575***
J2 The managers give positive feedback	3.76	3.14	32.988***
J3 The managers pass the employees' ideas to the upper levels of the organisation	3.76	3.39	10.922***
J4 The managers participate in ideation and development	4.10	3.76	10.619***

*$0.01 < p \leq 0.05$; **$0.001 < p \leq 0.01$; ***$p \leq 0.001$

questions, aside from the questions "there is an opportunity for flexible working and working hours in our organisation", "our organisation has measures for evaluating development", and "exploitation of external (customers, competitors etc.) knowledge is evaluated or measured in our organisation". As work wellbeing and expertise are drivers for the different dimensions of innovation capability (e.g., Viitala 2005; Salojärvi 2006), these issues should be enhanced to keep the perception of the employees on a more satisfactory level. Significant differences between the executives and employees concerning the different aspects of innovation capability -related measurement were also found in most of the questions. The most worrying result of this issue is that the perception of the executives was only on a neutral or negative level, and the perception of the employees was mainly on a negative level. Hence, there is a major challenge for both academics and practitioners to develop new practices and methods for the measurement and evaluation of the issues related to innovation capability.

As a summary of this section, it can be stated that the perceptions regarding the different dimensions of innovation capability, employee expertise, work wellbeing, and measurement of issues related to innovation capability seem to be dependent on the organisational position of the respondent. On the other hand, the size of the organisation (both the amount of revenue and the number of employees) or the industry do not have an effect on the perceptions.

Table 21.8 Comparison of the means depending on the position of the respondent

Research question	Mean		F-value
	Exec.	Empl.	
Expertise (know-how)			
O1 All employees have a possibility for education	3.82	3.44	6.966**
O2 We have instructions and responsible persons for work orientation	3.80	3.21	17.762***
O3 The employees are encouraged to be multi-skilled	4.18	3.57	42.477***
O4 Voluntary learning and development of expertise are supported in our organisation	3.86	3.52	8.403**
O5 There are practices for transferring tacit knowledge	3.25	2.68	16.920***
O6 In our organisation, learning is an investment, not an expense	3.91	3.45	14.515***
Work wellbeing			
T1 The employees prosper in our organisation	4.08	3.77	11.728***
T2 The employees are treated equally	4.10	3.26	53.729***
T3 The employees are appreciated for their work	3.96	3.36	42.998***
T4 The number of working tasks is suitable	3.39	3.00	7.864**
T5 The quality, demands and responsibility of tasks are suitable	3.79	3.57	3.953*
Measurement			
M3 The functionality of internal processes is evaluated or measured in our organisation	3.54	3.14	7.625**
M5 Leadership practices are evaluated or measured in our organisation	2.96	2.51	9.356**
M6 The employees' ideas are evaluated or measured in our organisation	3.04	2.66	7.067**
M7 The employees' work wellbeing is evaluated or measured in our organisation	3.75	3.00	26.454***
M8 The employees' expertise is evaluated or measured in our organisation	3.55	3.25	5.235*
M9 The measurement information is used for developing the actions and operations of our organisation	3.55	2.80	29.467***

*$0.01 < p \leq 0.05$; **$0.001 < p \leq 0.01$; ***$p \leq 0.001$

21.7 Summary and Conclusion

The current literature highlights the role of organisations' innovation capability as crucial, as innovations play a key role in the survival and growth of organisations, as well as in acquiring sustainable competitive advantage. The expertise and work wellbeing of employees are emphasised as drivers for innovation capability. Also the measurement and evaluation of innovation capability are seen as essential when focusing on the development of innovation capability.

The study focused on the current state of innovation capability, expertise, work wellbeing, and measurement in Finnish SMEs. It was concluded that the size of the organisation and the industry did not have a significant effect on innovation capability, expertise, work wellbeing, and measurement. The most important insight of the study was that the organisational position of the respondent seemed to have a significant effect on the different dimensions of innovation capability, employee expertise, work wellbeing, and measurement of issues related to innovation capability. The most significant differences between executives and employees concerning innovation capability were related to culture, leadership, and innovation structures. The entirety of measurement and the evaluation of issues related to innovation capability were on a poor level in comparison to the other issues studied. This can be explained by the intangible nature of innovation capability, which appears as difficulties in evaluation, and especially in measurement. As the sample covered 2,400 out of around 8,000 SMEs in Finland, and 311 valid responses were achieved, it can be stated that the results describe very well the current state in Finland of the issues studied.

For the future, both academics and practitioners should focus on the development of new methods and practices for enhancing innovation capability, and especially the measurement of issues related to it. From the viewpoint of the Doing, Using and Interacting (DUI) mode of innovation and practice-based innovation, it is essential to reduce the gap between the perceptions of executives and employees, and to get the perception of the employees to a more satisfactory level.

References

Alegre, J., & Chiva, R. (2008). Assessing the impact of organizational learning capability on product innovation performance: An empirical test. *Technovation, 28*(6), 315–326.

Amabile, T. M. (1997). Motivating creativity in organizations: On doing what you love and loving what you do. *California Management Review, 40*(1), 39–58.

Berg Jensen, M., Johnson, B., Lorenz, E., & Lundvall, B. (2007). Forms of knowledge and modes of innovation. *Research Policy, 36*(5), 680–693.

Bettencourt, L. A. (2004). Change-oriented organizational citizenship behaviors: The direct and moderating influence of goal orientation. *Journal of Retailing, 80*(3), 165–180.

Bontis, N. (2001). Assessing knowledge assets: A review of the models used to measure intellectual capital. *International Journal of Management Reviews, 3*(1), 41–60.

Burt, R. S. (2004). Structural holes and good ideas. *American Journal of Sociology, 110*(2), 349–399.

Çakar, N. D., & Ertürk, A. (2010). Comparing innovation capability of small and medium-sized enterprises: Examining the effects of organizational culture and empowerment. *Journal of Small Business Management, 48*(3), 325–359.

Calantone, R. J., Cavusgil, S. T., & Zhao, Y. (2002). Learning orientation, firm innovation capability, and firm performance. *Industrial Marketing Management, 31*(6), 515–524.

Capaldo, G., Iandoli, L., Raffa, M., & Zollo, G. (2003). The evaluation of innovation capabilities in small software firms: A methodological approach. *Small Business Economics, 21*(4), 343–354.

Carayannis, E. G., & Provance, M. (2008). Measuring firm innovativeness: Towards a composite innovation index built on firm innovative posture, propensity and performance attributes. *International Journal of Innovation and Regional Development, 1*(1), 90–107.

Cavusgil, S. T., Calantone, R. J., & Zhao, Y. (2003). Tacit knowledge transfer and firm innovation capability. *Journal of Business & Industrial Marketing, 18*(1), 6–21.

Cohen, W. M., & Levinthal, D. A. (1990). Absorptive capacity: A new perspective on learning and innovation. *Administrative Science Quarterly, 35*(1), 128–152.

Dobni, C. B. (2008). Measuring innovation culture in organizations: The development of a generalized innovation culture construct using exploratory factor analysis. *European Journal of Innovation Management, 11*(4), 539–559.
Francis, D., & Bessant, J. (2005). Targeting innovation and implications for capability development. *Technovation, 25*(3), 171–183.
Gopalakrishnan, S. (2000). Unraveling the links between dimensions of innovation and organizational performance. *The Journal of High Technology Management Research, 11*(1), 137–153.
Granovetter, M. (2005). The impact of social structure on economic outcomes. *Journal of Economic Perspectives, 19*(1), 33–50.
Harmaakorpi, V., & Melkas, H. (Eds.) (2008). *Innovation policy between systems.* Acta series no. 200. Helsinki: Association of Finnish Local and Regional Authorities & Lappeenranta University of Technology. (In Finnish.)
Kujansivu, P., Lönnqvist, A., Jääskeläinen, A., & Sillanpää, V. (2007). *Intangible success factors of trading. Measure, develop and lead.* Helsinki: Talentum. (In Finnish.)
Lampikoski, K., & Emden, J. B. (1999). *Managing innovatively: Exploit creative resources.* Porvoo: WSOY. (In Finnish.)
Lawson, B., & Samson, D. (2001). Developing innovation capability in organisations: A dynamic capabilities approach. *International Journal of Innovation Management, 5*(3), 377–400.
Lin, R.-J., Chen, R.-H., & Chiu, K. K.-S. (2010). Customer relationship management and innovation capability: An empirical study. *Industrial Management & Data Systems, 110*(1), 111–133.
Marr, B. (2007). Measuring and managing intangible value drivers. *Business Strategy Series, 8*(3), 172–178.
Nahapiet, J., & Ghoshal, S. (1998). Social capital, intellectual capital and the organizational advantage. *Academy of Management Review, 23*(2), 242–266.
Otala, L. (2003). *Wellbeing to workplace – outcome to the action.* Helsinki: WSOY. (In Finnish.)
Paalanen, A., Kujansivu, P., & Parjanen, S. (2009). Measuring the effects of an innovation-focused intervention. Proceedings of the XX ISPIM Future of Innovation Conference, Vienna, Austria.
Panayides, P. (2006). Enhancing innovation capability through relationship management and implications for performance. *European Journal of Innovation Management, 9*(4), 466–483.
Prajogo, D. I., & Ahmed, P. K. (2006). Relationships between innovation stimulus, innovation capacity, and innovation performance. *R&D Management, 36*(5), 499–515.
Romijn, H., & Albaladejo, M. (2002). Determinants of innovation capability in small electronics and software firms in southeast England. *Research Policy, 31*(7), 1053–1067.
Sáenz, J., Aramburu, N., & Rivera, O. (2009). Knowledge sharing and innovation performance – A comparison between high-tech and low-tech companies. *Journal of Intellectual Capital, 10*(1), 22–36.
Salojärvi, S. (2006). Expertise, work wellbeing and creativity – A positive spiral. In P. Vesterinen (Ed.), *Work wellbeing and leadership* (pp. 49–58). Helsinki: WSOYpro. (In Finnish.)
Shan, W., & Zhang, Q. (2009). Extension theory and its application in evaluation of independent innovation capability. *Kybernetes, 38*(3/4), 457–467.
Skarzynski, P., & Gibson, R. (2008). *Innovation to the core: A blueprint for transforming the way your company innovates.* Cambridge, MA: Harvard Business School Press.
Subramanian, A., & Nilakanta, S. (1996). Organizational Innovativeness: Exploring the Relationship Between Organizational Determinants of Innovation, Types of Innovations, and Measures of Organizational Performance. *International Journal of Management Science, 24*(6), 631–647.
Teece, D. J., Pisano, G., & Shuen, A. (1997). Dynamic Capabilities and Strategic Management. *Strategic Management Journal, 18*(7), 509–533.
Tidd, J., Bessant, J., & Pavitt, K. (2005). *Managing innovation: Integrating technological, market and organizational change.* West Sussex, England: John Wiley & Sons.
Tura, T., Harmaakorpi, V., & Pekkola, S. (2008). Breaking inside the black box: Towards a dynamic evaluation framework of regional innovative capability. *Science and Public Policy, 35*(10), 733–744.

Viitala, R. (2005). *Lead expertise! Leading the expertise from theory to practice*. Helsinki: Inforviestintä Oy. (In Finnish.)

Wan, D., Ong, C. H., & Lee, F. (2005). Determinants of firm innovation in Singapore. *Technovation, 25*(3), 261–268.

Wang, C. L., & Ahmed, P. K. (2003). Structure and structural dimensions for knowledge-based organizations. *Measuring Business Excellence, 7*(1), 51–62.

Yang, S.-B., & Choi, S. O. (2009). Employee empowerment and team performance. Autonomy, responsibility, information, and creativity. *Team Performance Management, 15*(5/6), 289–301.

Yang, C. C., Marlow, P. B., & Lu, C.-S. (2009). Assessing resources, logistics service capabilities, innovation capabilities and the performance of container shipping services in Taiwan. *International Journal of Production Economics, 122*(1), 4–20.

Yukl, G. A. (1998). *Leadership in organizations* (4th ed.). Upper Saddle River, NJ: Prentice Hall.

Chapter 22
Epilogue: Two Modes of Practice-Based Innovation

Vesa Harmaakorpi and Helinä Melkas

> We often think of grand innovations, like the invention of the light bulb, as what drives economic growth. But equally important, and perhaps more important, are the 1,001 small innovations that regular business managers and line workers do every day at their jobs. If we can find more effective ways of sharing those micro-innovations with one another so that each person doesn't have to reinvent the wheel ... then we're much more likely to be able to get a faster, more steady pace of economic growth–and improved competitive advantage for the companies that make that easy. [...] What we're going to see in the coming decade are companies whose whole culture is based on continuous improvement and experimentation–not just of specific processes, but of the entire way the company runs. I think this revolution can be fairly compared to the scientific revolution that happened centuries ago.
>
> Erik Brynjolfsson, MIT (2010)

22.1 Some Lessons Learned

This book contains many different kinds of views concerning practice-based innovation and various issues related to it. It touches upon many 'levels' in the discussion of innovation. This was a deliberate choice; it reflects our view and experience that innovation policies and activities suffer from a lack of bridging of structural holes and of utilisation of distances. The chapters in this book thus range from philosophical considerations concerning pragmatism to very concrete examples of practice-based innovation in, for instance, a care home for elderly people. Innovation is still often treated as something 'noble', but without bridging various structural holes between innovation researchers, practitioners, and authorities, inter alia, true understanding concerning means, challenges, and

V. Harmaakorpi (✉) • H. Melkas
Lahti School of Innovation, Lappeenranta University of Technology, Lahti, Finland
e-mail: vesa.harmaakorpi@lut.fi

H. Melkas and V. Harmaakorpi (eds.),
Practice-Based Innovation: Insights, Applications and Policy Implications,
DOI 10.1007/978-3-642-21723-4_22, © Springer-Verlag Berlin Heidelberg 2012

opportunities related to innovation is lacking. This is especially important in the context of practice-based innovation. In the Introduction (Chapter 1), we defined practice-based innovation processes as being *triggered by problem-setting in a practical context and conducted in non-linear processes utilising scientific and practical knowledge production and creation in cross-disciplinary innovation networks*. These processes essentially require combination of knowledge from theory and practice, as well as knowledge from different disciplines.

In the chapters in this book, practice-based innovation is understood by the authors in various ways. This is usual; concepts are incorporated into discussions in various forums already before their contents are clearly defined. The concept of practice-based innovation is a good example. In the preceding chapters, it is used more or less as a synonym for user-driven or employee-driven innovation, for instance. On the other hand, as noted in Chapter 2, the arguments for practice-based innovation are not only practical but also philosophical. The author aptly points out that to some, it may seem that practice-based innovation allows for ignorance, incoherence, or even inconsistency. Pragmatism, on the other hand, has deep roots as a respectable line of thought in philosophy, recognising that application is part of a learning process where ideas change in their application and yield new ideas, so that application is part of discovery. Thus, according to the analysis in Chapter 2, practice-based innovation has a firm foundation in the embodied cognition line in cognitive science and in the pragmatic tradition in philosophy. In our view, the concept of practice-based innovation – as defined in the Introduction and understood in various ways in the other chapters – is useful also because innovation researchers and practitioners, as well as funding organisations, for instance, do not necessarily (and unfortunately) acquaint themselves with philosophical works. The concept may then serve as a bridge between various 'possible worlds'.

The book aimed at variety, presenting a fraction of issues related to practice-based innovation and identifying also needs for further research as well as practice. One lesson learned is that all levels from the macro-level to the individual level, and both basic concepts – such as innovation and practice-based innovation – as well as 'circling' them with other central concepts require researchers and practitioners' attention. It is not sufficient if innovations are discussed at the upper level of national science and technology policy. High-level strategic documents, for instance, may provide quite a narrow view that does not benefit practical actors in companies or public organisations. Multi-actor innovation processes require consideration of other levels, too (regional, network, organisational, work community, and individual levels). Achieving and keeping a balance between these levels is vital. This book hopefully contributed to building up a stronger foundation for the concepts as well as for future studies.

Having emphasised the need to combine different levels and concepts, it is also necessary to note the danger of the innovation concept becoming downgraded and of novel innovation policies dissipating (cf. Edquist et al. 2009). When concepts are transferred back and forth between the academic and the public policy sphere, their character may become even more problematic and vague than they might otherwise be (cf. Miettinen 2002). But what is the overall aim? In innovation research that is, in our view, not that simple a question. Is the aim to avoid downgrading, dissipation and vagueness of concepts? That is certainly worthwhile, but if we want to find out how to combine various types of innovation activities into effective and sustainable wholes at various levels, the aim has to be different, and we have to be willing to accept vagueness, too.

Another lesson learned is that diversity is needed in innovation activities and policies. Policies to promote competitiveness should thus allow for diversity and divergence in order to enhance innovativeness, but manage to link diversity into a power that promotes sustainability. Diversity at all levels – from, for instance, individual innovation tools to diverse types of knowledge and to organisations, industries, and sectors – is important in this regard (cf. Harmaakorpi et al. 2011). The need for diversity is also an equality issue at various levels, which should not be forgotten when discussing (practice-based) innovation. This issue as such is not investigated in detail in this book, but in, for instance, studies in which well-being, learning, and innovativeness are focused on, the discussion and/or the data should be disaggregated by gender. Such disaggregation is crucial in order to make differences in women and men's views, needs, and experiences visible and thus give an informed and sustainable basis for reflection and action related to innovation. Otherwise, the relevance of the results may be questionable, and the results may be significantly biased – hiding information that should be available when deciding about organisational development efforts or policies at larger levels.

This problem of questionable relevance and biased information also applies to existing innovation policy diagnostics that tend to provide data at levels of aggregation that do not take into account the diversity of the situations and needs of firms, industries, and regions, as discussed in Chapter 5. The authors note that innovation-related arguments are typically formulated at a very high level of aggregation that prevents policy-makers from deriving policy prescriptions that would take into account the diversity of situations and needs of different types of organisations in different sectors, industries, regions, and countries. National innovation diagnostics may enable reasonable benchmarking of innovation performances across countries. The implementation of practice-based innovation policies, however, requires disaggregated data and evidence. Such shortcomings need to be addressed by researchers and policy-makers in the future. We also need combinations of qualitative and quantitative studies at different levels.

The results and insights included in this book cover a large spectrum, so it is not appropriate to repeat all of them in this chapter. By bringing up just a few here, we did not wish to be unfair. In the remainder of this chapter we present a framework with two modes of practice-based innovation and explore their characteristics as compared to science-based innovation. It summarises many of the issues discussed in previous chapters and combines them with a view to informing future comprehensive innovation policies and strategies – their planning, implementation, and evaluation – and to promoting their long-term effectiveness and sustainability.

22.2 Outlining the Pieces of the New Puzzle

22.2.1 Background

Lundvall (2007) emphasised that in the current era, there is a need for both strengthening the science-base and promoting experience-based learning; "This is absolutely fundamental when it comes to linking the analysis of national innovation systems to economic development". He suggested the following four new directions for research (on innovation systems):

1. It is necessary to develop a better understanding and more efficient analytical techniques to study institutional 'complementarity' and 'mismatch' in innovation systems.
2. There is a need to deepen the understanding of the production, diffusion, and use of knowledge. In this connection the focus should be on interactive learning processes and upon how 'social capital' evolves as a basis for interaction within and across organisational boundaries.
3. There is a need to understand and develop indicators of how and to what degree workplaces function as learning sites in different national systems.
4. A promising research strategy is to link organisational learning, mobility of people, and network formation. Networks will always involve interaction between people, and the specific career will have an impact on with whom and how agents interact.

Our thinking departs from Lundvall's (2007; cf. Lorenz and Lundvall 2006) arguments. He strongly supported the idea that understanding processes of experience-based learning is a key to the understanding of the specificities of national innovation systems. His analysis thus focused on the following:

- Understanding knowledge and learning;
- The co-evolution of the division of labour, interaction and cooperation;
- Firms as sites for employee learning; and
- The weak correlation between strength of the science-base and economic performance.

We operationalise Lundvall's arguments into our framework[1] as follows:

- As the most central factors of innovation activities ('the corner pieces of the puzzle') we look into innovation types, modes of knowledge production, innovation models, and proximities and distances.

[1] For a more detailed description, see Harmaakorpi et al. (forthcoming).

- As 'the other pieces' we look into economic logics, innovation capital, innovation processes, innovation methods, origins of innovation, fields of expertise, types and conversion of knowledge, knowledge bases, innovation environments, knowledge transfer mechanisms, as well as institutions (target organisations and educational organisations).

Lists like this may contain many shortcomings. For instance, Edquist (2005) listed ten *activities* ('functions') that should be studied in a systematic manner in terms of their respective 'causes and determinants'. The list encompasses quite disparate elements including, for instance, forms of knowledge creation and learning, organisational forms, market demand and public policy instruments. Lundvall (2007), again, criticised listing of a number of 'activities' and studying each of those separately. Listing is potentially useful if the aim is to establish a checklist for managers and policy makers (Rickne 2000), but in terms of theoretical understanding, it represents a step backwards since much of what we already know about the innovation process is neglected. Lundvall (2007) thus argued that it is not obvious how studying different activities separately would lead to more rigorous theory. In addition to – or instead of – our 'pieces', many others could be studied, such as organisational set-up of firms, interaction among firms, national specificities such as national education, labour markets, and welfare regimes, or network positioning of firms, inter alia. Our list is aimed to be both a theoretical and a practical framework. The concepts included in it originate from different levels.

22.2.2 The 'Corner Pieces'

We will not go into detail here concerning the various concepts, such as innovation types, as that is beyond the scope of this Epilogue. It is easy to get lost in the jungle of various innovation types that have been identified by scholars. In practice, categorising innovations into types is often very difficult, if at all possible. Various innovation types usually co-exist. Dividing innovation into types is about looking at innovation from different perspectives, and it may contribute to distinguishing cause-effect relationships and co-impacts of various innovations (Pekkarinen and Melkas 2010). It is also beneficial with regard to finding new policy options and directions. Policies at different levels need to contribute to and build up skills that cover the entire spectrum of various innovations and – very importantly – build up skills to define which skills and policies are needed.

22.2.2.1 Modes of Knowledge Production in Innovation

Gibbons and his colleagues (1994) defined two different processes of knowledge production: Mode 1 and Mode 2. Mode 1 is usually a hierarchical process, during which knowledge tends to preserve its form. Mode 2 is a more heterarchical process, transient by nature. One of the key contrasts between them is that in Mode 1, problem-solving is usually carried out following codes of practice relevant to a particular discipline, and it departs from a homogeneous theoretical basis, while in Mode 2, knowledge production activity is organised around a particular application and is more diffuse by nature. Mode 2 combines heterogeneous knowledge interests in a multidisciplinary manner – often in very practical environments.

Gibbons et al. (1994) claimed that Mode 2 knowledge production started emerging from the mid twentieth century, and it is context-driven, problem-focused and interdisciplinary. It involves multidisciplinary teams brought together for short periods of time to work on specific problems in the real world. Traditional research represents Mode 1, which is academic, investigator-initiated, and discipline-based knowledge production. Gibbons et al. already noted a shift in knowledge production from Mode 1 to Mode 2. This does not diminish the significance of Mode 1 knowledge production. Furthering Mode 1 knowledge production is mainly a task for science policy. In practice-based innovation activities, again, Mode 2 is likely to be the mainstream, but tools to support it are not yet very well developed, and until recently, the whole Mode 2 has not been highly valued as compared to Mode 1.

According to Gibbons et al. (1994), in line with Mode 2, quality control also becomes more context- and use-dependent. In a more dispersed institutional space, quality control takes on more transient and temporary forms and fluid norms. Essentially, success is defined differently in Mode 2. It includes additional criteria to the traditional one of scientific excellence, such as efficiency or usefulness that are defined in terms of the contributions the work has made to the overall solution of transdisciplinary problems (Gibbons et al. 1994). This is taken into account very poorly or not at all in today's measures and criteria of innovation. There is a clear need for new ways to measure results and identify them and new ways to conceive innovation information and knowledge (Lundvall 2007).

The notion of Mode 2 knowledge production has attracted considerable interest, and also criticism. Also Mode 3 knowledge has been put forward (e.g., Groundwater-Smith and Mockler 2009). According to Carayannis and Alexander (1999; 2004), Mode 3 for knowledge creation, diffusion and use is a multi-lateral, multi-nodal, multi-modal, and multi-level systems approach to the conceptualisation, design and implementation of government–university–industry public-private research and technology development cooperative partnerships. According to Jiménez (2008), Mode 3 knowledge production shares some of the properties of Mode 2 research, but with the distinctive characteristic of being closely linked to current societal needs. Mode 3 is bottom-up initiatives, while Mode 2 is top-down (Jiménez 2008).

The criticism concerning the concept of Mode 2 has focused, for instance, on its empirical validity, conceptual strength, and political value (Hessels and Van Lente 2008). Etzkowitz and Leydesdorff (2000) argued that Mode 2 was the material base of science already before its academic institutionalisation, so it is not new. Rip (2002), again, argued that the internal coherence of Mode 2 is questionable; multi-disciplinary, application-oriented research does not always show organisational diversity or novel types of quality control. Descriptive and normative elements are mixed in Mode 2 discussion, so it has been claimed to be rather a political ideology than a descriptive theory (Godin 1998; Shinn 2002).

Perhaps Mode 1 and 2 could be seen as a descriptive theory on policy. Gibbons et al. (1994) reminded that Mode 1 and 2 interact with each other. According to them, specialists trained in interdisciplinary sciences tend to enter Mode 2 knowledge production. This leads to the notion that if only or overwhelmingly Mode 1 is supported by public innovation (and economic and science) policies, the contribution to work life of all these specialists in interdisciplinary sciences is undermined and overlooked. This group of people is quite large in the society.

Gibbons et al. (1994) further noted that socially distributed knowledge is at the core of Mode 2. It is, in fact, linked to the ways in which IT is revolutionising innovation (Brynjolfsson 2010; also Gibbons and his colleagues referred to that already) as well as to the rise of virtual networks in work life. Mode 1 and Mode 2 knowledge production are at the core of the research concerning practice-based innovation, and we will claim that there is a need to reconsider this categorisation.

22.2.2.2 STI and DUI Models in Innovation

Mode 1 and 2 (as well as Mode 3) are related to DUI (doing, using, interacting) and STI (science, technology, innovation) innovation activities. Lundvall (2007) noted that already Adam Smith (1776) distinguished two different modes of innovation. One refers to industrial districts where the focus is on experience-based learning (DUI) and the other refers to the national system of research (STI). Lundvall claimed that in international organisations, as in national governments, the strong position of expertise based upon standard economics has contributed to a narrow interpretation of the national system of innovation. Innovation indicators reflect outputs such as number of patents or inputs that are easy to measure such as R&D expenditure. As to indicators of knowledge, there is a strong bias in favour of knowledge that is explicit. The know-how built up through learning by doing, using and interacting is much more difficult to measure. Human capital measurements may register formal investment in education, but what people learn at the workplace or as consumers is not easy to capture through standard measurements. The absence of indicators makes the area less visible for policy makers, and this contributes to a bias in innovation policy toward promoting STI activities rather than DUI activities. In recent studies based on a combination of survey and register data for Danish firms it has been demonstrated that firms that engage in R&D without establishing organisational forms that promote learning, and neglect customer interaction are much less innovative than firms that are strong both in terms of STI and DUI learning (Jensen et al. 2007; Lundvall 2007).

The fact that science and codified knowledge become increasingly important for more and more firms in different industries – including so-called low-technology ones – does not imply that experience-based learning and tacit knowledge have become less important for innovation (Lundvall 2007). To bring innovations – including science-based innovations – to the market, organisational learning, industrial networks as well as employee participation and competence building are more important than ever (DUI learning). Lundvall argued that the distinction between STI and DUI is fundamental, when it comes to analysing modern innovation systems – and also when it comes to designing management strategy as well as public policy. Although DUI is becoming better known, concrete strategy models are few. The DUI model is certainly not a competitor of the STI model, but supports rooting of knowledge produced by the STI model, too. An important role for DUI is thus to function as follow-up to STI.

22.2.2.3 Proximity and Distance in Innovation

Since the 1990s, there has been an increasing interest towards the notion of proximity in the context of economic development in general – and innovation in particular. As such, the notion of proximity is a sort of an 'umbrella' concept (Boschma 2005a) consisting of different dimensions. The general idea is that proximity, in whatever form, somehow reduces the uncertainty of economic activity, contributes to solving the problem of coordination between different actors, and facilitates interactive learning and innovation (Boschma 2005b).

When analysing the logic and dynamics of innovation, at least four functions of proximity have been identified. First, being close to each other helps companies to develop an efficient division of labour and coordinate their actions, facilitating the development of a core of specialised suppliers and partners. Second, there are externalities of proximity available to all within a region. In particular, these externalities are related to the localised human resources (workforce) and know-how (Feldman 1994; Torre and Gilly 2000). Third, there is evidence that when companies of the same industry are located close to each other, it forces them to innovate by creating an environment where companies compete, in a positive sense, with each other (see, e.g., Porter 1990). Fourth, and perhaps most importantly, proximity is relevant for the appearance of knowledge spillovers and learning processes between the actors (Audretsch and Feldman 2003; Malmberg and Maskell 2002). It is not clear, however, what the role of physical proximity is in contrast to the other forms of proximity in innovation activities. In his discussion on the relevance of proximity for learning, Boschma (2005b) argued that geographical proximity per se is neither a necessary nor sufficient condition for learning. Moreover, as Audretsch and Feldman (2003) argued in the context of knowledge spillovers, empirical evidence on the correlation between physical proximity and learning gives us no understanding about the ways in which knowledge spillovers really occur between relevant actors.

From the point of view of innovation, proximity in its different forms has contradictory effects. There are the apparent positive effects of proximity, but on the other hand, it has been persuasively argued that there is a phenomenon of 'having too much proximity', that is, a negative side of proximity. These negative effects relate to different forms of lock-ins, involuntary spillovers, a certain lack of flexibility, and a possibility of opportunism (see, e.g., Boschma 2005b; Tura and

Harmaakorpi 2005; Adler and Kwon 2000). The possibility of negative effects of proximity suggests that there are equally important conditions of innovation connected to the idea of physical, cognitive or functional *distance* or diversity between actors. Such models of innovation emphasise the generation and maintenance of social, cultural and cognitive diversity as basic functions of innovation policy. Boschma (2005b) identified two basic features of distance that are relevant for innovation processes: (1) by enhancing physical, cognitive, social and structural *openness*, distance enables new ideas and information to be triggered and thus fosters renewal of the regional knowledge base, and (2) distance enables the implementation of new ideas by the formation of an efficient combination of control and flexibility.

When discussing the roles of proximity and distance in innovation policy, we thus face a critical dilemma: on the one hand, there is a need for mechanisms for enhancing physical, social and cognitive proximity between the relevant actors of innovation processes. This is the basic idea behind most of the modern innovation policies and, for example, behind the different models of technology centres and science parks. On the other hand, innovation policy has to include mechanisms for enhancing social and cognitive diversity, openness of innovation networks and the ability of an innovation network to connect itself to the wider national and global knowledge base – that is, mechanisms for ensuring sufficient distance between the actors. This perspective is not unfamiliar to innovation policy makers, but it has not attracted similar attention as its opposite one. In particular, when focusing on the practical models and methods of innovation policy, this imbalance is striking. The main intention of innovation policy models appears to be the generation of different forms of closeness. However, if the above argument is accepted, the central challenge is not to maximise proximity, but to create an efficient *balance* between the contradictory purposes of enhancing proximity and distance.

The interplay between proximity and distance is a very challenging task to manage with regard to supporting practice-based innovation. Practice-based innovation is often hybrid by nature, and its effectiveness may become visible only after quite some time has passed. Therefore, methods and approaches to managing the interplay between proximity and distance successfully require considerable attention also in future research (Harmaakorpi et al. 2011).

22.3 The Need for Two New Sub-Categories

On the basis of the multi-faceted research presented in this book and its synthesis, we argue that it is necessary and fruitful to divide Mode 2 knowledge production into two sub-categories to understand the prerequisites of practice-based and broad-based innovation activities, and to support them in practice:

- Sub-category 2a contains intellectual cross-fertilisation, for instance in innovation sessions, in which scientific and practical expertise are combined with the help of various ideation and creative methods; such sessions may aim at, for instance, a concrete product or process innovation.

- Sub-category 2b, again, contains more heterogeneous development of organisations, the effectiveness of which becomes visible more slowly. Such development may be conducted with the help of, for instance, applied community-based theatre methods and learning by doing. In this kind of development that aims at, for example, organisational and social innovations, every employee – and customer – is an expert.

Our recategorisation of the modes including sub-categories 2a and 2b is presented in Table 22.1. The points of view on the left include the 'corner pieces' and the 'other pieces' of the 'puzzle'. In this Epilogue, we have briefly gone through the corner pieces only; our work is presented in greater detail in Harmaakorpi et al. (forthcoming). The footnotes related to Table 22.1 point to just a few of the relevant references. Most of the concepts have been discussed also in the preceding chapters.

When looking at Table 22.1, it needs to be kept in mind that the categorisation is simplified, and the borders between the cells are not always clear or self-evident. The categorisation, however, helps in understanding the differences in the modes and the resulting needs for innovation policy. The modes are mutually complementary, but each of them needs to be managed. In our view, this should be taken into account in future evaluations of innovation policy. For instance, if Mode 1 innovation policies are concentrated on at the national level, it is not sufficient. Such shortcomings are typical – for instance, many international seminars are pure Mode 1, while Mode 2a and 2b gain much less or non-existent attention in innovation policy. Moreover, in international cooperation projects, Mode 2a and 2b may be much more relevant and sustainable to develop than Mode 1.

22.4 Discussion and Conclusions: It Takes Three to Tango

The conclusion, thus, is that three different kinds of innovation policies are also needed. Things get mixed up, and even false assumptions and negative effects may ensue, if the Mode 1 approach or something related to it is utilised when, for instance, developing public sector services and organisations. The three different innovation policy wholes need to 'dance seamlessly together' (see also Fig. 22.1).

Critical elements in this joint dance are, for instance, the importance of brokerage of information and knowledge, and creation of arenas for combinations of the different modes (knowledge conversion mechanisms and *ba*s). Modes 2a and 2b require different types of expertise and brokerage skills, which also have to be focused on. In addition to being more and more often practice-based, many innovation processes are more and more open. Innovations are increasingly boundary-breaking creations that require knowledge and skills that are hardly found even in one big company. Even giants like Nokia are developing innovation systems that integrate customers. The customer-driven approach has become more prevalent,

Table 22.1 Modes 1, 2a and 2b, and innovation policy[a]

Point of view	Innovation policy types		
	Science-based innovation (STI, Mode 1)	Practice-based innovation (DUI, Mode 2a)	Practice-based innovation (DUI, Mode 2b)
Most typical innovation types	Radical technological innovations and related concepts	Radical concept innovations – technological system innovations	Organisational innovations – social innovations – service innovations
Most typical fuels of innovation	Proximity	Distance	'Near distance'
Most typical logics	Agglomeration – clusters – economies of scale	Related variety – innovation platforms	Developing innovation capability – breaking 'silos' and preventing bottlenecks
Most typical capital	Intellectual capital – financial capital	Social capital – institutional capital	Social capital – structural capital
Most typical innovation processes[b]	Analytical	Interpretative	Interpretative
Most typical innovation methods	Scientific methods	Methods of intellectual cross-fertilisation (also virtual)	Problem-based learning (e.g., culture-based methods)
Most typical origins of innovations	Science and related expertise	Networks – serendipity – customers	'Normal' staff – customers
Most typical fields of expertise	World class scientific expertise in narrow fields	Brokering – general ability to build possible worlds	Brokering – general ability to build possible worlds
Most typical types of knowledge	Explicit knowledge	Self-transcending knowledge[c]	Tacit knowledge
Most typical knowledge bases[d]	Analytical	Synthetic	Symbolic
Most typical logics of knowledge production	Homogeneous knowledge production	Heterogeneous knowledge production	Heterogeneous knowledge production
Most typical innovation environments	World class scientific centres	Arenas of intellectual cross-fertilisation in value networks	Arenas of developing organisational innovation capability

(continued)

Table 22.1 (continued)

Point of view	Innovation policy types		
	Science-based innovation (STI, Mode 1)	Practice-based innovation (DUI, Mode 2a)	Practice-based innovation (DUI, Mode 2b)
Most typical knowledge transfer mechanisms	Technology diffusion for the firms of cluster	Scanning and absorbing technology and market signals	Organisational learning
Most typical target organisations	Big companies – technology gaselles	SMEs, big companies	Big companies – SMEs – public and third sector
Most typical educational organisations	Universities	Universities – polytechnics	Polytechnics – colleges – vocational education

[a] For further information and descriptions of the background and the concepts, see Harmaakorpi et al. (forthcoming).

[b] Lester and Piore (2004) divided innovation processes into analytical and interpretative ones. The goal of interpretative innovation is to discover new definitions. This process of sense-making is understood to be a fragmented, ongoing, open-ended (and multi-voiced) process of dialogue that emphasises interaction and communication. In an interpretative innovation process, incompleteness and distance need to be tolerated, and participants have to be willing to withstand multiple viewpoints and a lack of universal truths – as there may be no single 'answer', rather multiple suggestions and proposals.

[c] See, e.g., Scharmer (2001); Harmaakorpi and Melkas (2005).

[d] Asheim and Coenen (2005; 2006; see also, e.g., Asheim et al. 2005) made a distinction between three types of regional knowledge bases: analytical (science-based), synthetic (engineering-based), and symbolic. These types indicate different mixes of tacit and codified knowledge, codification possibilities and limits, qualifications and skills, required organisations and institutions involved, as well as specific competitive challenges from the globalising economy, which have different implications for different sectors of industry, and, thus, for the kind of innovation support needed.

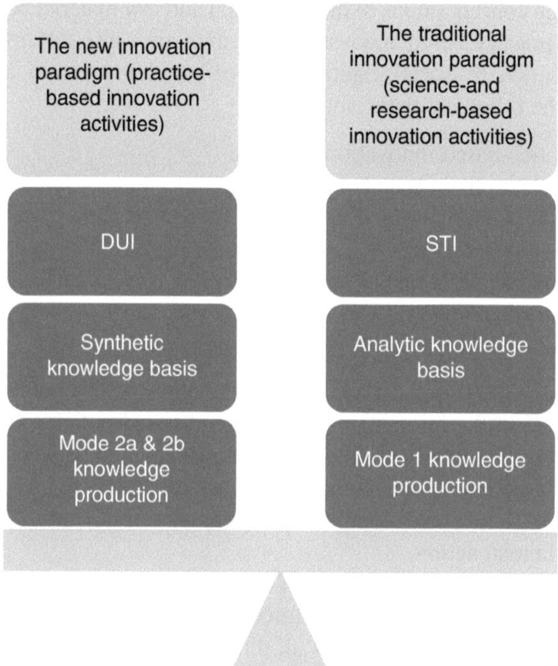

Fig. 22.1 A simplified demonstration of the need for balance between certain conceptual pairs related to innovation

but the impact of information and knowledge produced by customers on innovation processes has in practice not necessarily increased. The different forms and impacts of customer involvement also have to be focused on carefully.

Also in the future, contributions to science, technology and centres of expertise are just as important as earlier. If, however, increasing investments to furthering the practice-based mainstream of innovations and to combining practice- and science-based approaches are not made, much of the innovation potential hiding in innovation networks may fail to be utilised (Harmaakorpi 2010). The same applies to individuals and organisations.

By definition (see the Introduction), practice-based innovation does not mean that only practice-based knowledge would be combined, but the essential thing is problem-setting in a practical context, and impacts and consequences of this on innovation processes. Innovation processes often utilise highly scientific knowledge, but in the case of practice-based innovation, problem-setting is practice-based, and often knowledge from different scientific fields is combined. This is a significant difference as compared to science-based innovation, in which problem-setting takes place on the conditions of theoretical knowledge. Thus it requires a different kind of investigation and attention (Harmaakorpi 2010).

Practice-based innovation activities are typically based on employees, customers, or partner networks in daily operations. While science-based innovation activities are strongly based on concentrations of mental capital and raw material born in those, the core issues in practice-based innovation activities are quite different and mainly based on utilisation of social capital. Practice-based innovation activities place challenges on the following fields, inter alia:

- Finding practice-based innovation potential
- Developing absorptive capacity
- Furthering knowledge production in Mode 2a and 2b
- Forming social capital
- Forming expertise and roles of actors
- Enhancement of creativity
- Development of practice-based innovation policy

Edquist et al. (2009) noted concerning innovation policies and strategies that caution is needed in the implementation of broad-based policies so as not to add complexity to the support system. If we adopt the distinction between Modes 2a and 2b in practice-based innovation, complexity is actually likely to increase. However, what is the goal – to avoid complexity or enhance innovation? Complexity is inherent in all human action.

What we need is a skilful combination of innovation policies to work together – sometimes even visibly against market forces – this is part of the essence of Modes 2a and 2b. Modes 2a and 2b do not at all diminish the importance of Mode 1 and STI. Mode 1 knowledge production may be a basis for knowledge-combining innovation processes in addition to the more traditional tasks. Mode 1 knowledge production is enhanced particularly by means of science policy, whereas in practice-based innovation activities, Modes 2a and 2b are the mainstream. In the beginning of this chapter, in the quotation from Brynjolfsson (2010), what might be called a major 'innovation revolution' was anticipated. New ways are needed to support this revolution by means of innovation activities, policies and strategies, and that is where we claim Modes 2a and 2b may be of help.

References

Adler, P., & Kwon, S.-W. (2000). Social capital: The good, the bad and the ugly. In E. L. Lesser (Ed.), *Knowledge and social capital: Foundations and applications* (pp. 89–115). Boston, MA: Butterworth-Heinemann.

Asheim, B. T., & Coenen, L. (2005). Knowledge bases and regional innovation systems: Comparing Nordic clusters. *Research Policy, 34*(8), 1173–1190.

Asheim, B., & Coenen, L. (2006). Contextualising regional innovation systems in a globalising learning economy: On knowledge bases and institutional frameworks. *The Journal of Technology Transfer, 31*(1), 163–173.

Asheim, B., Coenen, L., Moodysson, J., & Vang, J. (2005). *Regional innovation system policy: A knowledge-based approach.* CIRCLE Electronic Working Papers, No. 2005/13. Lund: CIRCLE, Center for Innovation, Research and Competences in the Learning Economy, Lund University. http://swopec.hhs.se/lucirc/abs/lucirc2005_013.htm. Accessed 29 December 2010.

Audretsch, D. B., & Feldman, M. P. (2003). Knowledge spillovers and the geography of innovation. In J. Vernon Henderson & J. Thisse (Eds.), *Handbook of urban and regional economics: Cities and geography* (Vol. 4, pp. 2713–2739). Amsterdam: North Holland Publishing.

Boschma, R. (2005a). Role of proximity in interaction and performance: Conceptual and empirical challenges. *Regional Studies, 39*(1), 41–45.

Boschma, R. (2005b). Proximity and innovation: A critical assessment. *Regional Studies, 39*(1), 61–74.

Brynjolfsson, E. (2010). The four ways IT is revolutionizing innovation, an interview with Erik Brynjolfsson. *MIT Sloan Management Review*, 1 April. http://sloanreview.mit.edu/the-magazine/articles/2010/spring/51330/it-innovation-brynjolfsson-article/. Accessed 29 December 2010.

Carayannis, E., & Alexander, J. (1999). Winning by co-opeting in strategic government-university-industry (GUI) partnerships: The power of complex, dynamic knowledge networks. *Journal of Technology Transfer, 24*(2/3), 197–210.

Carayannis, E., & Alexander, J. (2004). Strategy, structure and performance issues of pre-competitive R&D consortia: Insights and lessons learned from SEMATECH. *IEEE Transactions of Engineering Management, 51*(2), 226–232.

Edquist, C. (2005). Systems of innovation: Perspectives and challenges. In J. Fagerberg, D. Mowery, & R. R. Nelson (Eds.), *The Oxford Handbook of Innovation* (pp. 181–208). Norfolk: Oxford University Press.

Edquist, C., Luukkonen, T., & Sotarauta, M. (2009). Broad-based innovation policy. In *Evaluation of the Finnish national innovation system – full report* (pp. 11–69). Helsinki: Taloustieto Oy. www.evaluation.fi. Accessed 5 November 2009.

Etzkowitz, H., & Leydesdorff, L. (2000). The dynamics of innovation: from National Systems and 'Mode 2' to Triple Helix of university-industry-government relations. *Research Policy, 29*(2), 109–123.

Feldman, M. P. (1994). *The geography of innovation*. Boston, MA: Kluwer.

Gibbons, M., Limoges, C., Nowotny, H., Schwarzman, S., Scott, P., & Trow, M. (1994). *The new production of knowledge*. London: Sage.

Godin, B. (1998). Writing performative history: The new new Atlantis? *Social Studies of Science, 28*(3), 465–483.

Groundwater-Smith, S., & Mockler, N. (2009). *Teacher professional learning in an age of compliance*. Berlin, Heidelberg: Springer.

Harmaakorpi, V. (2010). The 'Regional Development Platform Method' as a tool for innovation policy. In A. Eriksson (Ed.), *The matrix – post cluster innovation policy* (pp. 23–34). Vinnova report 2010:10. Stockholm: VINNOVA. http://www.vinnova.se/en/Publications/Products/The-Matrix/.

Harmaakorpi, V., & Melkas, H. (2005). Knowledge management in regional innovation networks: The case of Lahti, Finland. *European Planning Studies, 13*(5), 641–659.

Harmaakorpi, V., Melkas, H., & Uotila, T. (Forthcoming). Broad-based innovation and the emergence of two modes of practice-based innovation. Journal manuscript.

Harmaakorpi, V., Tura, T., & Melkas, H. (2011). Regional innovation platforms. In P. Cooke, B. Asheim, R. Boschma, R. Martin, D. Schwartz, & F. Tödtling (Eds.), *The handbook of regional innovation & growth*. Cheltenham, UK: Edward Elgar. In press.

Hessels, L. K., & van Lente, H. (2008). Re-thinking new knowledge production: A literature review and a research agenda. *Research Policy, 37*(4), 740–760.

Jensen, M. B., Johnson, B., Lorenz, E., & Lundvall, B.-Å. (2007). Forms of knowledge, modes of innovation and innovation systems. *Research Policy, 36*(5), 680–693.

Jiménez, J. (2008). Research socially responsible: May we speak of a Mode 3 knowledge production? *RECIIS – Electronic Journal of Communication, Information & Innovation in Health, 2*(1), 48–56.

Lester, R. K., & Piore, M. J. (2004). *Innovation: The missing dimension*. Cambridge, MA: Harvard University Press.

Lorenz, E., & Lundvall, B.-Å. (Eds.). (2006). *How Europe's economies learn*. Oxford: Oxford University Press.
Lundvall, B.-Å. (2007). *Innovation system research: Where it came from and where it might go*. Globelics Working Paper Series No. 2007–01. The Global Network for Economics of Learning, Innovation, and Competence Building System. http://dcsh.xoc.uam.mx/eii/globelicswp/wpg0701.pdf. Accessed 29 December 2010.
Malmberg, A., & Maskell, P. (2002). The elusive concept of localization economies: Towards a knowledge-based theory of spatial clustering. *Environment and Planning A, 34*, 429–449.
Miettinen, R. (2002). *National innovation system: Scientific concept or political rhetoric*. Helsinki: Edita.
Pekkarinen, S., & Melkas, H. (2010). Safety alarm systems and related services: From potholes to innovation opportunities. *International Journal of Service Science, Management, Engineering, and Technology, 1*(4), 53–70.
Porter, M. E. (1990). Cluster and the new economics of competition. *Harvard Business Review*, November–December, 77–90.
Rickne, A. (2000). *New technology-based firms and industrial dynamics: Evidence from the technological system of biomaterials in Sweden, Ohio and Massachusetts*. Göteborg: Chalmers University of Technology.
Rip, A. (2002). Science for the 21st century. In P. Tindemans, A. Verrijn-Stuart, & R. Visser (Eds.), *The future of science and the humanities* (pp. 99–148). Amsterdam: Amsterdam University Press.
Scharmer, C. O. (2001). Self-transcending knowledge: Organizing around emerging realities. In I. Nonaka & D. Teece (Eds.), *Managing industrial knowledge: Creation, transfer and utilization* (pp. 68–90). London: Sage Publications.
Shinn, T. (2002). The Triple Helix and new production of knowledge: Prepackaged thinking on science and technology. *Social Studies of Science, 32*(4), 599–614.
Smith, A. (1776/1904). *An inquiry into the nature and causes of the wealth of nations* (E. Cannan, Ed.). 1904, 5th ed. London: Methuen and Co., Ltd.
Torre, A., & Gilly, J. P. (2000). On the analytical dimension of proximity dynamics. *Regional Studies, 34*(2), 169–180.
Tura, T., & Harmaakorpi, V. (2005). Social capital in building regional innovative capability. *Regional Studies, 39*(8), 1111–1125.

MIX
Papier aus verantwortungsvollen Quellen
Paper from responsible sources
FSC® C105338

If you have any concerns about our products,
you can contact us on
ProductSafety@springernature.com

In case Publisher is established outside the EU,
the EU authorized representative is:
Springer Nature Customer Service Center GmbH
Europaplatz 3, 69115 Heidelberg, Germany

Printed by Libri Plureos GmbH
in Hamburg, Germany